Problems in the Behavioural Sciences

GENERAL EDITOR: Jeffrey Gray
EDITORIAL BOARD: Michael Gelder, Richard (
    Christopher Longuet-Hi;

The psychology of fear and stress

Problems in the Behavioural Sciences

The psychology of fear and stress

SECOND EDITION

Jeffrey Alan Gray

Professor of Psychology, Institute of Psychiatry, London

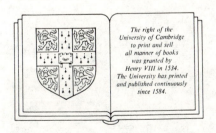

*The right of the
University of Cambridge
to print and sell
all manner of books
was granted by
Henry VIII in 1534.
The University has printed
and published continuously
since 1584.*

CAMBRIDGE UNIVERSITY PRESS
Cambridge

New York   New Rochelle   Melbourne Sydney

Published by the Press Syndicate of the University of Cambridge
The Pitt Building, Trumpington Street, Cambridge CB2 1RP
32 East 57th Street, New York, NY 10022, USA
10 Stamford Road, Oakleigh, Melbourne 3166, Australia

First published by Weidenfeld and Nicolson 1971
Second edition published by Cambridge University Press 1987

Printed in Great Britain

British Library cataloguing in publication data
Gray, Jeffrey A.
The psychology of fear and stress.—2nd ed.—(Problems in the behavioural
sciences; v.5)
1. Fear
I. Tittle   II. Series
152.4   BF375.F2

Library of Congress cataloging-in-publication data
Gray, Jeffrey Alan.
The psychology of fear and stress.
(Problems in the behavioural sciences)
Bibliography
Includes index.
1. Fear.   2. Stress (Psychology)   3. Environmental psychology.
4. Neuropsychology.   5. Social psychology.
I. Title.   II. Series.   [DNLM: 1. Fear.   2. Stress,
Psychological.   WM 178 G779p]
BF575.F2G73   1987   156'.2432   86–33387

ISBN 0521 24958 9 hard covers
ISBN 0521 27098 7 paperback

MU

Contents

Foreword

The space in which science lives and grows has some curious features. It is natural in some ways to imagine a tree diagram: a thick trunk (philosophy?) divides first into a few major branches (physicochemical, biochemical, social sciences), then into somewhat thinner branches (physics, for example, now separating from chemistry), and so on repeatedly until the twigs that display the latest, developing buds ramify in all their recondite glory. The trouble with this image is that it leaves no room for the fusions between long-separate branches which make up one of the most powerful driving forces in contemporary science. If biology and chemistry once split off from each other as branches that aim at different places in the sun, how can we put them together again in our tree diagram to symbolise that modern prodigy, biochemistry?

Another image, avoiding this problem, is that of the bicycle wheel: a central hub (philosophy?) with spokes radiating in all directions, the distance between spokes being proportional to the distance between the sciences they stand for. Now, to celebrate the birth of biochemistry, we merely have to slot in another spoke, equidistant between the parent disciplines. And, to preserve the necessary metaphor of growth that is integral to the tree diagram, we can make our spokes expand proportionally to the maturity of the subjects they represent. But it is a consequence of this image that spokes get further and further from each other as they grow: the quickest route between two disciplines is always through the philosophical hub; and, as subjects mature, they become more isolated, more idiosyncratic. To be sure, this feature of 'bicycle wheel' space is, to some extent, veridical: bear witness the latest journal devoted entirely (I invent, but barely) to the eating habits of the *ob/ob* mouse after lesions to the ventromedial nucleus of the hypothalamus. But, from another point of view, it is the growing points of science that are the closest together, the circumference of the wheel that is shorter than the hub. To take an example close to our interests, it is those workers who push the frontiers of brain science forward most vigorously whose research may be transformed overnight by developments in physics, chemistry, or linguistics; and the linguists' subject may be revolutionised by advances in brain science.

The bicycle wheel is ill-designed for these paradoxes. Our series, *Problems in the Behavioural Sciences*, in contrast, is designed precisely to house them. Its intention is to provide a forum in which research workers on any frontier that relates to psychology *de facto*, even if not *de jure*, can communicate their discoveries, their questions, and their problems to their peers in other disciplines, and to a new generation of students in their own. The central role played by psychology in this series guarantees that it will be multi-disciplinary, for psychology already draws on advances in fields as diverse as biophysics and biochemistry, logic and linguistics, social anthropology and sociology. Conversely, the increasing use of behavioural methods – often, alas, ill-understood and worse-applied – by scientists in other fields of biological or sociological enquiry means that they too need to keep a wary eye on how the battle rages at the frontiers of psychology. They will find in *Problems in the Behavioural Sciences* the latest war reports.

The Psychology of Fear and Stress fits naturally into such a multi-disciplinary framework. There is nothing, I suppose, about fear and avoidance behaviour that makes them essentially more multi-disciplinary than other things; but in practice these forms of behaviour have probably been investigated from more different angles than any other aspect of psychology. Genes and environment, biological and social sciences, the theory of learning and the complexities of drug action, hormones and neurons, personality traits and psychopathology – all these have both thrown light on, and themselves been elucidated by, the study of fear; and this book faithfully reflects this diversity of influences, while yet attempting to construct from them a coherent whole. The result is a synthesis that combines concepts drawn from the behavioural and the neural sciences in a manner that may sometimes surprise those unaware of the rapid progress that is currently being made in unifying these hitherto separate disciplines.

There is a second way also in which this book fits naturally into our series. Previous volumes have dealt with the major principles of animal learning theory (Dickinson), simple motivational systems such as hunger (Le Magnen) and thirst (Rolls & Rolls), and the interaction between motivational systems (Toates). *The Psychology of Fear and Stress* takes us to the next level of complexity: that of the emotions. Learning theory and motivation are the twin pillars upon which the psychology of emotion rests. Thus anyone who has read these earlier volumes will find that they have contained much that is relevant to the topics covered here. At the same time, he will find pointers towards other fields that have until now figured less prominently in our series, in particular, those of personality, psychiatric disorder, and behavioural treatment. Later volumes will build further upon these new beginnings.

Preface

It will come as no surprise to a working psychologist who reads this book that I started my graduate career in Hans Eysenck's department at the Institute of Psychiatry, London, or that my supervisor at that time was Peter Broadhurst: the deep impression that these two scientists have made upon my thinking will be obvious. The first edition was largely written in 1968–69 while I held a Medical Research Council Travelling Fellowship in Neal Miller's laboratory at the Rockefeller University, New York; this gave me the opportunity to experience at first hand the beneficial influence which until then I had known only through Dr Miller's writings. The book first appeared in 1971, in the World University Library (published by Weidenfeld and Nicolson in the U.K. and by McGraw-Hill in the USA). I have preserved the original style of the first edition and much of the original content, although this second edition has undergone a considerable expansion to bring it up to date. I largely wrote the second edition during two happy stays in Lexington, Virginia, the first (in the autumn of 1982) in the Department of Psychology at Washington and Lee University, the second (in the autumn of 1985) in the Department of Psychology and Philosophy at the Virginia Military Institute, where I held the Mary Moody Northen Chair as Visiting Eminent Scholar. I am extremely grateful to my erstwhile colleagues at each of these three universities (Rockefeller, Washington and Lee, and the Virginia Military Institute) for their hospitality and help. I am also deeply indebted to Eileen Gibson (of the Rockefeller University) for her invaluable secretarial help in preparing the manuscript for the first edition; and to Shirley Chumbley (at the Institute of Psychiatry, London) for the patience and skill with which she has prepared the manuscript for the second. Geraldine Flanagan's painstaking criticism of an early draft of the first edition was invaluable in at least turning it into something better.

Preservation of the style of the first edition has had the consequence that in certain respects the format of this book departs from that of others in *Problems in the Behavioural Sciences*. In particular, I have not given formal name-and-date references, although the appropriate reference will often be clear from the text. If it is not, the reader should consult the list of sources for each chapter which precedes the

References. If, nevertheless, no specific reference can be found for a point of fact made in the text, then it is to be found in one of the major sources given for the chapter concerned.

Acknowledgements

The author would like to acknowledge the following sources for the figures (for full details of the publication, see References): 2.1 Tinbergen (1951); 2.2 Jersild & Holmes (1935); 2.3 Tinbergen (1948); 2.4 Hunt & Otis (1953); 2.5 Guttman & Kalish (1956); 3.1 Izard (1971); 3.2 Marler & Hamilton (1966); 3.3 Andrew (1965); 4.1 Broadhurst (1960*a*); 4.2 Cattell (1950); 4.3 Cattell (1965); 4.4 Hall (1951) and Broadhurst (1960*a*); 4.6 Broadhurst (1960*a*); 4.7 Estes & Skinner (1941); 5.1 Langley (1965); 5.2 Morgan (1965) and Langley (1965); 5.3 Brown & Barker (1966); 5.4 Langley (1965); 5.6 Eccles (1965); 5.7 Morgan (1965); 5.8, 5.9 Axelrod & Reisine (1984); 6.2 Christian (1959); 7.1 Broadhurst (1960*a*); 7.2 Whishaw & Vanderwolf (1973); 7.3 James *et al.* (1977); 7.4, 7.5 Drewett *et al.* (1977); 7.6, 7.7 Levine (1966); 8.1 Denenberg (1964); 8.3, 8.4 Ward & Ward (1985); 8.5 Gray, Lean & Keynes (1969); 8.6 Levine & Mullins (1966); 8.7 Haltmeyer, Denenberg & Zarrow (1967) and Levine *et al.* (1967); 8.8 Levine (1962*b*); 9.1 Moynihan (1955); 9.2, 9.3 Miller (1959); 9.4 Sidman (1966); 9.5 Hearst (1965); 9.6 McNaughton & Gray (1983); 9.7 Fowler & Miller (1963); 9.9 Church, Wooten & Matthews (1970); 10.1 Gray & Smith (1969); 10.2 Strongman (1965); 10.3 Gray (1967); 10.4 Brown (1961); 10.5 Brown & Wagner (1964); 10.6, 10.7 Gray (1969); 10.8 Feldon *et al.* (1979); 10.9 Davis *et al.* (1981); 10.10 Gray (1969); 11.1 Kamin (1956); 11.2 Bolles, Stokes & Younger (1966); 11.4 Morris (1975); 11.5 Starr & Mineka (1977); 12.2 Gray & Smith (1969); 12.3 Mackintosh (1983); 13.1 Starzl *et al.* (1951); 13.2 Grzanna & Molliver (1980); 13.3 Livett (1973); 13.4, 13.5, 13.6, 13.7 Melzack & Wall (1983); 13.8 Sokolov (1960); 13.9 Douglas (1967); 13.10 Smythies (1966); 13.11 Haefely *et al.* (1985); 13.12 Haefely (1984); 13.15 O'Keefe & Nadel (1978); 13.17, 13.18 Gray (1982); 13.19 Cooper *et al.* (1982); 13.20 Soubrié (1986); 13.21 Rawlins *et al.* (1980*a*) and Owen *et al.* (1982); 13.22 De Molina & Hunsperger (1962); 13.24 Melzack & Wall (1983); 13.25 Olds & Olds (1965); 14.1 Fuster (1980); 14.2 Eysenck (1955); 14.3 Eysenck (1967); 14.4, 14.5 Eysenck & Rachman (1965); 14.6 Gray (1970*a*); 14.7 Roth *et al.* (1976).

1 Introduction

This book is written in the belief that human behaviour is influenced in innumerable ways by fundamental biological mechanisms which may well go back to the roots of our mammalian heritage or beyond. While this statement is generally applicable, it carries particular force when our concern is with emotional behaviour. Which of us has not felt, in moments of great emotion, that his behaviour and his feelings are due to forces of which he has little understanding and over which he has still less control? If we are ever to achieve such understanding – and only understanding will bring control – it is essential that we subject our behaviour, and the behaviour of our animal relatives, to the same kind of systematic observation and experiment which has paid such huge dividends in the physical sciences.

In carrying out such a programme of scientific research, there are, broadly speaking, two kinds of approach one can adopt. One can consider the historical *origins* of the behaviour Man displays today, or one can investigate the actual *organisation* of this behaviour. These two approaches are by no means mutually exclusive: on the contrary, they supplement each other, the one kind of investigation often supplying valuable clues as to the correct interpretation of the other. In this book, we shall adopt both approaches, often simultaneously; but, by and large, the first half of the book (up to Chapter 8) is more concerned with the origins, both phylogenetic (affecting the evolution of species) and ontogenetic (affecting the development of the individual), of fearful behaviour, and the second half is more concerned with its organisation.

The organising mechanisms of behaviour are in the brain (the central nervous system) and the rest of the nervous system, and in the network of hormonal pathways known as the 'endocrine system'; these function so closely together that we do best, in fact, to speak of a single 'neuro-endocrine system'. The evolutionary origins of the neuro-endocrine system, and of the behaviour it controls, are to be found in the processes of natural selection which were first seen clearly by Darwin a century ago and which have been studied continuously since.

This gives us two good reasons why a book which is concerned with fear, usually thought of as a subjective emotion confined to the

1

human psyche, is full of information obtained from experiments on the behaviour of animals (though our own species is by no means neglected). In the first place, it is only with animals that it is permissible to conduct experiments which submit the role of the nervous and endocrine systems in the control of behaviour to scientific scrutiny. In the second place, when we study animals we are studying the origins of our own behaviour; and, if we choose species which are relatively close to us on the evolutionary scale (such as monkeys, cats, rats, and the other mammals), the behaviour we study may not be so different from our own.

Two kinds of question

Scientific investigation starts with questions. Suppose you know someone who is afraid of snakes, or heights, or women. There are two big questions you might wish to ask about this: (1) why is he *afraid*? – i.e. what are the conditions which gave rise to his fear; and (2) why is *he* afraid? – i.e. what is it that predisposed this person in particular to become frightened under these conditions. If we generalise these questions, we get: (1) what are the conditions which give rise to fear? (2) what are the conditions which affect susceptibility to fear, or 'fearfulness'? If we generalise still further, to cover all such states, and not just fear alone, we get the two very large and important branches of experimental psychology known as 'motivation' and 'personality'. In this book we shall be concerned with both these kinds of question, as they relate to fear and fearful behaviour.

But what kind of behaviour is it that we study when we study fear? And what do we mean by 'fear' anyway? The long answer to these questions is to be found by reading the book; though even then it will inevitably be inadequate. The short answer is that we are concerned with the same kind of behaviour that the man in the street would consider 'fearful' or 'frightened'; and that I mean by 'fear' roughly what he does.

The man in the street would probably regard fear as a state of mind, or a state of feeling, having certain causal antecedents in the environment and leading to certain causal consequences in behaviour. With this view I have no quarrel; except that I would prefer to regard fear as a state, not of the mind, but of the neuro-endocrine system. To be sure, this state has subjective concomitants, but these are known only to the one person experiencing fear and, as such, are of little use for scientific enquiry. The brain, by contrast, is an eminently suitable object for scientific experiment; and there can be no doubt that behaviour is the result of events in the brain. Furthermore, while the connection between brain-and-behaviour, on the one hand, and subjective experience, on the other, remains mysterious, we can at least be sure that the connection is an intimate one. It seems reasonable to proceed, therefore, by treating fear as *a hypothetical state of the brain, or neuro-endocrine system, arising under certain conditions and eventuating in certain forms of behaviour.*

Under what conditions? And eventuating in what forms of behaviour? These are the questions with which we shall be concerned.

The nature of the emotions

Fear is usually listed among the emotions. If one looks down such a list, including states like love, hate, anger, joy, shame, guilt, etc., it is extremely difficult to see anything which they have in common with one another, but not with states not called emotions. Psychologists have not so far done much to illuminate this obscurity. But recently the lines of a general theory of the emotions have begun to emerge. Roughly speaking, such a theory would hold that the common element binding the emotions into a class is that they all represent some kind of reaction to a 'reinforcing event' or to signals of impending reinforcing events. 'Reinforcing events' amount to rewards and punishments, including (among the punishments) the removal of a reward or the failure of an expected reward to occur and (among the rewards) the removal of a punishment or the failure of an expected punishment to occur. The specific quality of a particular emotion then results from two things: (1) the particular type of reinforcing event involved; and (2) the person's knowledge of these events. Viewed in this light, fear is one form of emotional reaction to the threat of punishment, where a 'punishment' may be operationally defined as any stimulus which members of the species concerned will work to terminate, escape from, or avoid.

So much, then (stated in the most general way), for the conditions which give rise to fear. What of the behaviour which then occurs? Again limiting ourselves to the most general statement possible, we may say that a frightened animal is most likely to try one of the three Fs – freezing (keeping absolutely still and silent), flight, or fight – when he is faced with a punishment or the threat of a punishment; or he may learn something quite new which will terminate the danger or keep him out of the dangerous situation in the future. That most interesting of the animals, Man, behaves in much the same way.

Enough of definitions and generalities: in the immortal words of J. Alfred Prufrock,

> Oh, do not ask, 'What is it?'
> Let us go and make our visit.

(T. S. Eliot, 'The Love Song of J. Alfred Prufrock')

2   Fears, innate and acquired

The Behaviourist Revolution was announced by J. B. Watson in the United States in 1913. (Actually, the Russians had been Behaviourists since 1863, when the well-known physiologist I. M. Sechenov published his influential book *Reflexes of the Brain*.) Watson's famous article 'Psychology as the behaviourist views it' was, as a Marxist might say, progressive at the time. It was a healthy reaction to the bankruptcy of the introspectionist methods of studying psychology then prevalent. Watson's insistence that the only thing the psychologist can *observe* is behaviour (not the mind, nor sensation, nor yet emotion) – *methodological* behaviourism – is now accepted beyond dispute. But many of the scientists who followed Watson's banner went further than this and maintained not only that we should observe only what is observable, but also that we should *talk* only about what is observable. In this, they were conforming to the generally positivist temper of the times, for very similar movements were afoot in other sciences and in philosophy. But the consequences were in some ways specially harmful for psychology, which saw a whole generation of scientists who, for fear of the sin of mentalism, dared not speculate even about what goes on in the brain.

Watson's theory of fear

For some reason, Watson and his followers felt that learning is a more hard-headed principle than instinct, that it commits one to less cerebral (or mental, depending on one's point of view) furniture. This belief can be seen clearly in the theory of fear proposed by Watson in 1924, which held that the innate stimuli for fear (i.e. those which are capable of causing fear without special learning that they are dangerous or noxious) can be limited to loud noise, sudden loss of support, and pain. All other stimuli which can be seen to produce fear are then supposed to acquire this power as the result of a form of learning known as 'classical conditioning', which we shall be looking at later in the chapter. A curious feature of this theory, and one which is very characteristic of Behaviourist thought, is that it treats all species as equal: according to this theory only one statement of the stimuli for fear appeared sufficient to cover them all.

Science likes to start with simple ideas. Even so, Watson's theory of fear seems excessively simple. And, seen in a general biological context, the principles upon which he relied are extremely flimsy. There is nothing more hard-headed about learning than about instinct as a principle with which to explain the establishment of a mode of behaviour. Just as natural selection leads to the evolution of such physiological mechanisms as the circulation of the blood, so it can lead to the evolution of innate behavioural mechanisms, if these aid the survival of the species. And there is no reason why the same mechanisms should evolve in all species. So it will be no surprise to find that a number of quite complex stimuli besides those listed by Watson are innately capable of causing fear, or that these stimuli may be quite different in different species.

Many of the discoveries made in this field are due to the work of the 'ethologists'. An ethologist is a zoologist who has chosen to study behaviour. Under the leadership of such distinguished scientists as Konrad Lorenz and Niko Tinbergen, ethology grew up on the continent of Europe and in Great Britain in virtual isolation from experimental psychology, though sharing with it an identical subject matter. Better versed in general biology and zoology than their psychological counterparts, the ethologists have avoided some of the traps that the latter have fallen into; but they have fallen into a number of their own.

Where the psychologist sees learning, the ethologist, often with as little reason, is nudged by his own preconceptions into seeing instinct. As a result, the two groups of scientists have frequently joined battle in the so-called 'nature–nurture' controversy. This controversy concerns the relative extents to which behaviour is determined by influences proceeding from the environment, including those of learning, and by the hereditary influences transmitted via the genes. In relation to fear it must concern us in three different ways. First, there is the question of which *stimuli* can innately arouse fear. Secondly, there is the question of what forms of *behaviour* innately occur in states of fear. Thirdly, there is the question of the extent to which an individual's particular degree of *susceptibility to fear* is determined by heredity or environment. In this chapter we trace the skirmishes which have been fought around the first of these questions.

Innate fears

One such series of skirmishes has been concerned with the existence (or otherwise) in certain birds of an innate mechanism for the recognition of the hawks which prey on them. In his book, *The Study of*

Instinct, Tinbergen suggested that ducks and geese are able to distinguish hawks flying overhead from more harmless birds because they come into the world equipped with a special mechanism for the detection of the hawk's short neck. In testing this suggestion, he and Lorenz made use of a model so shaped that when it was pulled in one direction, it resembled the silhouette of a hawk, while, pulled in the other direction, it resembled a goose (Fig. 2.1). Experiments using this and other models did indeed show that ducks and geese are more alarmed by the hawk shape than by the goose shape when these 'fly' overhead. However, the interpretation of these observations remains very much in doubt. In the first place, Tinbergen's birds had been reared in a natural environment and had seen geese fly overhead much more often than hawks. Now novelty is itself an important stimulus for fear, so Tinbergen's hawk shape might have been more frightening simply because it was more novel. And, secondly, other investigators, working under more controlled conditions, have failed to repeat Tinbergen's findings.

More substantial evidence for the existence of innate mechanisms for the recognition of predators can be found in the story of the owl and the chaffinch. Many birds react to owls, as well as to other (especially novel) stimuli with a form of behaviour known as 'mobbing'. The British ethologist Robert Hinde, in an important paper published in 1954, describes this behaviour in the chaffinch. This bird approaches the owl for a certain distance, calling 'chink, chink'. At a certain point it ceases advancing, confining itself to short flights towards the predator and short flights back again. Hinde has analysed

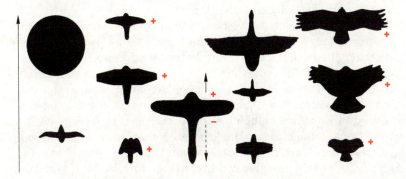

Fig. 2.1. Tinbergen's 'hawk–goose' shapes. Silhouettes that resemble hawks, when pulled over birds subject to hawk predation in nature in the direction of the arrow on the left, elicit escape (+). The same model (centre) may elicit escape when pulled in this direction but not when pulled in the opposite direction.

this behaviour in some detail, and has shown that the chaffinch is in conflict between a tendency to approach the predator and a tendency to flee from it. His experiments, and those published by Hartley in the same year, made use of live and stuffed owls, as well as models ranging from near-realistic to abstract shapes, to analyse the critical features which cause (or 'elicit', as the jargon has it) mobbing in a variety of passerine birds. These appear to be: an owl-like silhouette, with a big head, a short neck, and a short tail; a solid contour; a patterned surface, with spots, streaks, or bars; a brown or grey colour or combination of these colours; and a beak and a pair of frontal eyes. Moreover, Hinde's evidence suggests that at least some features of this highly complex percept are innate in the chaffinch, yellow bunting, and reed bunting, for hand-reared members of these species, when tested with a stuffed tawny owl, showed flight or mobbing.

Maturation of fear

Hinde's experiments also contain observations of the important phenomenon of 'maturation'. If one observes the responses of an animal to a particular stimulus at different ages and finds that it shows no signs of fear at an early age, but all the signs at a later age, the temptation is to assume that the animal *learnt* in the interval that the stimulus is dangerous. But this need not be the case at all. In many such instances what happens is that the neural mechanisms mediating the behaviour are not fully functional at the early age, but undergo further development (maturation) by the later age. (This is what happens, for example, when a child 'learns' to walk, an accomplishment in which virtually no learning is involved at all.)

In Hinde's experiments the hand-reared birds displayed flight or mobbing if they were faced with the stuffed tawny owl when they were more than about a month old, but not when tested at an earlier age. On its own, this observation is not sufficient to tell us that the processes of stimulus recognition had undergone maturation, for the birds might simply have been unable to carry out the movements involved in flight and mobbing at the earlier age. However, control experiments conducted by Hinde showed that other stimuli (loud noise and capture by the experimenter) did make the young birds try to flee. Our most reasonable conclusion, therefore, is that these birds have an innate mechanism which enables them to detect owls and respond to them in an appropriate fashion; and that this mechanism reaches a mature level of functioning only some time after birth.

The owl and the chaffinch experiments, then, demonstrate an apparently innate mechanism for the detection of a complex set of stimuli characteristic of a predator. A more familiar example of the same kind of thing is the fear of snakes. This appears to be quite common in human beings, both adults and children. The Freudians have seen sexual significance in this fact, regarding snakes as symbolic of the penis. No doubt they would find support for this interpretation in the fact that women report significantly more fear of snakes than do men. (They also claim to be more frightened of worms, rats, mice, and spiders.) However, the fear of snakes is also found among our primate cousins, so we must seek a less extravagant explanation for this intriguing difference between the sexes in Man (unless we are to conclude that the chimpanzee – who is, admittedly, a very clever animal – is capable of symbolic sexual fears).

The evidence from humans suggests that the fear of snakes is innate, but does not develop until the child is several years old. In 1928 Jones & Jones set a large, active, but harmless snake free in an enclosure with children of different ages. There were no signs of fear before the age of two; between three and four the children showed signs of cautiousness, and at four and over there were definite signs of fear which increased with age. In fact, the fear displayed increased in frequency and intensity up to the age of about seventeen in people who had never been harmed by a snake. All this strongly suggests a maturation process; but, of course, the results obtained may be due to the subjects' having read or heard about snakes. It is fortunate, therefore, that the question may be resolved by experiments on the chimpanzee, an animal which has no natural language (though he can be specially trained to use one). These experiments were carried out by the well-known Canadian psychologist D. O. Hebb in 1946.

Hebb presented a number of objects to each of 30 chimpanzees and recorded their efforts to withdraw from the stimulus, as well as such other signs of fear as hair erection and screaming. Among the objects was 'a painted wax replica of a coiled 24-inch snake'. This produced signs of fear in 21 of the animals tested and was surpassed in effectiveness only by 'the skull of a 5-year-old chimpanzee with movable jaw controlled by a string'. Since nine of the animals tested had been born and bred in captivity and their behaviour did not differ from that of the others, we can conclude that the fear of snakes in the chimpanzee is innate. Maturation of the fear of snakes does not seem to have been studied systematically in the chimpanzee, though there is evidence that it increases to a maximum at about 6–7 years.

The similarity between the chimpanzee's emotional life and Man's extends well beyond the fear of snakes, as Hebb makes clear in a penetrating discussion. It includes fear of dead or mutilated bodies and fear of strangers. Hebb's chimpanzees sometimes displayed 'paroxysms of terror' when exposed to models of a human or chimpanzee head detached from the body, to the chimpanzee skull described earlier, to a 'painted human eye and eyebrow', or to an anaesthetised chimpanzee. Moreover, these fears too were subject to maturation: 'Young infants showed no fear, increasing excitement was evident in the older (half-grown) animals, and those adults that were not frankly terrified were still considerably excited.' According to other observations made by Hebb, the chimpanzee develops a fear of strangers at about 4 months of age. The same fear develops in the human infant at about 8 months of age, which fits reasonably well with the general pattern that the human takes about 1½ times as long as the chimpanzee to reach a corresponding level of development. That this maturation of the fear of strangers is innate is shown by its appearance in animals reared in the laboratory. That it depends, however, on prior learning to recognise *familiar* individuals is shown by another experiment of Hebb's, in which chimpanzees which had been reared in darkness to an age at which they would normally show fear of strangers failed to do so when first exposed to the light.

Another type of fear which appears to be subject to maturation is the fear of the dark shown by children. The English psychologist C. W. Valentine carried out extensive observations on his own children during the 1920s. He first noted signs of fear of the dark at about 2 years of age, and it continued to grow in intensity until about five. Similar observations must have been made by many a parent. The importance of Valentine's observations is that his detailed records preclude the possibility that the appearance of the fear was due to some unpleasant experience occurring during the dark and leading to the learning ('conditioning') of fear. Again, however, it is possible that the children could have read or heard something about darkness being frightening. But the regularity with which fear of the dark develops at this age in children seems more in keeping with maturation than with suggestion. Jersild & Holmes collected data on the relative frequency of various fears in children of different ages, and it can be seen from their results (Fig. 2.2) that, as in Valentine's observations, fear of the dark is still rising at 5 years. Although I have been unable to find similar data for the chimpanzee, the German psychologist Köhler observed that young animals of this species show a violent avoidance of solitude, a condition which resembles darkness in consisting of a lack of stimulation.

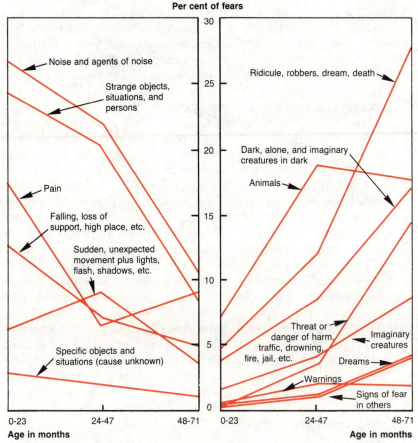

Fig. 2.2. Relative frequency of fears of different kinds shown by children of different ages who were observed by parents or teachers.

Valentine also offers some valuable observations about fear of animals in children. This (Fig. 2.2) rises to a maximum between about 2 and 4 years. Valentine notes how fear of dogs developed in one of his children at the age of 20 months, although the child was unafraid of dogs before this age and nothing had happened which would be likely to arouse fear of them. Other evidence which suggests that neither fear of the dark nor fear of animals is due in any great degree to learning is Jersild's finding that only about 2% of a group of children named darkness or animals as being involved in the 'worst happenings' that had befallen them, whereas 14% said that they were afraid of both these things.

A final example of maturation in children concerns the fear of heights. As any mother knows, very young babies have remarkably little appreciation of the danger of falling from a height; left to themselves, for example, they will crawl right over the edge of the bed. Gradually, however, a proper caution emerges. It is possible to test the fear of heights with an apparatus known as the 'visual cliff'. This consists of a transparent glass sheet placed over a floor with two levels, one flush beneath the glass, the other several feet lower. The different depths are made visually striking by covering the floor with a black-and-white checkerboard pattern. When a baby is first able to crawl, it will gaily cross the 'cliff' to join its mother, who has been placed by the cunning experimenter on the deep side. But at about 9 months of age the baby abruptly stops doing this, instead seeking a detour or halting altogether. Questioning of the babies' mothers fails to reveal any correlation between the emergence of fear of the deep side of the cliff and experiences of falling from a height in the home. Other experiments show that the baby can perceive the difference between the shallow and deep side of the cliff some time before it begins to keep away from the deep side. If lowered towards the glass sheet, for example, the baby will stretch out its arms for support appropriately as it approaches the shallow side, but not as it approaches the deep side. So it is the *fear*, not the perception of heights, that is subject to maturation. This process of maturation appears to be secondary to the development of crawling, since Bertenthal, Campos & Barrett have shown that fear of the deep side of the visual cliff first appears at a relatively constant period (about 10–40 days) after crawling begins, rather than at a fixed chronological age. Whether this perfectly ordinary fear of heights is related to cases of more severe height phobias in adults is not known.

Social interaction

Another important source of stimuli for fear lies in the behaviour of the members of one's own species, or 'conspecifics'. Stimuli arising in this way, often called 'sign-stimuli', have been studied in some detail by the ethologists. Tinbergen, for example, has described the reactions of a favourite ethological fish, the three-spined stickleback, to the series of models depicted in Fig. 2.3. In the spring, the male of this species develops a 'nuptial dress of whitish-blue on the back and brilliant red on the throat and belly'. At the same time, he occupies a territory which he prepares to defend against other males, builds a nest there, and entices a female to enter it and spawn. When Tinbergen introduced his models into an occupied territory, the resident male

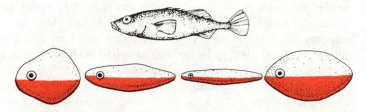

Fig. 2.3. Tinbergen's stickleback models: only those with a red belly acted as effective deterrents to a live stickleback, even when they were much cruder than the uncoloured model.

invariably attacked them, and those models which had a red belly were attacked most fiercely. In a second experiment he removed the owner of a territory. As soon as this was done, a neighbouring male always tried to move in. But if Tinbergen substituted a model for the previous occupant of the territory, the invader was much slower and more hesitant; only models with a red belly acted as effective deterrents in this way. Incidentally, the red breast of the robin serves a very similar function.

Of course, sign-stimuli need not consist of distinctive colouring or other markings actually present on the body surface. Frequently they consist of ritual movements or gestures made by the threatening animal. Tinbergen's sticklebacks, for example, carried out the following movements during boundary disputes: 'Standing nearly vertically with his head pointing downwards, his broad side turned to the rival and (often) the ventral spine on that side erected, he thrusts his snout into the bottom with abrupt jerks.' Tinbergen was able to imitate these movements with his models and found that, irrespective of the colour of the model, this threat posture had more effect than a horizontal presentation.

All these effects could, of course, be due to learning. The evidence that they are not derives from one male reared by Tinbergen in isolation and tested with the models before it had ever seen another stickleback. The same effects were obtained, suggesting that recognition of the threatening significance of these stimuli from the stickleback's conspecifics is innate.

Similar findings have been reported in species that are much closer to our own than is the stickleback. Sackett reared rhesus monkeys in total isolation from other monkeys, but with a slide projector which either the experimenter or the monkeys themselves (by pressing a bar) could use to show pictures. Pictures of monkeys with threatening expressions on their faces elicited clear signs of fear and withdrawal in the isolated animals. This response was subject to maturation and reached a peak at about 3 months of age. At that time, the rate of bar-pressing to show slides of threatening monkeys also declined, although before then the isolated animals had been as eager to inspect pictures of this kind as to inspect other pictures of their conspecifics.

Animals respond appropriately to signs of fear in their conspecifics, as well as to threat. In some species the critical signals are odours emitted by the frightened animal. Rats, for example, can use the sense of smell to detect places in which other rats have experienced an electric shock to the feet, and consequently to avoid such places. They can similarly detect and avoid places in which other rats have been 'frustrated'. To demonstrate this, Collerain & Ludvigson trained a group of 'donor' rats to find food reward in a box and then withheld the reward (so 'frustrating' them). Other rats now avoided the box in which the donor rats had been frustrated. This was probably because the donor rats emitted a 'frustration' odour. There is reason to believe that the rat's frustration odour is very similar to its fear odour; as we shall see in Chapter 10, there are many other similarities between fear and frustration. Whether such odour cues (or 'pheromones') are involved in human social interaction is not known, but the possibility certainly cannot be ruled out.

Classical conditioning

As we have seen, many fears appear to be largely innate. But learning too is important in the genesis of fear. A central role in such learning is usually attributed to the process of classical conditioning.

Classical conditioning was first investigated at the turn of the century by the Russian Nobel prize winner Ivan Petrovich Pavlov, and we can do no better than describe a typical experimental situation used in his laboratory. The experimental subject (in Pavlov's case, a dog) is repeatedly exposed to a sequence in which an initially neutral stimulus (e.g. a tone or a flash of light) is followed shortly after by a stimulus of some biological significance to the animal, such as food. The latter, or 'unconditioned stimulus' (UCS), elicits without prior learning a range of 'unconditioned reponses' (UCR). As the pairings of stimuli are continued, it is found that the initially neutral but now 'conditioned stimulus' (CS) comes to elicit on its own some part of the total pattern of responses formerly elicited only by the UCS; this is known as the 'conditioned response' (CR), and in Pavlov's case was most often the response of salivating. In common-sense terms we may say that the dog has come to expect food after the CS and salivates in anticipation.

Classical or Pavlovian conditioning has to be distinguished from a second, equally important learning process, known as 'instrumental' or 'operant' conditioning. This depends, not upon the relationship between two stimuli as does classical conditioning, but upon the relationship between a response and a stimulus that follows the response. The response is typically chosen for its experimental convenience, e.g. running in an alley or pressing a bar if the subject of the experiment is a rat, pecking at a key if it is a pigeon. The stimulus that is made to follow upon the response is known as a 'reinforcer'. Reinforcers may act as rewards (e.g. food if the animal is hungry, water if it is thirsty) or as punishments (e.g. a painful stimulus such as electric foot-shock); the former, 'appetitive' stimuli increase the probability of future emission of the response upon which they are made contingent (e.g. bar-pressing); the latter, 'aversive' stimuli decrease that probability. We shall have more to say about instrumental conditioning later in the book, especially in Chapter 11; for the moment it is classical conditioning that holds the centre of the stage.

As we have seen, particular emphasis was laid by Watson on the classical conditioning of fear as a way of accounting for the very wide range of stimuli to which both animals and Man may respond with fear. And, although we have widened the class of innate stimuli for fear beyond the limits set by Watson, it is clear that some such mechanism is still required.

That at least one index or symptom of fear can be classically conditioned is shown by an experiment conducted by Hunt & Otis. The response they used as their index was that of defecation, which, as we shall see in Chapter 4, is one of the best validated measures of fear available to the experimenter. They presented a rat with a flashing light for 3 minutes, at the end of which time an electric shock was delivered to the animal's feet. As shown in Fig. 2.4, the result of this pairing was that defecation, which at first occurred primarily when the shock was delivered, came increasingly to occur during the presentation of the light. When, in a further phase of the experiment, light was presented without the following shock, the tendency for the light to elicit defecation gradually faded, producing a typical 'extinction' curve. There are many other experiments showing that other responses symptomatic of fear (changes in heart rate, changes in respiration, etc.) can be classically conditioned in a similar manner. There can, indeed, be no doubt that Watson was right in his belief that classical conditioning is of major importance in the acquisition of the capacity

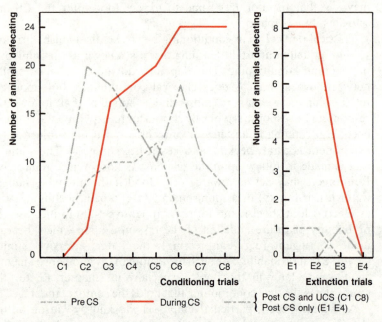

Fig. 2.4. The formation and extinction of a conditioned fear response: the number of rats (out of 24 during conditioning trials and 8 during extinction trials) defecating before, during, or after presentation of a warning signal (CS) followed, during conditioning, by an electric shock (UCS).

to elicit fear by initially neutral stimuli, and this process may play an important role in the formation of certain neurotic symptoms in Man (Chapter 14).

One of Watson's own experiments in this field has acquired some fame. It was conducted with an 11-month-old infant known as 'little Albert'. Albert was first tested and found to show no fear of animals, whereas he did show signs of fear when a steel bar was struck loudly. A white rat was now put in front of him and at the same time the bar was struck, causing him to start violently. After half-a-dozen of these combinations, little Albert would cry at the sight of the rat and do his best to get away from it. He also displayed a phenomenon known as 'stimulus generalisation'. This describes the tendency for stimuli similar to the original CS to elicit the CR, the magnitude of the response diminishing as the similarity of the two stimuli decreases, and it is a universal property of learnt responses (see Fig. 2.5). In Albert's case, the generalisation extended to rabbits, a fur coat and, to a smaller extent, a dog, all of which now made him appear frightened.

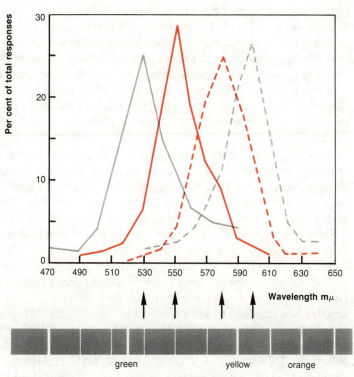

Fig. 2.5. The phenomenon of stimulus generalisation appears to be a universal property of learnt responses. In the particular case illustrated, four groups of pigeons were rewarded for pecking at a disc on to which light of the four different wavelengths indicated by the arrows was projected. When the wavelength of light was changed, there was an orderly decline in the pigeons' rates of response: the greater the stimulus change, the greater was the decline.

However, the interpretation of these results was questioned by Valentine in an excellent paper. He conducted very similar experiments to Watson's, using his year-old daughter as subject. First he showed the little girl some opera glasses and sounded a loud whistle as she reached for them. She showed no sign of fear of either the whistle or the combination, glasses–whistle. She was now shown a 'woolly caterpillar', at which 'she repeatedly turned away with a shrug of the shoulders or slight shudder'; no great fear here, either. Valentine now sounded the whistle when next the little girl turned to look at the caterpillar, with the result that she gave a loud scream and turned away. Valentine quotes several other observations of a similar nature. They all go to show that it is much easier to transfer fear (by an apparent classical conditioning procedure) to some stimuli (e.g. caterpillars) than to others (e.g. opera glasses). Given what we know (but Watson did not) about the maturation of certain fears, especially of animals, it is easy to accept Valentine's conclusion that certain stimuli may elicit a 'lurking fear in the background' (owing to innate factors) so that an added disturbance can bring out a full-blown fear reaction.

In spite of the undoubted importance of classical conditioning, then, these observations suggest that we should be cautious in attributing all the fear-arousing properties of a particular stimulus to this form of learning.

Another peculiar feature of the classical conditioning of fear comes out when we consider one particular response which has often been used as an index of conditioning, namely, 'freezing'. Many species display this pattern of tense and silent immobility in time of danger. In the Hunt & Otis experiment described above, for example, the flashing light used as CS came to evoke freezing as well as defecation. Now it has often been said that classical conditioning consists of 'stimulus substitution', by which it is meant that the CS comes to elicit exactly the same responses which were originally elicited by the UCS. It has long been clear that this view of conditioning is not strictly accurate: the pattern of responses elicited by the CS is probably never an exact copy of the reaction to the UCS. However, it is true to say that the CS does not normally produce a change in behaviour unless the UCS produces a change of the same kind in the same direction (though not necessarily to the same degree). Conditioned freezing, however, constitutes a clear-cut exception to this rule. The rat normally responds to an electric shock with a great increase in activity, either directed towards finding an exit from the situation, or attacking some feature of the environment, or consisting of frantic scampering and jumping; yet a stimulus followed by shock comes to elicit the reverse of this – immobility.

There are two ways (not mutually exclusive) out of this dilemma. One is to suppose that the rat (and many other species) comes equipped with an *innate* response (that of freezing) to stimuli which have been followed by unconditioned fear stimuli and only needs to learn (by conditioning) *which* originally neutral stimuli he should respond to. As we have seen, this kind of admixture of learning and innate behaviour would not be without parallel; and it makes good adaptive sense to respond to, say, pain inflicted by a predator by running and struggling but to stimuli which *warn* of impending danger by remaining still and perhaps escaping the predator's attention. Furthermore, there is evidence (considered in Chapter 13) that separate brain systems control freezing and more active forms of escape and avoidance behaviour.

The second way out of the dilemma is to abandon the general view that, in classical conditioning, what gets conditioned to the CS is a response at all. Instead, many have supposed that the CS comes, as a result of conditioning, to evoke some kind of changed state in the nervous system. This view has its dangers. If stated loosely, it may be much less amenable to rigorous experimental test than the response-conditioning alternative. If it is to be adopted at all, then, the properties of the changed state which the CS is supposed to evoke must be stated precisely. Much of the rest of this book is devoted to this task.

Pain and fear

Watson's theory held that pain is one of the few stimuli that are innately capable of eliciting fear. But this tenet runs into difficulty with the facts considered in the previous section. If a painful UCS and a CS which has been paired with the UCS elicit different – indeed, in some respects diametrically opposed – forms of response, is it possible to attribute both forms to the same underlying state of fear? Surely not. and, if one must choose between the two, common sense dictates that one should name 'fear' that state which is elicited by stimuli that give *warning* of pain, not the state elicited by the painful stimulus itself. Fear, in other words, is due to the *anticipation* of pain.

In agreement with this conclusion, the last decade has seen an accumulation of evidence that the central states of pain and fear are clearly different. We have just seen some of this evidence: painful stimuli elicit movement, conditioned stimuli associated with pain elicit freezing. Another difference between the responses evoked by the two kinds of stimuli concerns aggression. Provided there is another animal in close proximity, a painful UCS is a highly reliable way to elicit fighting and attack; but it is very difficult to condition this type of behaviour to a CS associated with pain. As we shall see in Chapter 13, this behavioural difference is paralleled by differences in the brain. There is little overlap between the brain structures which control, on the one hand, freezing and other kinds of fearful behaviour and, on the other, fighting and attack. Pharmacological evidence similarly points to a separation between the physiological mechanisms subserving fear and reactions to pain, respectively. Analgesics, such as morphine, reduce the response to a painful stimulus (e.g. they make a rat or mouse slower to move away from a hot floor or an electrified grid), yet they do not weaken responses to a CS which has been associated with such a painful stimulus. Conversely, drugs which reduce fear have no analgesic properties. (These drug studies are considered further in Chapters 10 and 12.)

Studies of human emotional expression (Chapter 3) confirm these inferences. The expressions characteristic of fear and pain are quite different from each other. Furthermore, in agreement with the animal studies mentioned above, Izard has demonstrated a close relationship between pain and anger (the equivalent in emotional terms of aggressive behaviour). The expression elicited in babies by a painful stimulus (an inoculation injection) gradually changed over the first 1½ years of life from one of distress to one of anger. At no time did the inoculation give rise to an expression of fear. This observation is quite contrary to Watson's theory.

Recent evidence suggests not only that the central states of pain and fear are different, but also that when they interact it is to inhibit rather than enhance each other. Anecdotes abound relating how, during periods of intense emotional excitement, one may experience no pain from even severe injury: it is after the battle that the soldier most feels his wounds. The physiological basis of this effect has been clarified by the recent discovery that the brain and spinal cord possess their own analgesics, to which synthetic drugs like opium or morphine are closely related chemically. These so-called 'endogenous opiates' or 'endorphins' (short for 'endogenous morphine-like substances') apparently function both as hormones (released from the pituitary gland; Fig. 5.2) and neurotransmitters (Fig. 5.6).

There is now considerable evidence that the brain and pituitary react to a variety of stressors by releasing endorphins, which then dampen transmission in neural pathways that subserve pain (see Chapter 13). Fanselow & Baackes demonstrated an effect of this kind when rats were presented with a fear CS (established by prior association with shock in a standard classical conditioning paradigm). Reaction to pain was measured by the time the animal spent licking its paw after local application of an irritant (formalin). Exposure of the animal to the fear CS reduced this reaction, as did exposure to odours released by a second rat that had been given foot-shock. These effects of the fear CS and the 'fear odour' were themselves blocked by injecting the animal with a drug (naloxone) which prevents endorphins from acting. The implication of these findings is clear: during fear, endogenous opiates are released, reducing the intensity of pain. The adaptive value of this mechanism is also clear: the time to lick your wounds is when the danger has passed.

The classification of fear stimuli

It is by now obvious that Watson's theory was *too* simple. Worse, it was wrong, in at least three respects. The list of unconditioned fear stimuli (pain, loud noise, and sudden loss of support) was too narrow; it omitted a whole range of apparently innate, species-specific fears, of which we have considered several examples in this chapter. The list was also too broad in that it included pain. And too great a role was allotted to classical conditioning. But it remains true that science likes simple ideas. Can we offer an alternative scheme which, while doing justice to the facts, retains a degree of simplicity?

It would appear to be possible to subsume all fear stimuli under one of five general principles: intensity, novelty, 'special evolutionary dangers' (a phrase which will be elaborated in a moment), stimuli arising during social interaction, and conditioned fear stimuli.

Two of Watson's original suggestions – pain and loud noise – are fairly obvious examples of *intense* stimuli. We have had to delete pain from the list of fear stimuli. But there is a case for continuing to include intense stimuli that impinge upon distance receptors (i.e. visual, auditory, and olfactory stimuli emanating from a source with which the animal is not in physical contact). J. S. Myer has pointed out that animals respond to such stimuli with the same kind of inhibition of movement that is seen in response to conditioned fear stimuli (see above); whereas, in response to 'proximal' stimuli (i.e. those that are effective only when bodily receptor surfaces are actually in contact with the source of stimulation), the animal engages in active escape or aggressive behaviour, as we have seen in the case of electric shock. Phenomenologically, as Myer also points out, this distinction corresponds with the difference between painful and non-painful aversive stimulation, and with the common-sense view that fear is a response to danger signals (distal stimuli) rather than to pain itself. Also in line with this distinction is the fact that (as we shall see in Chapter 13) the brain structures most intimately concerned with the emotion of fear (notably, the septo-hippocampal system) receive major projections from sensory systems concerned with the analysis of distal stimuli; whereas somaesthetic sensations (initiated by stimuli impinging upon bodily surfaces), including pain, are processed by brain systems that are closely related to the organisation of aggressive defense and unconditioned escape behaviour.

The fear of *novelty* is apparent in a number of observations discussed in this chapter, of which the most obvious is the fear of strangers. Hebb also describes how a chimpanzee may be frightened by the sight of a familiar attendant wearing the equally familiar coat that is normally worn by another attendant. Working with Hebb in Montreal, Melzack noted how dogs which were used to seeing an open umbrella were frightened when he closed it. Watson's 'sudden loss of support' is probably a mixed case of stimulus novelty and stimulus intensity (for the more sudden a stimulus, the more intense is the neural message it initiates). Suddenness is in general an important factor in determining how effective a stimulus is in eliciting fear; Fig. 2.2, for example, includes the effects on children of sudden, unexpected movement.

One rather puzzling class of situations which elicit fear are those which consist of a *lack* of stimulation. Some members of this class may be special instances of novelty. An anaesthetised chimpanzee could be described as a normal chimpanzee with the added novelty of 'no movement'; solitude could be the novelty of 'no companions'. This is not simply quibbling with words; for there is very good evidence (see Chapter 13) that the failure of a stimulus to occur at a point in time or space where it usually occurs acts like any other kind of novel stimulus. However, the intensity of the fear evoked by the sight of a dead or mutilated body is so much greater than that evoked by more ordinary forms of novelty that we perhaps ought to seek an alternative explanation for the effects of this stimulus. Fear of the dark is also difficult to account for in terms of novelty, since by the time this fear matures darkness is no less familiar than the light.

We turn to our third principle – that of the *'special evolutionary danger'* – to account for the development of the two latter fears. This principle is the one by which we would seek to explain the special fears of predators considered earlier in the chapter. It can be stated as follows: where a particular situation is repeatedly responsible for the death of a significantly large proportion of the members of a species over a sufficiently large (on an evolutionary scale) span of time, the individuals of that species may be expected to develop an innate fear of some of the stimuli characteristic of that situation and to avoid them. The increased likelihood of survival of a species which evolves such innate fears should be obvious. This principle is almost certainly responsible for the fear of snakes displayed by chimpanzee and Man alike, as well as for other fears of predators in many species. It no doubt operates as well in the fear of dead or mutilated bodies described by Hebb: it is after all likely that a place littered with the bodies of your conspecifics is an unhealthy place for you. And it may well be at work in the human fear of the dark (which is a dangerous circumstance for a species relying so heavily on vision); in the fears of enclosed spaces and of very open spaces common both in Man and in many other species (for one is easily trapped by a predator in the one case and easily spotted by a predator in the other); and in the fears of heights common in our own species and easily observed, for example, in the laboratory rat.

The fourth class of fear stimuli – those arising from *social interaction with conspecifics* – is the result of a more complex evolution, involving the development of mutually interacting forms of behaviour (such as those characteristic of dominance and submission), of the sign-stimuli produced during these forms of behaviour, and of the mechanisms for recognition of these sign-stimuli. We shall have more to say about this important topic both in the next chapter and in Chapter 6.

The fifth and final class of fear stimuli consists of those that are formed by learning, and especially by the process of classical condition-ing. Although this kind of learning is not the Super-glue for fear that Watson believed it to be, it still has a very important role to play. We shall examine this role in some detail later in the book.

We should beware, however, of the trap of supposing that behaviour is *either* learned *or* innate. On the contrary, learning normally *co-oper-ates* with innate mechanisms to ensure that organisms are well adapted to the environments in which they must survive. Moreover, the blend between these two modes of adaptation to the environment varies from a strong predominance of the one to an equally strong predomi-nance of the other (with all shades in between), depending on the evolution of the particular form of behaviour in the particular species.

A recent experiment with blackbirds, for example, has shown that mobbing (which in Hinde's experiments with chaffinches seemed to be largely innate) is subject to cultural transmission from one gener-ation to the next. Young birds were allowed to observe adults (of a different but related species) mobbing targets not usually mobbed by blackbirds. The young blackbirds rapidly acquired this mobbing response. Similarly, Susan Mineka (in contrast to Hebb's work with chimpanzees) found hardly any sign of an innate fear of snakes in laboratory-reared rhesus monkeys; but her animals very rapidly acquired an intense and persistent fear of snakes simply as a result of observing their wild-reared parents behave fearfully in the presence of one. Since nearly all the wild-reared monkeys (captured in India many years before) showed a strong fear of both real and toy snakes without laboratory training of any kind, it is likely that they too had benefitted from similar observational learning during their own infancy. The many poisonous snakes in India would otherwise take a heavy toll of a species which learned to avoid this predator only after personal experience of the danger – often fatal – that it represents.

The American psychologist Martin Seligman has used the term 'prepared' to describe fear stimuli that are related to special evolution-ary dangers. He has suggested that such stimuli are not completely innate in their capacity to elicit fear, but rather that they are prepared (by Darwinian natural selection acting on an evolutionary time scale) to enter into an association with other dangerous or painful stimuli in a particularly rapid and persistent manner. One example of such preparedness is perhaps the rapid observational learning of fear of snakes in Mineka's rhesus monkeys; another is the 'lurking fears' described by Valentine in his daughter.

Seligman's suggestion has been investigated experimentally by A. Öhman's group in Uppsala, Sweden. These workers used an electric shock to the hand as the UCS to condition fear in human subjects. The CR was measured as an increase in the electrical conductance of the skin on the palm of the hand; this is the 'galvanic skin reflex' or 'skin conductance response'. (The skin conductance response appears to be due to the release of sweat, which gives rise to the clammy hands that you will no doubt recognise from your own moments of anxiety; it forms the basis of the lie-detector used by the police.) Pairs of stimuli were used as CSs, of which one was followed by shock and the other was not. Öhman reports that if the stimulus pair consisted of slides of snakes or spiders (prepared stimuli, according to Seligman), the conditioned response was more resistant to extinction (when shock was discontinued) and less open to cognitive control (when the subject was told there would be no more shocks) relative to a control condition in which the pair of slides consisted of pictures of flowers or mushrooms (unprepared stimuli). That this had to do with our distant evolutionary past rather than any objective evaluation of the danger represented by snakes and spiders was shown by a further experiment using pictures of rifles and revolvers. Although these objects are clearly more dangerous to modern Man than snakes or spiders, the CR that they elicited was no more persistent than the response to flowers or mushrooms.

Öhman and Seligman believe that these observations demonstrate that prepared stimuli are susceptible to particularly effective classical conditioning; and they attribute the genesis of most human phobias to this process. A closer look at Öhman's data, however, suggests that very little if any conditioning is involved in the special effectiveness of picture of snakes and spiders that he observed. For almost identical results were obtained in a procedure in which the subject was exposed to shock or the threat of shock while inspecting slides – but with no consistent temporal relationship between shocks and the appearance of a particular picture. This type of result, known as 'sensitisation' (of the response to snakes or spiders by the shocks), is usually taken to *exclude* classical conditioning, for this depends upon a precise temporal relationship between CS and UCS, allowing the subject to predict something from the former about the likely occurrence of the latter. Thus it would be more accurate to interpret Öhman's results as showing that in an existing state of fear (aroused by the threat of shock) one is more likely to react also with fear to other, prepared stimuli (snakes and spiders, but not flowers and mushrooms); and specific classical conditioning to those stimuli is *not* necessary for this to happen.

Watson allotted a central role to classical conditioning in the genesis of fears and phobias. Seligman's concept of preparedness reduces that role, by having the outcome of conditioning depend critically on the nature of the stimulus that serves as CS. And, if my interpretation of Öhman's data is correct, the role of classical conditioning in the genesis of phobic responses is reduced still further. In other words, 'prepared' stimuli are simply those that I have previously termed 'innate stimuli for fear', though they may require some additional source of emotional disturbance before the 'lurking' fear (Valentine's term) that they elicit is transformed into a full-blooded fear reaction.

It seems, then, that we can classify the stimuli for fear into those which are distal and intense; novel; characteristic of special evolutionary dangers; arising during social interaction with conspecifics; or conditioned by fear. Given this classification, it is of some interest to look again at Fig. 2.2. There it can be seen that all the stimuli which might be classed under 'intensity' and 'novelty' – noise, strange objects and persons, falling and loss of support, and sudden unexpected movement – fall in the left half of the figure: that is, they all diminish fairly rapidly with age. This is hardly surprising in terms of our analysis, since it is in the nature of an intense stimulus that, with repetition, it becomes less intense (a phenomenon known as 'adaptation'), and in the nature of a novel stimulus that it becomes familiar with repetition (a phenomenon known as 'habituation'). In contrast, as we have seen throughout this chapter, the special fears of predators and conspecifics often take time to mature, especially in advanced animals. This makes quite good biological sense: at an age at which an animal's defence must depend primarily on its parents, it can confer little or no selective advantage to be frightened of other animals. In this connection, it is worth noting that in Fig. 2.2 fear of the dark shows a gradual *rise* with age, making it more plausible to class this among the special fears of potentially dangerous situations, as suggested above.

The situations and objects feared by different people – let alone those feared by individuals from a range of different animal species – present at first sight a bewildering picture of diversity. But the principles of classical conditioning, together with the principles by which we have sought to classify the innate stimuli for fear, seem to be able to impose a degree of order on this diversity; and these two sets of principles are of considerable importance in the attempt to understand the phenomena of human neurosis, as we shall see in our final chapter.

3   The expression of fear

Psychologists are constantly faced with their own particular chicken-and-egg problem: which should come first, stimulus or response? The stimuli for fear (or for any other form of behaviour) can only be so classified in virtue of the behaviour which they evoke; but the totality of behaviour patterns displayed by a particular species can only be broken down into classes of behaviour ('fearful', 'hungry', and so on) in virtue of the stimuli which evoke them. There is, in other words, an ineluctable interdependence between the classification and definition of a set of stimuli and the classification and definition of the set of responses elicited by those stimuli. So, if we chose to begin in the previous chapter with a consideration of the *stimuli* for fear, this was an arbitrary decision; and we were able to do so only by begging the question of the way in which we recognise the fear which such stimuli evoke. In this chapter we must begin to remedy the omission.

One possible way of going about this is to see whether the experimental subject *looks* afraid. Though this may sound vague and unscientific, it is a perfectly serious suggestion. The work of the ethologists has demonstrated the importance of emotional sign-stimuli in regulating the social organisation of a variety of species. Now, to be efficient, sign-stimuli must be easily recognised by the members of the species. May we not learn, then, to recognise them too? In pursuing this line of enquiry – an enquiry into *the expression of the emotions* – we shall be interested not only in the signs of fear and related forms of behaviour (e.g. fearful threat, submission), but also in the signs of those complementary forms of behaviour to which fearful behaviour is a common social response (e.g., anger, aggression, confident threat, dominance).

As we saw in the previous chapter, animals do indeed seem to be good at recognising and reacting to the sign-stimuli emitted by their conspecifics. Recall, for example, Sackett's socially isolated infant monkey responding with fear to slides of threatening adults of its own species; or Mineka's monkeys learning from observation of their parents' reactions that a snake is something to be avoided. Recently, Seyfarth and his collaborators have made similar observations in the wild. They observed that vervet monkeys characteristically make three different alarm calls depending on which kind of a predator they have spotted – a leopard, eagle, or snake. Furthermore, in response to these calls, vervets display three different kinds of behaviour. They take to the trees in response to a 'leopard' call, they look up when they hear an 'eagle' call, and they look down when they hear a 'snake' call. The appropriateness and survival value of these reactions is clear; but they might each be triggered by sight of the actual predator rather than the alarm call. To show that this was not so, Seyfarth recorded the different calls and played them from hidden loudspeakers in the vervets' natural habitat. Even when there was no predator around, the monkeys that heard the broadcast call responded in a manner appropriate to the predator it signalled. Thus they can tell not only that the calling monkey is alarmed, but also what alarmed it.

A similar expertise in decoding emotionally expressive signals has been demonstrated in laboratory studies of 'co-operative conditioning' in rhesus monkeys by R. E. Miller. These experiments involve a 'receiver' monkey, trained to avoid foot-shock by performing a simple manual response upon receipt of a signal warning of shock delivery; this signal is transmitted by a second, 'sender' monkey, the only one of the pair allowed to perceive a primary signal of shock provided by the experimenter. In some of these experiments Miller showed that the receiver monkey readily detected signs of fear in the sender's reaction to the primary signal of shock and responded appropriately. In others the receiver monkey was trained to respond to the sight of the sender monkey in a calm state. The avoidance responding of the receiver monkey was then extinguished (by omission of all shocks). Now, for the first time, a shock was delivered to the sender monkey (and not to the receiver). When the receiver monkey saw the sender's reaction to the painful shock, he immediately re-commenced the extinguished avoidance responding. Similar results were obtained when Miller used (in place of live sender monkeys) pictures of monkey faces on a slide projector. Thus monkeys are highly skilled at detecting fear in the facial expressions of members of their own species.

These experiments show that rhesus monkeys can recognise when another monkey is frightened, and that they treat this percept as similar to a stimulus that has been associated with shock. A comparable demonstration has been made with human subjects by Orr & Lanzetta. A tone was first paired with shock in a classical conditioning paradigm, the emotional response to the tone being measured by the skin conductance response. After conditioning was complete the response to the tone was extinguished by presenting it without shocks. During extinction trials tones were always accompanied by a second stimulus, a slide in which faces were pictured with various kinds of expression.

Pictures of faces with fearful but not other kinds of expressions retarded extinction. These results suggest that, for people as well as monkeys, fearful facial expressions intensify previously acquired fear responses.

For this kind of emotional communication to have much effect it is obviously essential that we should be good at recognising each other's emotions. At one time, the experimental evidence appeared to indicate the converse, namely, that people do badly when asked to classify human emotional expressions. However, a series of studies by two groups of psychologists, one directed by Cal Izard and the other by Paul Ekman, have met with more success. Izard, for example, presented subjects with a set of photographs of facial expressions of eight basic emotions. Some of the photographs used for fear and anger are shown in Fig. 3.1. The American sample of subjects on which the photographs were originally tried out agreed in their classification of the four pictures representing fear on average about 76% of the time; the corresponding figure for anger was even higher – 89%. Even more encouraging, very similar results were obtained with other groups of judges (but the same pictures) in England, Germany, Sweden, France, Switzerland, and Greece, although the average agreement dropped to about 57% for a group of Japanese subjects and to about 50% for a group of Africans. Thus, provided the culture of the judge of emotion is not too different from that of the actor of the emotion, we seem to be quite good at recognising the significance of facial expressions.

Whether the signs of human anger act like the stickleback's red belly as a sign-stimulus for fear is still an open question. But our common experience strongly suggests that they do. This impression is supported by an experiment conducted by Öhman & Dimberg. These investigators compared classical conditioning to pictures of faces with angry and neutral expressions, respectively, as the CS, and an electric shock to the hand as UCS. The skin conductance response to pictures of angry faces was substantially more persistent than the response to neutral faces.

Darwin's principles of emotional expression

The main criterion for interpreting the signs of emotion in animals is the behaviour by which they are followed. The foundations for the scientific study of this topic were laid by Charles Darwin in 1872, in his book *The Expression of the Emotions in Man and Animals*. Darwin

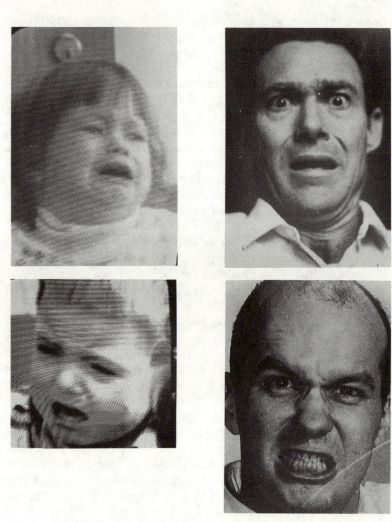

Fig. 3.1. Photographs of adult subjects (to the right) and video snap shots of children (to the left) with expressions of fear (above) or anger (below). Using photographs of this kind, Izard showed good agreement between American and European subjects in judgements of the emotion expressed by the models. However, the same judgements were only made about 50% of the time by groups of Africans and Japanese.

suggested that emotional expression could be understood in terms of three principles.

The first of these he called the '*principle of serviceable associated habits*', and much the same idea has been invoked more recently by

Tinbergen and other ethologists under the title 'intention movements'. As the English zoologist Richard Andrew writes:

Comparing the facial gestures of various primates and other mammals, Darwin concluded that their origins could be traced to primitive activities such as drawing back the lips for biting. The original acts evolved into displays that conveyed a derivative meaning (such as threat) to other animals. . . . In the course of evolution facial expressions changed in form (through changes in the facial muscles) and sometimes in meaning. Endowed, however, with the function of communication, they survived from one stage of animal evolution to the next and thus were passed along the evolutionary ladder to man. (Andrew, 1965)

Herbert Spencer (quoted by Darwin in illustration of his thesis) had written in a similar vein that 'the destructive passions are shown in gnashing of the teeth and protrusion of the claws and these are weaker forms of the actions that accompany the killing of prey'. Another example, considered in some detail by Andrew, is the flattening of the ears shown by many mammals when they are startled, in danger of attack, or about to launch a defensive attack themselves. Originally a protective response, preserving the ears from injury, this is known by anyone who owns a horse to be a reliable sign of impending trouble.

Darwin's second principle is that of *antithesis*, according to which two behavioural dispositions opposite in kind (e.g. aggressive and friendly) are expressed in ways which are also the opposite of each other. A very common example, found among fish, birds, and mammals (including Man) is the opposition between a direct stare (signifying dominance and perhaps the imminence of a confident attack) and an averted gaze (signifying submission). Examples of this principle of antithesis are shown in Fig. 3.2.

The third of Darwin's principles is that of *the direct action of the nervous system*. Under this heading he appeared to have mainly in mind the physiological adjustments which are involved in a given emotional state and which are preparatory to the undertaking of an appropriate course of action. Most of these adjustments involve the autonomic nervous system and the endocrine system, whose role in emotional behaviour we shall consider in Chapter 5.

The study of emotional expression in animals, besides being of importance in its own right, throws considerable light on the origin of the forms of emotional expression used by Man. Not unnaturally, the most interesting observations in this connection are those made with primates. The frown, for example, as a warning or indication of annoyance is used by baboons and chimpanzees, as well as Man; it is suggested by Andrew that the function of the frown is to emphasise

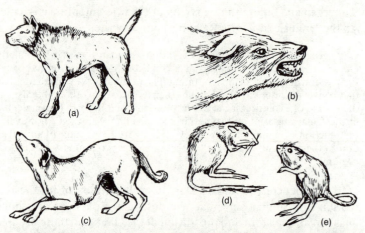

Fig. 3.2. Darwin's principles of emotional expression. The threat postures (a) and (b) include both 'serviceable associated habits' (e.g., withdrawal of the lips from the teeth) and 'movements due to direct action of the nervous system' (e.g., erection of the hair on the back). The submissive posture (c) is the reverse of threat behaviour in many respects, illustrating Darwin's principle of antithesis. A similar contrast is seen between the aggressive and submissive postures of the kangaroo rat (d) and (e).

the fixed stare which is used in confident threat. This would have come about as a result of an initial association between staring and the lowering of the eyebrows during fixed attention to a visual stimulus. The latter function is, of course, still with us in the 'frown of concentration'. Another example is the yawn. It is a common observation that a man may yawn under conditions of extreme tension, as during an examination. Monkeys appear to do the same thing. It seems likely that the significance of the yawn as an emotional display is that of exposing the teeth in fearful threat. The human pout of indignation probably had its origin, in a similar manner, in the use of this gesture by other primates as part of the expression of confident threat. A final example is that of the grin. This is used by many primates as an expression of fearful threat (as the baring of the teeth might lead us to expect). Its use by a subordinate animal to signify 'I am not hostile' and thereby ward off the attack of a more dominant animal may be at the root of the transformation of the grin and the smile into a friendly greeting in Man, and indeed it is also used in this way by some other primates (Fig. 3.3). However, Andrew points out that these expressions still retain their original function in Man, as when, for example, 'a man maintains a fixed smile in response to the verbal attacks of a superior'.

Fig. 3.3. Social greeting, with grin, by a gelada baboon.

The study of emotional expression in both animals and Man, then, has yielded some interesting and important results. It seems that we can recognise instances of a given class of emotional behaviour with some degree of accuracy. But in the final analysis, an emotional expression can only be interpreted in the light of the behaviour by which it is most often accompanied or followed. Thus, at some stage, we must consider what this behaviour is.

One simple move is to equate fear with efforts to escape or avoid the feared object. But this move begs the whole question. Certainly escape and avoidance behaviour provide one index of fear, and perhaps the most important one. We would hesitate to say that someone was afraid of a particular object or situation if he *never* made any attempt to get away from it. But we must beware of saying that the *absence* of avoidance behaviour indicates an *absence* of fear, for many other forms of behaviour may occur in response to a feared object. An animal may attack it; he may remain silent and immobile; or he may be 'paralysed with fear' (which is perhaps the same thing). Thus avoidance behaviour is *one* criterion of fear, but certainly not *the* criterion of fear. Indeed, the attempt to define fear as avoidance behaviour is usually preparatory to the conclusion – very welcome to the radical Behaviourist, who wants nothing to intervene between stimulus and response – that we have no need for the concept of fear at all, since we can always talk of avoidance behaviour instead.

It is, of course, true that if there were a single piece of behaviour which was perfectly correlated with our use of the term 'fear', and if this were the only form of behaviour correlated with it, then 'fear' *would* be redundant. It would be equally redundant if 'fear'were said to result always and only from the presentation of an independently denumerable list of stimuli. In either case, the radical Behaviourist could get on quietly with the job of describing such perfect stimulus–response relationships.

In fact, the behaviour of organisms is not so neat. Experimental psychology – as well as common sense – has been forced to invent the hypothesis of a complex psychological state, 'fear', precisely in order to make sense out of the otherwise shifting and imprecise relationships observed between stimuli and responses. But having invented this hypothetical state, it is essential that we should reliably recognise instances of its occurrence. However, this in itself is insufficient. If we are to discover the way in which the brain is involved in fear, or the way in which fear can lead to the learning of new behaviour or to the disruption of existing behaviour, or the extent to which a person's susceptibility to fear is a product either of his environment or of his heredity – if we are to do any of these vital things, we must be able to estimate the *degree* of fear. Can we do this? Can we measure fear? In the next chapter we shall see that this is indeed possible.

4   The inheritance of fear

It is a common-place observation that some people are more easily and more intensely frightened than others. Is this due to their heredity or their upbringing?

If the nature–nurture controversy is stated this way, as an either–or question, it is a pseudo-problem. All behaviour is the result of an *interaction* between genes and environment. But there is still the problem of estimating in each case exactly *how much* effect each of these sources of variation has had.

As soon as we start asking questions about 'how much', we are forced beyond the mere recognition of states of fear into the much harder task of measuring the intensity of fear. The problem of measurement is one that has constantly faced the experimentalist in all branches of science. It may sound like a purely technical matter: how to construct a foot ruler. In reality, measurement is inextricably part of the general business of theory construction. We can use a mercury thermometer to measure temperature only because of our theoretical understanding of the nature of heat and of its effects on the expansion of materials. Theoretical understanding of this depth is absent from virtually all branches of experimental psychology, and the study of fear is no exception. Thus, in what follows, we shall need to consider simultaneously the claims made for particular methods of measuring fear and the kinds of general theoretical arguments which can be used to support or refute these claims.

To say that some aspect of behaviour or of the physiological changes associated with behaviour may be used as a measure, or 'index', of fear is to make the following linked hypotheses: (1) that the species we are investigating displays a class of behaviour which can be meaningfully and usefully distinguished from its other classes of behaviour; (2) that this class of behaviour results from a particular state ('fear') of the individual animal's neuro-endocrine system; and (3) that the value assumed by the proposed index is a valid measure of the occurrence and intensity of this state.

To see how this works, let us take one particular index of fear, widely used by experimental psychologists, and consider how it has been used and how its use has been justified: the degree to which an animal, usually a small rodent, defecates under mildly stressful conditions. This measure of fear has played a prominent part in studies of the inheritance of fearfulness.

A natural question to ask is, why choose defecation in the first place? One good reason is purely anthropomorphic: it is a matter of universal experience that strong emotion, especially fear, can cause involuntary excretion, and our own culture is not the only one to have celebrated this fact in certain idiomatic expressions. Questionnaire studies have reported that during World War II the strain of battle resulted in involuntary defecation in 21% of the sample studied and involuntary urination in 9%. That this sign of fear is not limited to the human species will be vouched for by anyone who has worked in an animal laboratory.

Now this of course is only the beginning. We must provide better reasons than this in support of our hypotheses. The distinction of a class of fearful behaviour implies that we shall be able to find a number of different items of behaviour which co-vary, in the sense that conditions which make one item in this class more likely to occur also make other items in the class more likely. Thus, in order to test the validity of the hypothesis that defecation is a measure of fear we shall need to develop as well other measures or 'tests' of fear. Since these other presumed measures of fear will be as invalid as the defecation measure which we are trying to validate, it might appear that we are entering on an infinite regress. Very similar problems have arisen in the realm of intelligence testing. There a way out of the difficulty has been found by using a 'real-life' criterion of intelligence (e.g. scholastic success) for the validation of intelligence tests. However, no such criterion appears to be available for the measurement of fear.

Fortunately, the danger of an infinite regress is more apparent than real. What we must do is to develop a series of tests of fear, preferably as diverse as possible, each of which stems from some part of our common-sense beliefs about fear or from some more precisely stated theory of fear which will, no doubt, in turn spring from common sense. We must then show that: (1) the measures we take from our various test of fear co-vary; (2) these measures do not, on the whole, co-vary with the results of presumed tests of other classes of behaviour (such as general activity, aggression, sexual behaviour, etc.); (3) a measure of defecation, in particular, correlates highly with other measures of fear, but does not correlate highly with tests of other classes of behaviour.

Now there are essentially two ways of showing that different tests correlate with one another and therefore measure the same thing. In either case the degree of correlation may be expressed quantitatively in the form of a 'correlation coefficient'. This can take any value between +1 and −1. The value +1 indicates a perfect positive relationship; that is, the two measuring devices from which the two sets of scores have been obtained give exactly equivalent results and a high score on the one corresponds to a high score on the other. The value −1 also indicates perfect correspondence, but now a high score on the one test corresponds to a low score on the other. If there is no correspondence between the two measuring devices, the correlation coefficient is zero.

One way of demonstrating correlation is to apply different tests to a number of samples of behaviour under varying conditions but from the same individuals. One might thereby discover that *conditions* which make it highly probable that a rat will defecate also make it highly probable that it will run away, that its heart will beat faster, that it will try to bite the experimenter, and so on. A second way is to apply the tests to a number of different individuals under standard conditions. In this way, one might discover that a *rat* which is very likely (compared with other rats) to defecate is also very likely to have a high heart rate, to run away, and to bite the experimenter. This distinction is, of course, the one which we first met in Chapter 1: the distinction between motivation and personality. In what follows, experiments on both motivation and personality will be cited in evidence; but we shall have more to say about personality, that is, about fearfulness.

The Open Field test

The link between the study of fearfulness and the use of defecation as a measure of fear is due to Calvin S. Hall, an American pioneer in the field of 'psychogenetics', i.e. the investigation of the inheritance of behaviour. He developed a standard situation (known as the Open Field test) for the measurement of emotional defecation and used it during the 1930s in an attempt to selectively breed animals for differing levels of fearfulness. His work has been repeated and extended in more recent years by an English psychologist, P. L. Broadhurst. A photograph of Broadhurst's Open Field apparatus is shown in Fig. 4.1. It consists of an arena, nearly 3 feet in diameter, above which are mounted a battery of photoflood lamps and loudspeakers through which the animal is exposed simultaneously to bright light and loud noise. A typical procedure is to place the experimental subject (a rat) for 2 minutes at a time in this arena on 4 consecutive days. The

Fig. 4.1. A rat in the Open Field test as standardised by Broadhurst. The compartments marked out on the floor of the arena are used to calculate the amount the animal moves about (the 'ambulation score').

defecation response is measured simply as the number of fecal boluses dropped by the animal. The experimenter also keeps a record of the animal's movements and from this he is able to calculate the distance it travels (the 'ambulation' score).

Here then is our test situation. Our main interest is in knowing whether the scores it yields can be used as measures of fear. But we would be wise first to ask a different question about them: are they reliable enough to measure anything at all? In other words, do different investigators using the test (or a single investigator using it on different occasions) under similar conditions obtain similar results? If this question had been asked of certain other proposed measures of psychological attributes – the Rorschach ink blot test is a notorious example – much time, money, and false hopes would have been spared; for when (many years too late) their reliability has finally been assessed, it has turned out to be negligible.

There are various ways of assessing the reliability of a measuring device, but they all result in a correlation coefficient describing the extent of agreement between the different observers or the different occasions of measurement. It is generally reckoned that a reliability coefficient of about 0.8 is sufficiently high for a test to be useful. The values obtained for the Open Field defecation and ambulation scores are usually of this order.

Unlike Rorschach ink blots, then, the Open Field test can at least measure something. It is far harder to show that it measures fear.

As a first move it is necessary to show that defecation occurs as a response to a whole variety of situations which one would expect to elicit fear. We would obviously want to reconsider our use of defecation as a measure of fear if it were elicited only by auditory stimuli or only by tweaks on the tail. Parker looked at this problem in 1939. He recorded the amount of defecation elicited in 200 rats by six different situations: the Open Field test, a buzzer, sudden dropping, a tilting box which caused the rat to slide down an inclined plane, forced swimming, and immobilisation. The correlations obtained between the scores on every pair of tests were consistently high and positive, ranging from 0.6 to 0.9. The results were further analysed by a mathematical technique known as factor analysis. This replaces the matrix of correlations by a description in terms of a number of axes (or 'factors') to which each test may be related (Fig. 4.2). In this way it is possible to judge the extent to which a series of tests are measuring the same or different things (i.e. the same or different factors). It is factor analysis, for example, which was used to isolate the factor of general intelligence ('g') which is measured by the IQ tests. Parker's results, when analysed in this way, similarly indicated a general factor which all the tests were measuring. Since his experiment was concerned with the personality side of the personality–motivation dichotomy, we may call this factor 'fearfulness'.

Having shown that defecation is a fairly general response to a wide variety of frightening situations, our next concern is to demonstrate that the correlations between defecation scores and other features of the animal's behaviour fit the hypothesis that defecation is the result of fear. Some of the experiments which have sought for evidence of this kind have examined the correlations between defecation in the Open Field and other behaviour recorded simultaneously in this situation. Hall, for example, argued that a frightened animal should not eat in a strange environment, and there should therefore be a negative correlation between defecation and readiness to eat in the Open Field; and this is the result that he obtained.

We might similarly expect exploratory behaviour and fear to be negatively correlated with each other. There is good reason to suppose that the ambulation score obtained in the Open Field test is in part a measure of exploratory behaviour; and it is a frequent observation that the defecation and ambulation scores are negatively correlated.

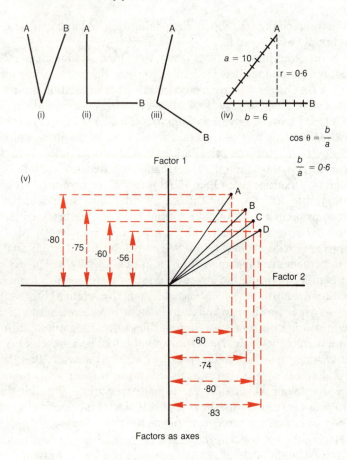

Fig. 4.2 Geometric representation of correlation (i–iv, *above*) and of factor analysis (v, *below*). The closer the correlation, the more acute is the angle between A and B: in (i) the acute angle represents a positive correlation; in (ii) the right angle represents a complete lack of correlation; and in (iii) the obtuse angle represents a negative correlation. The convention is that the cosine of the angle is made equal to the correlation coefficient, as in (iv). In (v) a set of correlations between four different measures (A, B, C, and D) are pictured together as a sheaf of arrows set out according to these conventions. The total set of correlations may be simplified by referring each measure to the axes shown in (v) as 'Factor 1' and 'Factor 2'. In the simple case shown, the intercorrelations between A, B, C, and D allow all four arrows to lie in the same plane, so that the projections of the four points on the two axes, or factors, give an accurate description of their position. In more complex cases, the pattern of intercorrelations would push one or more of the arrows into the third or higher dimensions, and three or more axes would be needed. The values obtained for the projections on the axes (dotted red arrows) are referred to as the 'loadings' of the measures on the factors.

This general interpretation of the relation between fear and exploration, and of the significance of the ambulation score, receives strong support from a factor-analytic study reported in 1967 by Whimbey & Denenberg. They found that ambulation was strongly and positively related to a factor of exploratory behaviour and (except on the first day of Open Field testing) negatively related to a factor of fearfulness. This general pattern – high fearfulness being accompanied by high levels of defecation and low levels of ambulation – has emerged (with differences in detail) in a dozen factor-analytic studies of Open Field behaviour in both rats and mice recently reviewed by Joseph Royce.

Changes in the defecation score itself are consistent with the hypothesis that it is a measure of fear: as we would expect, it goes up if we increase the physical intensity of the light and sound to which the animal is exposed, and it goes down as the animal gets used to the arena over the successive days of testing.

As well as correlating defecation with other responses in the Open Field, one may also investigate the relation between defecation in one situation and some other form of fearful behaviour in a different situation. An example of this approach is an experiment carried out by E. E. Anderson in 1938. He measured defecation both in the Open Field and while the rat was forced to wade in deep water, and also the speed with which the animal would emerge into an open space from the home cage or from the cover provided by a stove pipe (this technique is often called an 'emergence test'). In line with the hypothesis that fear may disrupt exploratory behaviour, the speed of emergence was greatest in those animals which defecated least.

The heredity of fearfulness

But it is from studies of the heredity of fearfulness that the most impressive studies relating defecation in the Open Field to other forms of behaviour have come. In fact, these studies have been so closely bound up with the Open Field test that one can regard them equally well either as establishing the extent of genetic control of fearful behaviour, or as validating the use of defecation as a measure of fear.

As we noted earlier, the problem facing the psychogeneticist is not to show whether a particular form of behaviour is 'due to' genes or 'due to' the environment, for all behaviour is due to both; but to estimate how much each of these kinds of causal influence has contributed to the final behavioural product. But it is important to note that there can never be a completely general answer to the question, 'how much?' The answer obtained must depend on the particular sample of individuals and the particular set of environmental conditions studied.

This is easy to show. Suppose your population consisted entirely of genetically identical individuals; then it *must* be the case that the total variation observed in their behaviour is due to environmental factors. Conversely, suppose you manage to provide each individual with an absolutely identical environment; then the total variation observed in their behaviour *must* be due to genetic factors. We see, then, that the more you are able to hold environmental factors constant, the more the remaining variability will be due to heredity; and the more you are able to hold genetic factors constant, the more the remaining variability will be due to the environment.

The fact that we can get no absolutely general answer to the question, 'how much is due to heredity, how much to environment?' does not mean that we should not study the problem at all. It will still be the case that for a given range of conditions in a given population, some characteristics of behaviour will be more dependent on genes or environment than others, and this may constitute important information, both from a practical point of view and in allowing one then to proceed to an analysis of the causal chain producing the behaviour in question. However, the problem of disentangling genetic and environmental effects from one another is a formidable one. It will help to begin with a clear picture of the diverse points at which the environment can modify the effects of the genes. Fig. 4.3 has been prepared to this end. The only part of this chain which can be called

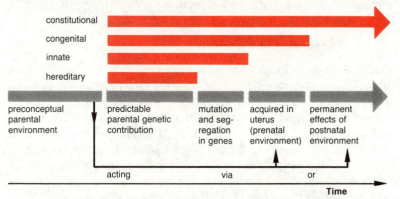

Fig. 4.3. Interaction between genes and environment in the determination of the behaviour of the adult animal. Only the segment labelled 'predictable parental genetic contribution' is a purely genetic effect. The environment may interfere at any other point in the chain, though the form taken by the environmental effects is in turn determined by the inherited nature of the system on which they are acting. The various kinds of environmental effect shown in the figure are taken up in detail in Chapter 8.

'hereditary' is the segment labelled 'predictable parental genetic contribution'; the environment may interfere at any or all of the remaining links.

Selective breeding

Both Hall and later Broadhurst have made use of a method known as selective breeding in their efforts to throw light on the heredity of fearfulness. What they did was to obtain defecation scores in the Open Field from a population of rats and then mate together high-scoring males with high-scoring females on the one hand (to form a high-scoring line called by Hall 'Emotional' and by Broadhurst 'Maudsley Reactive'), and low-scoring males with low-scoring females on the other (to form a 'Non-emotional' or 'Maudsley Nonreactive' line). In subsequent generations the procedure is repeated, high scorers within the high-scoring line and low scorers within the low-scoring line being again mated. If there is a steady separation between the mean scores obtained by each line as a function of generation of selective breeding, this is presumptive evidence that the behaviour in question is determined to some extent by the genes handed down from parents to offspring. That this separation occurred in both Hall's and Broadhurst's experiments can be seen from Fig. 4.4.

These results, however, are not sufficient by themselves to show that Open Field defecation is under some degree of genetic control. Remember that an animal with Reactive parents and therefore Reactive genes is also carried in a Reactive womb and given Reactive mothering. Thus it may differ from an animal with Non-reactive parents because of environmental influences transmitted to it by its mother either prenatally or postnatally. In order to control for this possibility we must separate the genetic mother from the prenatal mother and from the postnatal mother. This can be done for the postnatal mother by the technique of cross-fostering: at birth Reactive and Nonreactive litters are simply interchanged, so that genetically Reactive rats are provided with Nonreactive mothering, and *vice versa* (Fig. 4.5). This experiment was carried out by Broadhurst during the first six generations of selection, with the result that the animals' behaviour upon testing in adulthood in the Open Field always resembled that of their genetic parents, not their foster-mothers'.

To control for the effects of the *pre*natal maternal environment, an equally simple technique known as a 'reciprocal cross' may be used (Fig. 4.5). If a Reactive male is mated with a Nonreactive female and, conversely, a Reactive female with a Nonreactive male, the offspring of the two crosses should not, on average, differ in the genes

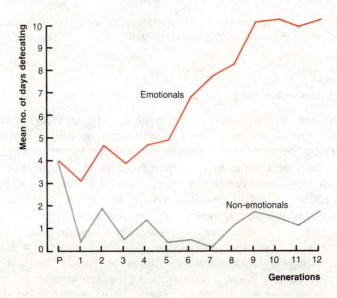

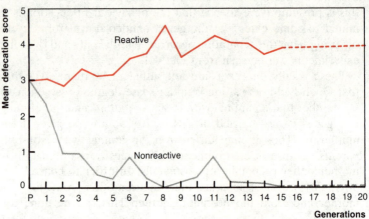

Fig. 4.4. Results of Hall's (*above*) and Broadhurst's (*below*) selective breeding experiments. Both experiments started with a 'parental' generation of rats (the points marked 'P' on the abscissa) which were tested in the Open Field (Fig. 4.1.). Subsequent generations were obtained by mating together high-defecating animals (to form the strains whose scores are shown as red lines in the figure) or low-defecating animals (black lines). As a result, the red and black lines steadily diverge. In Broadhurst's experiment, selection was relaxed between generations 15 and 20, i.e. animals were mated within existing strains but not according to their own defecation scores. The stability of the differences established by the first 15 generations of selective breeding is shown by the fact that scores for Reactive and Nonreactive strains remain as far apart at generation 20 in spite of this relaxation of selection.

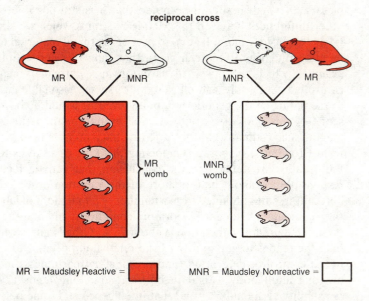

Fig. 4.5. Cross-fostering (*above*) controls for the effects of the *postnatal* maternal environment by interchanging at birth litters which are of different genetic constitution and have had different prenatal environments. The reciprocal cross (*below*) controls for the effects of the *prenatal* environment by depositing animals of the same genetic constitution (obtained here by mating Reactive males with Nonreactive females *or* Nonreactive males with Reactive females) in different sorts of womb (depending on whether the mother is Reactive or Nonreactive).

they receive from their parents. But they do differ in their pre- and postnatal environments, the former receiving the Nonreactive variety and the latter the Reactive kind. (Fathers play no part in the rearing of young rats.) Now we have already excluded postnatal maternal environmental effects, so, if differences are observed between the reciprocal crosses, this is presumptive evidence for the existence of a prenatal environmental influence on the behaviour concerned. Broadhurst carried out this experiment with the sixth, seventh, and eighth generations of the Maudsley strains as parents and, taking the results for the three generations together, there appeared to be little or no effect of the intra-uterine environment on the Open Field defecation score. It is worth noting, however, that there *was* an effect on the ambulation score, offspring of mothers of the high-ambulating Nonreactive strain themselves ambulating more than offspring of mothers of the low-ambulating Reactive strain.

The evidence is good, then, that the changes in Open Field behaviour produced by selective breeding have been due to genetic factors. (Though it should be noted that it is still possible that the other differences between the Maudsley strains which we shall now go on to consider might be due to pre- or postnatal maternal environmental influences.) One can go further than this and estimate the *extent* to which the observed variation in a particular measure is controlled genetically, as well as to make some reasonable guesses as to the nature of the genetic system involved. It would be beyond the scope of this book to go into the theory and methods involved in making such estimates. Suffice it to say that a number of experiments from Broadhurst's laboratory applying this kind of analysis to Open Field data have tended to the conclusion that genetic factors are responsible for about half or more of the variation observed in the defecation scores obtained from a number of populations of laboratory rats reared under standard laboratory conditions; and that the corresponding figure for ambulation is about 75%. Other figures which will help the reader to get an idea of the relative importance of heredity in the control of defecation (and presumably fearfulness) in the rat are that adult body weight in both rat and Man tends to have a heritability of about 75%; and that intelligence and some personality traits in Man have a heritability of about 50%.

Fearful and fearless rats?

As a result of these breeding experiments, then, we have two strains of rats, one which obtains a high Open Field defecation score, the other obtaining a low one. If defecation is a valid test of fear, these strains should differ in a consistent manner in a variety of other ways. Does the evidence bear this out?

If we first look at behaviour in the Open Field itself, we find an interesting change in the ambulation scores (which formed no part of the selection criterion) in the two Maudsley strains (Fig. 4.6). Both strains have shown an increase in ambulation as a function of generation of selection. This may be due to the one variable which is operating in the same direction for both strains, namely, inbreeding. More important for our present purposes, there is now a clear-cut difference between the two strains, Nonreactives ambulating much more than Reactives, in line with the disruptive effects of fear on exploration discussed earlier.

The performance of both Hall's two strains and the two Maudsley strains has also been investigated in a variety of situations outside the Open Field. The range and kind of differences between emotional and non-emotional strains which have been found place it beyond all reasonable doubt that the selective breeding has succeeded in affecting *some* general attribute of behaviour. What is far more difficult to prove is that this attribute is best described as fearfulness, although this is certainly the most plausible hypothesis.

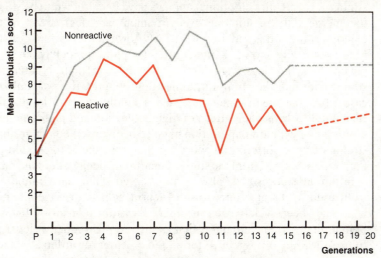

Fig. 4.6. Changes in the Open Field ambulation score (Fig. 4.1) in the Maudsley Reactive and Nonreactive strains of rats as a function of generation of breeding for high or low defecation scores. The point marked 'P' on the abscissa represents the parental generation. Selective breeding was relaxed (dotted lines) between generations 15 and 20, without effect on the differing levels of ambulation recorded for the two strains.

Defecation itself is an unlearnt ('unconditioned') response to fear stimuli. Another mode of behaviour which seems to perform a similar function, especially in situations involving conflict, is that of grooming or preening. Willingham, for example, in a factor-analytic study of the mouse, found grooming to be one of the major modes of response to fear stimuli in this species; and Cohen & Price observed an increased rate of grooming in rats which heard a tape recording of another rat screaming. I have observed a higher rate of grooming in Maudsley Reactive rats compared to Nonreactive animals when both were placed in a novel environment, but, as our hypothesis would lead us to expect, only when that environment was made relatively stressful by the inclusion of a loud noise.

When stronger fear stimuli have been used, the high-defecating animals have again shown greater responsiveness. Thus the Maudsley Reactive strain has a faster speed of escape from electric shock and a faster speed of swimming under water towards an exit to the air than their Nonreactive counterparts.

Other forms of unlearnt behaviour which are not themselves direct responses to fear stimuli may be disrupted by fear. We have already come across the disruptive effects of fear on exploratory behaviour. This relationship has appeared both in Hall's and in Broadhurst's strains of rats. Hall's emotional animals have been shown to be slower to emerge into an unfamiliar environment and less active in the exploration of this environment. My own observations of the Maudsley strains showed a higher rate of exploratory behaviour (rearing up on the hind legs to sniff at the top of the cage) in the Nonreactive strain, and, like the strain difference in grooming, this was only the case when the environment included a mildly stressful loud noise. Incidentally, this same pattern of high defecation and high grooming scores going with low exploration has also been observed in pure-bred strains of mice, offering support for the cross-species generality of the results obtained from the study of emotional behaviour in the rat.

Strain differences have also been observed when fear stimuli have been used to disrupt some form of learnt behaviour. One situation which has been widely used in this way was introduced by Estes & Skinner in 1941. The animal is first trained to press a bar or lever (or to peck at a key if it is a bird) for a food or water reward in a 'Skinner box'. While it is responding in this way, it is presented with a stimulus which has in its past experience been followed by a noxious stimulus, such as an electric shock. The consequent disruption of the animal's behaviour is measured as a fall in the rate of bar-pressing ('conditioned suppression'; Fig. 4.7). As we would expect, this reduction in bar-pres-

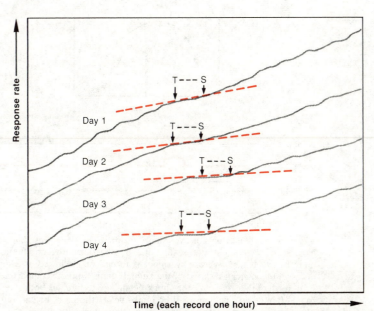

Fig. 4.7. The Estes–Skinner 'conditioned suppression' technique for measuring the intensity of fear. The animal is first trained to press a lever at a stable rate in order to obtain a food or water reward. It is then repeatedly presented with a warning signal (here a tone, marked 'T') followed by an electric shock ('S'). As a result, responding on the lever gradually ceases during the warning signal. The example shown is for a group of six rats over 4 consecutive days. By the fourth day there is almost complete suppression of response during the tone. The degree of response suppression may be expressed as a ratio of response rate during the warning to response rate in its absence, and thus provides a sensitive quantitative index of the intensity of the fear evoked by the warning.

sing rate has been found to be greater in Reactive than in Nonreactive animals. In a similar experiment investigating 'passive avoidance' behaviour (see Chapter 9), it was found that the speed with which the rat ran along a runway to a food reward was reduced by an electric shock in the goalbox to a greater extent in the Reactive than in the Nonreactive strain.

The disruptive effects of fear on learning and the performance of a learnt response can be observed even when the learning is itself based on the avoidance of punishment. Thus, although the Reactive animal is quicker to *escape* from an electric shock, it is poorer at undertaking anticipatory *avoidance* action in a 'shuttlebox' (Fig. 4.8) during the presentation of a warning signal. (We return to this finding in Chapter 12, where it is given a more detailed explanation.)

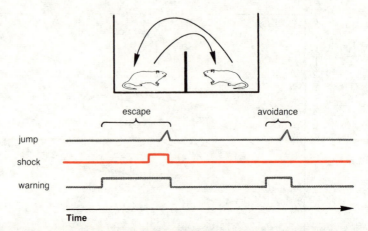

Fig. 4.8. A 'shuttlebox' for avoidance learning. This widely used apparatus consists of two compartments, separated by a barrier, with a grid floor which may be electrified. A warning signal is turned on, followed at a definite interval by an electric shock. If the animal jumps from the compartment he is in to the other one *after* the shock begins, he terminates both shock and signal (an 'escape' response). If he jumps *before* the shock begins, but after the signal is turned on, he terminates the signal and avoids the shock altogether (an 'avoidance' response).

In evaluating any proposed measure of a psychological attribute it is as important to take note of what does not correlate with it as it is to consider what does. Defecation, for example, might be heavily influenced by body weight (since heavier animals will have more faeces to dispose of) or by a tendency towards constipation; and, needless to say, if either of these possibilities were in fact the case, the usefulness of defecation as a measure of fear would be very restricted. Actually, they can both be disregarded: there is no correlation between Open Field defecation and body weight, nor between Open Field defecation and defecation in the home cage.

Of course, differences between the Maudsley strains in learning or problem-solving abilities would not be expected in a situation which does not arouse fear. This point can be brought out and given empirical support by describing two contrasting studies.

Das & Broadhurst used the Hebb–Williams maze, which consists of a number of problems designed explicitly on an analogy with human intelligence tests; these are presented to the animal only after it is fully used to the experimental situation and involve no noxious stimulation. As one would expect, there were no differences between Reactives and Nonreactives in this experiment. Furthermore, other workers have shown that two strains selectively bred for good and bad performance on the Hebb–Williams test do not differ in their Open Field behaviour. (These results should be compared with the absence of any relation between intelligence and the incidence of neurosis in Man, apart from a slight tendency for neurotic behaviour to be less common in the highly intelligent.)

Eysenck, on the other hand, imposed a very different kind of problem on the rats in his experiment. The animals were first trained to take a pellet of food from a trough when a buzzer sounded. A 'rule' was then introduced to the effect that, if the animal took the pellet immediately it dropped in the trough, it would be shocked, while if it waited for 3 seconds it could take the pellet with impunity. The adaptive solution to this problem, of course, is to wait 3 seconds and then take the food. Two kinds of maladaptive behaviour were also observed: taking the food more quickly and receiving the punishment, or not taking the food at all. Eysenck compares the latter kind of maladaptive behaviour with that shown by the human neurotic, suffering from, say, a phobia, who is seen in the psychiatric clinic; and the former kind to the behaviour of that other kind of neurotic, the criminal or 'psychopath'. Both kinds of maladaptive behaviour were more common in the Reactive strain of rats than in the Nonreactives.

This experiment must serve as the finale to the work we have discussed in this chapter; though it might serve equally well as an introduction to the complex – and vital – problems posed by the nature of human personality, problems which we shall take up again in our final chapter. For the present, it would seem reasonable to come to the following conclusions: (1) that the changes in Open Field defecation brought about by selective breeding are not just changes in responsiveness to this test situation, but indicate a more general change in a whole class of behaviour; (2) that this general change is a change in level of fearfulness; (3) therefore, that defecation under conditions of stress is a valid measure of fear; (4) that the changes produced by selective breeding are due to genetic factors; (5) therefore, that level of fearfulness is to an important extent under genetic control.

These are important conclusions.

5   The physiology of the emotions: fear and stress

Every science is based on certain fundamental assumptions which are not open to confirmation or disproof in the normal way, but are eventually justified, or discarded, in the light of the whole programme of research which the science sets itself. The postulates of Darwinism, for example, stand firm after a century of research not because they have been subjected to that process formidably called the 'hypothetico-deductive method', but because the programme of research which they inspired has vastly increased our understanding of the nature of biological evolution. Of course, this increase in our understanding has involved a large number of specific hypotheses and sub-theories which have been successfully submitted to the test of deduction and experiment. If this were not so, Darwinism would not be part of science at all, but some other occupation.

Few experimental psychologists would deny the assumptions that: (1) behaviour, feeling, and sensation are the results of specific physiological events within the organism; and (2) these events are themselves the result of a complex interaction between hereditary factors (working through genetically controlled biochemical systems) and environmental factors, past and present.

It is simple enough to state such a set of assumptions. The problem is to show, in detail, how the operation of the factors specified in them leads to particular instances of behaviour taking the form they do. Only when we have succeeded in doing this for a large number of different kinds of behaviour will our assumptions have proved their usefulness. The demonstration that fearfulness is in part due to the operation of genes, and that it may be measured to an important degree by the defecation response, has been a step along this long road.

The James–Lange theory of emotion

At the close of the nineteenth century the American philosopher and psychologist William James, and Carl Lange, a Dane, independently formulated a general theory of the relation between physiological events and emotional feelings. According to the James–Lange theory, as it is known, emotional feelings consist of the perception of the physiological changes initiated by the emotional stimulus. To feel afraid, on this view, is to feel one's heart pound, one's hair stand on end, one's breathing become faster, and so on – and nothing more. All these changes – like the defecation response – are due to the activity of the autonomic nervous system. If we include within the scope of the James–Lange theory (as did James, but not Lange) the physiological events controlled by the other great branch of the nervous system – the skeletal nervous system responsible for movements of the trunk and limbs – this theory may be stated at its most paradoxical: we do not run because we feel afraid, we feel afraid because we run.

Now, if the James–Lange theory is correct, there should be recognisable differences between the patterns of physiological change observed in the different emotions. Chapter 3 has already told us of one way in which this statement is likely to be true. We are quite good at recognising emotion in others, implying that there are characteristic expressions by which each emotion can be so recognised. Moreover, in line with the James–Lange theory, recent experiments have shown that felt emotion is influenced by just such emotional expressions. The subjects of these experiments have been requested (muscle by muscle, and without the use of emotionally loaded words) to set their faces into what is in fact (but unbeknown to them) an angry, happy, sad, or fearful expression. They are then asked to check off on a list of adjectives the ones that best describe their current emotional state ('sad', 'tense', 'worried', etc.). There is a definite bias towards checking off adjectives that correspond to the emotional expression that has been induced in the subject's face.

The autonomic and endocrine systems

In experiments that have studied the physiological events that are controlled by the autonomic nervous system and the endocrine system, however, the emotion-specific differences postulated by the James–Lange theory have proved hard to detect. To make clearer what we are talking about, Fig. 5.1 shows a map of the autonomic nervous system, subdivided into its 'sympathetic' and 'parasympathetic' branches: and Fig. 5.2 pictures the endocrine glands as they are situated in the human body, and the chief interrelations between the glands. An endocrine gland is an organ which secretes hormones into the internal environment of the body tissues. Hormones are

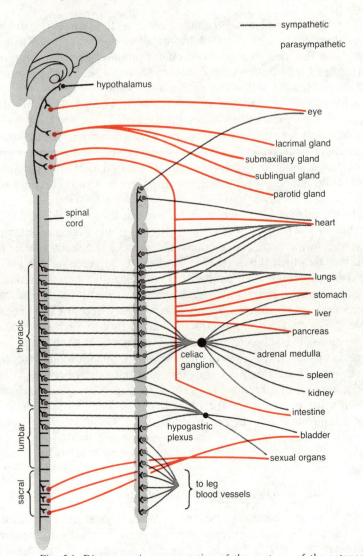

Fig. 5.1. Diagrammatic representation of the anatomy of the autonomic nervous system, divided into its sympathetic and parasympathetic branches.

specialised biochemicals which are able to regulate various functions of the bodily organs (often including other endocrine glands) which they reach by travelling through the blood stream.

The relations between the endocrine glands and the central nervous system are mediated particularly by the anterior pituitary gland (also known as the adenohypophysis). This is controlled by neurohormones (know as 'releasing factors') that are manufactured in the hypothalamus, a part of the brain situated immediately above the anterior pituitary (Fig. 5.3). These releasing factors are secreted by hypothalamic neurons into a separate, local segment of the blood stream (the portal stalk) which takes them rapidly to the anterior pituitary. This gland is thereby stimulated to secrete into the general blood supply various 'tropic' (or 'trophic') hormones, such as thyroid-stimulating hormone (TSH), adrenocorticotropic hormone (ACTH), etc. (Fig. 5.2), the particular tropic hormone released depending upon the particular releasing factors arriving from the hypothalamus. The outer part of the adrenal gland, or adrenal cortex (Fig. 5.4), the thyroid, and the gonads or sex glands (ovary or testis) are controlled by the anterior pituitary in this way.

The nervous and endocrine systems come into close contact at two other points. The posterior pituitary, or neurohypophysis, is connected to the hypothalamus by nerve-fibres (Fig. 5.3) and is therefore under direct nervous control. And the adrenal medulla (the central part of the adrenal gland: Fig. 5.4) is controlled by nerve-fibres coming from the sympathetic branch of the autonomic nervous system. The sympathetic nervous system (and, indeed, the whole of the autonomic nervous system) is itself under the direct control of the brain, of which the most important portion is again in the hypothalamus. It is significant that it is the hypothalamus which controls and coordinates the activities of the autonomic nervous system and the endocrine system, for this part of the brain is intimately concerned in the regulation of all forms of emotional behaviour.

The way in which these systems respond to stimuli of a kind causing pain, rage, or fear has been studied in considerable detail by physiologists and endocrinologists during this century. Two notable workers in this field were Walter B. Cannon, who concentrated on the description of the organism's immediate responses to such stimuli, and the late Hans Selye, whose 'general adaptation syndrome' includes, besides Cannon's immediate 'emergency reaction', the long-term adjustments called forth by prolonged stress.

The emergency reaction

The emergency reaction, as described by Cannon, is due to the sympathetic nervous system acting in conjunction with the hormones secreted by the adrenal medulla (known as 'adrenaline' and 'noradrenaline' in England, but as 'epinephrine' and 'norepinephrine' in America). Its function is to mobilise the body's resources for the swift action – 'fight or flight' – that may be needed. There is an increase

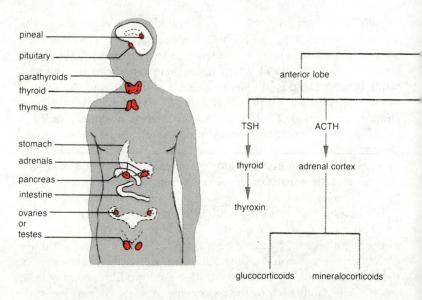

Fig. 5.2. The location of the endocrine glands in the human body and the functions of the pituitary. TSH, thyroid-stimulating hormone; ACTH,

in the rate and strength of the heart beat, allowing oxygen to be pumped round more rapidly; contraction of the spleen, releasing stored red blood cells to carry this oxygen; release of stored sugar from the liver for the use of the muscles; redistribution of the blood supply from the skin and viscera to the muscles and brain; deepening of respiration and dilation of the bronchi, to take in more oxygen; dilation of the pupils, perhaps to increase visual efficiency; an increase in the blood's ability to seal wounds by coagulating; and a rise in the supply of the special blood cells known as 'lymphocytes', whose function is to help repair damage to the tissues. All this takes place in a matter of seconds or minutes. Under modern conditions, it is often too late and in any case unnecessary, as the danger probably calls for neither fight nor flight. You have almost certainly experienced this if you have narrowly missed an accident while driving and your heart starts pounding several seconds after you have steered into safety.

Adrenaline and noradrenaline are monoamines, that is, substances that contain a single NH_2 group in their chemical composition. Two classes of monoamines are of particular importance in the endocrine and nervous systems: the catecholamines (including adrenaline and noradrenaline) and the indoleamines (of which the most important

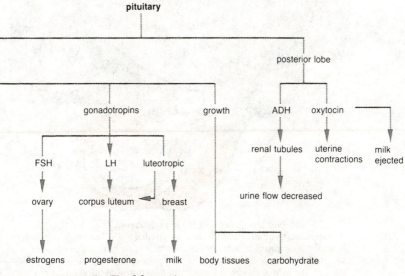

Caption for Fig. 5.2 (*cont.*)

adrenocorticotropic hormone; FSH, follicle-stimulating hormone; LH, luteinising hormone; ADH, antidiuretic hormone, also known as vasopressin.

representative is 5-hydroxytryptamine or serotonin, a substance we shall meet later). The biosynthetic pathways that produce the catecholamines are all linked; they start off with a simple amino-acid, tyrosine, that is absorbed from the diet. As shown in Fig. 5.5, the first steps lead to the synthesis of dopamine, which is a major transmitter substance (see below) in the brain. The critical enzyme that controls the rate at which dopamine is synthesised is tyrosine hydroxylase; since both noradrenaline and adrenaline are then synthesised from dopamine, this is also the rate-limiting enzyme for their synthesis. A second important enzyme is dopamine-β-hydroxylase, which controls the synthesis of noradrenaline from dopamine. Finally, the enzyme phenylethanolamine-*N*-methyltransferase (PNMT) is responsible for the synthesis of adrenaline from noradrenaline. These synthetic steps all take place within the cell concerned: different types of cells contain only the machinery for the production of dopamine (dopaminergic cells), or in addition the machinery that will produce noradrenaline (noradrenergic cells) or adrenaline (adrenergic cells).

Noradrenaline, as well as being released from the adrenal medulla, acts as a neurotransmitter in the sympathetic nervous system and brain. Nerve cells or 'neurons' are usually separated from their target

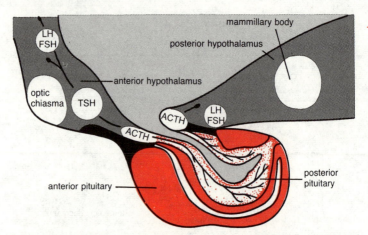

Fig. 5.3. The pituitary gland and the hypothalamus. The posterior pituitary is under direct neural control from the hypothalamus. The anterior pituitary is under the control of hormones (known as 'releasing factors') which are secreted by the hypothalamus and travel to the anterior pituitary in the blood stream. The role of the hypothalamus in the control of the anterior pituitary has been shown by experiments in which (a) electrical stimulation of areas of the hypothalamus causes the release of hormones from the anterior pituitary; or (b) lesions to the hypothalamus block their release. The areas of the hypothalamus where such results have been obtained are indicated in the figure by the abbreviations for the appropriate hormones: LH, luteinising hormone; FSH follicle-stimulating hormone; TSH, thyroid-stimulating hormone; ACTH, adrenocorticotropic hormone.

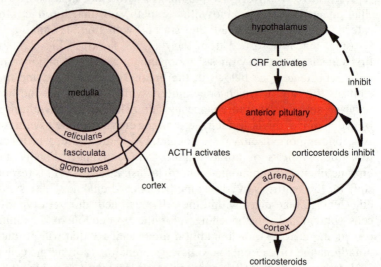

Fig. 5.4. Diagrammatic representation of the components of the adrenal gland and regulation of the adrenal cortex. CRF, corticotropin releasing factor; ACTH, adrenocorticotropic hormone.

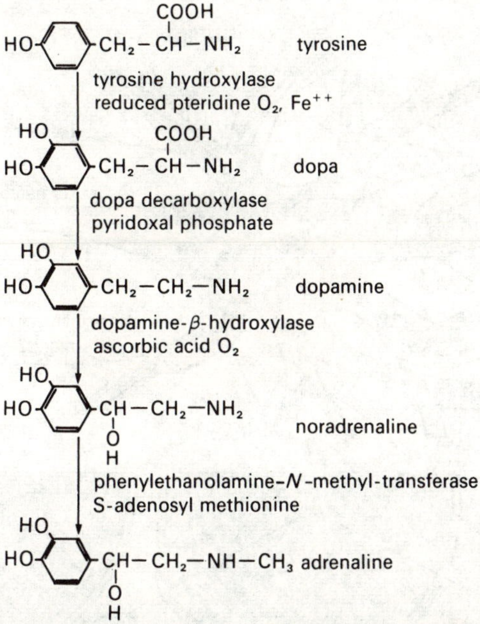

Fig. 5.5. The biosynthesis of adrenaline from tyrosine.

cells (be they muscle, endocrine, or other nerve cells) by a very small space or cleft called the synapse (Fig. 5.6). Thus, in the brain, which is made up of millions of discrete neurons laid out in ramifying networks of extraordinary complexity, each neuron is separated from the next by a synapse. Neurons operate or 'fire' by propagating a wave of electrochemical energy along their length, from the cell-body via a long shaft termed the 'axon' to their terminals at the synapse. A cell which has just fired releases into the synaptic cleft a small packet of a chemical substance (the neurotransmitter) which affects the firing of the next, or 'post-synaptic' neuron. Many different substances are known or suspected to act as transmitters in this way, including noradrenaline, dopamine, serotonin, and acetylcholine (which, among other things, is the major transmitter at the neuromuscular junction). Some transmitters are excitatory, that is, they increase the firing rate of the post-synaptic cells; others are inhibitory, decreasing post-synaptic firing rate; still others have more complicated effects, sometimes described as 'neuromodulation' rather than 'neurotransmission'; and many substances have different kinds of effects depending upon the

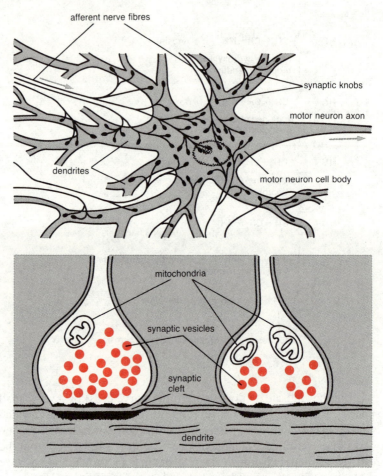

Fig. 5.6. Motor neuron cell body (*above*) and branches, called dendrites, are covered with synaptic knobs, which are the terminals of axons, or impulse-carrying fibres, from other nerve cells. The axon of ech motor neuron, in turn, terminates on a muscle fibre. The synaptic knobs (*below*) are designed to deliver short bursts of a chemical transmitter substance into the synaptic cleft, where it can act on the surface of the membrane of the nerve-cell below. Before release, molecules of the chemical transmitter are stored in numerous vesicles, or sacs. Mitochondria are specialised structures that help to supply the cell with energy.

cell in which they are located and the post-synaptic cell upon which they act. Whatever their effects, transmitters normally act upon specialised receptors located on the membrane that encloses the post-synaptic cell; these receptors are clustered particularly strongly in the synapse. In the sympathetic nervous system noradrenaline acts as the transmitter for the post-ganglionic nerves, that is, those which innervate the various target organs listed to the right of Fig. 5.1. The pre-ganglionic nerves (i.e. those that emerge from the spinal cord and synapse with the noradrenergic post-ganglionic nerves; Fig. 5.1) are cholinergic, that is, they use acetylcholine as their transmitter.

The general adaptation syndrome

For the more long-term responses described by Selye, the centre of attention passes from the adrenal medulla to the adrenal cortex and to the anterior pituitary which is responsible for activating the adrenal cortex. Selye has divided the hormones of the adrenal cortex into two classes: those which favour inflammatory processes ('pro-phlogistic') and which are also concerned with the metabolism of mineral substances (hence 'mineralocorticoids'); and those which hinder inflammatory processes ('anti-phlogistic') and are concerned with sugar metabolism ('glucocorticoids'). It is the latter class of adrenocortical hormones which is chiefly involved in the 'stage of resistance' said by Selye to follow the initial stage of the general adaptation syndrome ('alarm reaction') if stress is prolonged (Fig. 5.7). The

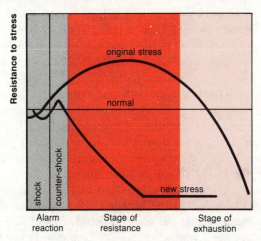

Fig. 5.7. Selye's general adaptation syndrome. After the initial 'alarm reaction', there follows a 'stage of resistance' during which resistance to the original stress (which continues to act) is increased, but resistance to other forms of stress is lowered. If the stress continues long enough, the stage of resistance is replaced by a final 'stage of exhaustion', with a catastrophic decline in resistance to all forms of stress.

release of glucocorticoids from the adrenal cortex is under the control of ACTH, while the release of ACTH from the anterior pituitary is itself largely controlled by corticotropin releasing factor secreted into the portal stalk by hypothalamic neurons. Corticotropin releasing factor also causes the anterior pituitary to release one of the major endogenous opiates, β-endorphin. This is probably one way in which conditioned fear stimuli and other forms of stress give rise to endogenous analgesia (see Chapters 2 and 9).

Recent work from Julius Axelrod's laboratory has demonstrated that while corticotropin releasing factor plays a central role in regulating the release of ACTH and β-endorphin, it is far from being the only substance involved. Indeed, the regulation of ACTH release turns out to be extremely complex, involving interactions between many different factors, including adrenaline and noradrenaline (both of which stimulate ACTH release at the level of the pituitary), several additional hypothalamic neurohormones (including vasopressin, which is secreted by hypothalamic neurons as well as by cells in the posterior pituitary), and a number of influences acting upstream to the hypothalamus. Axelrod's summary of the major findings in this field is illustrated in Fig. 5.8. Additional controls from higher brain centres are not shown in this figure. One important control of this kind involves the hippocampus (see Fig. 13.9). This structure contains large numbers of glucocorticoid receptors and appears to take part in the negative feedback loop by which high levels of circulating glucocorticoids inhibit further ACTH release. There is evidence that modification in the functioning of this loop may underlie some of the effects of early experience upon adult emotional behaviour (see Chapter 8).

The glucocorticoids include hydrocortisone, corticosterone, and cortisol. They are able to promote the transformation of non-sugars into sugars and to increase the deposition of sugar in the liver. In this way they continue the job, started during the alarm stage, of providing the body with rapidly mobilised sources of energy. They also facilitate the reactions of the blood vessels to adrenaline and noradrenaline, further increasing the potentiality for strenuous action if a second immediate danger is added to the background of continuing stress. It is much more puzzling that the glucocorticoids should *hamper* the inflammatory processes by which the body combats damage to its tissues, and *reduce* resistance to infection. Their anti-inflammatory powers may be familiar to you in such medical applications as the treatment of arthritis. In experimental animals, injection of these hormones has been shown to delay the growth of new tissue round

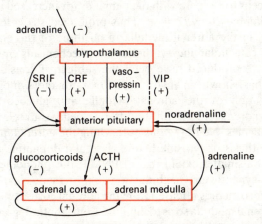

Fig. 5.8. Multihormonal control of ACTH release. Adrenocorticotropic hormone (ACTH) release from the anterior pituitary is regulated by a number of different hormones. Corticotropin releasing factor (CRF) and vasopressin, which are present in hypothalamic neurons, reach the anterior pituitary by a portal system and stimulate (+) ACTH release. The secretion of the releasing factors may be inhibited (−) by an adrenaline input to the hypothalamus. Hypothalamic vasoactive intestinal peptide (VIP) stimulates ACTH release from both human and mouse tumour cells, suggesting a possible role of this peptide as a CRF-like substance. Neuronal noradrenaline or adrenaline released from the adrenal medulla can also evoke ACTH release. ACTH stimulates the synthesis of glucocorticoids in the adrenal cortex. Glucocorticoids can stimulate the synthesis of adrenaline or inhibit ACTH release. Glucocorticoids can either act directly on the anterior pituitary to inhibit the formation of messenger RNA for ACTH or inhibit ACTH release. In addition, these steroids can act in the hypothalamus to effect CRF release (not shown). Somatostatin (SRIF) can block the evoked release of ACTH from tumour cells, suggesting a role for this peptide in ACTH secretion. From Axelrod & Reisine (1984).

a wound or a foreign body which has penetrated the skin; to inhibit the formation of antibodies; and to decrease the number of white blood cells (lymphocytes and eosinophils) and the weight of the thymus gland and other lymphoid tissues, all involved in fighting tissue damage and infection. A further frequent observation is the production of ulcers in the stomach and gastro-intestinal tract, though their mechanism of causation is obscure. We shall have more to say later about these odd ways of meeting stress.

During the resistance stage of the general adaptation syndrome important changes take place in the functioning of the biochemical pathways that control the synthesis of the catecholamines (Fig. 5.5). These changes appear to have the function of increasing the capacity of the relevant cells to keep up with the demands for increased release of the catecholamines they contain. Thus prolonged stress leads to increased levels in the adrenal medulla of all three of the enzymes responsible for catecholamine synthesis. The mechanisms by which these changes are produced have been the subject of considerable research. This has shown an involvement of both the sympathetic nervous innervation of the adrenal medulla and glucocorticoids released by the adrenal cortex, but to differing degrees in relation to the different enzymes. Neural impulses reaching the adrenal medulla increase the synthesis of tyrosine hydroxylase and dopamine-β-hydroxylase, and to a smaller extent that of PNMT; this phenomenon is known as 'trans-synaptic enzyme induction', a term made clear below. In addition glucocorticoids released from the adrenal cortex by ACTH act upon the adrenal medulla so as to inhibit the enzymatic degradation of PNMT and, to a lesser extent, of tyrosine hydroxylase and dopamine-β-hydroxylase. These mechanisms are summarised in Fig. 5.9.

Similar changes take place in noradrenergic neurons in the sympathetic nervous system. Chronic stress leads to an increase in the numbers of molecules of both tyrosine hydroxylase and dopamine-β-hydroxylase. This effect does not occur if the cholinergic pre-ganglionic nerves are first cut. So, here as in the adrenal medulla (Fig. 5.9), the effects of chronic stress upon these two enzymes are mediated by the neural input from the pre-synaptic cell (trans-synaptic enzyme induction). Phenomena of a similar kind have also been described in the brain, as we shall see in Chapter 13, and they may underlie certain kinds of behavioural adaptation to stress (Chapter 9).

Other changes produced by prolonged stress affect bodily growth and metabolism. There is a depression of activity in the thyroid gland, probably arising from a decrease in the secretion of thyroid-stimulating hormone by the anterior pituitary. There is also a marked inhibition of body growth. In rodents (though apparently not in dogs or primates) this effect is probably due to reduced secretion by the anterior pituitary of growth hormone (somatotropic hormone), itself due to the release of somatostatin, which exercises inhibitory control over growth hormone. The decreased activity of the thyroid and direct effects of adrenocortical hormones may also play a part in the inhibition of body growth seen under conditions of prolonged stress.

A very important part of the resistance stage of the general adaptation syndrome is the suppression of a number of bodily functions connected with sexual and reproductive behaviour. In the male, there is a fall in the production of spermatozoa; a reduction in the secretion

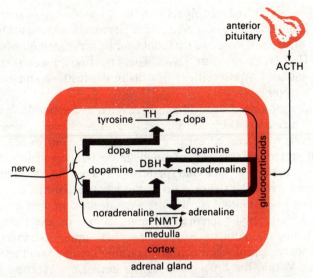

Fig. 5.9. Regulation of catecholamine biosynthesis in the adrenal medulla. The expression of catecholamines in the adrenal medulla is regulated at a number of different steps in the biosynthetic pathway. Tyrosine hydroxylase (TH) activity, the rate-limiting step in dopamine synthesis, is affected by nerve activity and to a minor extent by glucocorticoids of the adrenal cortex. Dopamine-β-hydroxylase (DBH) activity is regulated by nerve activity and glucocorticoids. Phenylethanolamine-*N*-methyltransferase (PNMT), the enzyme that converts noradrenaline to adrenaline, is predominantly regulated by glucocorticoids and to a small degree by nerve activity. Glucocorticoid synthesis is stimulated by ACTH which is released from the anterior pituitary. From Axelrod & Reisine (1984).

by the testis of testosterone, the male sex hormone or 'androgen'; and delay or complete suppression of puberty. In the female, there is disruption or suppression of the menstrual cycle in primate species or of the equivalent estrous cycle in sub-primates ('estrus' is the technical term for 'heat'); a decrease in the weight of the uterus; failure to ovulate or failure of the fertilised ovum to become implanted in the womb; an increase in the number of spontaneous abortions; and a failure in lactation, which can result in the death or slowed growth of the young. These drastic effects of prolonged stress on reproductive function have not yet been fully explained. They may in part be due to a reduction in the secretion by the anterior pituitary of gonadotropic hormones (follicle-stimulating hormone, FSH, and luteinising hormone, LH; Fig. 5.2) which stimulate the testis to secrete testosterone and stimulate the ovary to secrete estrogen and progesterone. A role may also be played by the anterior pituitary hormone, prolactin, which is rapidly and reliably released under conditions of stress. However, Collu *et al.* have observed a marked reduction in the testosterone levels of stressed male rats that could not be accounted for by changes in FSH, LH, or prolactin levels. This effect is apparently due to a stress-induced hyposensitivity of cells in the testis to stimulation by gonadotropins.

Under prolonged stress, then, there is a massive shutdown of those bodily activities which are directed towards growth, reproduction, and even resistance to existing infection, in favour of mechanisms which promote readiness for immediate high-energy action. And this general pattern – like the pattern observed in the immediate emergency reaction – is seen no matter what kind of stress is applied, whether it be physiological (surgery, injection of foreign protein, anaesthesia), environmental (extreme cold), or psychological (a threatening predator, intense competition among members of the same species, prolonged conflict, learning how to avoid an electric shock). Within the broad lines of the emergency reaction and the general adaptation syndrome, however, attempts have been made to show that, as the James–Lange theory would lead us to expect, specific patterns of physiological response are characteristic of particular emotions.

Physiological differences between fear and anger?

It has been claimed, for example, that fear is more dependent on adrenaline secretion and anger on the secretion of noradrenaline. Some evidence in support of this claim came from a group of heroic human subjects studied by Ax in 1953. They had been swathed in wires and electrodes used for the physiological recordings when Ax informed them – in order to induce a state of fear – that there was a 'high-voltage short-circuit' in the apparatus. Ax's method for inducing anger sounds equally effective: a technician came into the room and, on various slight pretexts, insulted everyone in sight, including the experimental subject. The physiological changes observed were different in the two cases: they were like those produced by adrenaline injections in the state of fear, while they resembled those produced by a simultaneous injection of adrenaline and noradrenaline when the subjects were angered.

Funkenstein obtained results similar to Ax's by using a drug called 'mecholyl'. This drug has different effects on the raised blood pressure produced by injections of adrenaline and noradrenaline respectively and may therefore be used to distinguish between high levels of blood pressure caused by the spontaneous secretion of the one hormone or the other. When it was administered to psychiatric patients whose predominant emotion appeared to be fear, the drop in blood pressure which occurred was similar to the response to mecholyl after *adrenaline* injections. In patients whose predominant emotion appeared to be anger, the response to mecholyl was like the response to this drug after *noradrenaline* injections.

More recent research, however, has failed to substantiate these claims. The Swedish psychologist Marianne Frankenhaeuser measured the urinary excretion of adrenaline and noradrenaline under a wide variety of experimental conditions. She found that both substances were excreted in many different emotional states, including both anger and fear, and also while the subjects were watching a funny film or playing a game of chance. Noradrenaline differed from adrenaline only in requiring a high level of emotional arousal for its release. Frankenhaeuser also found no difference between the emotional states elicited by infusions of adrenaline and noradrenaline, respectively, although the effects of noradrenaline were generally less intense than those of adrenaline. Nor could she find any difference in the excretion of either hormone as between predominantly anxious or predominantly aggressive individuals. This pattern of findings, then, suggests that adrenaline and noradrenaline are both released during *any* state of emotional arousal.

Nonetheless, Funkenstein's experiments with mecholyl suggest that there are different ways in which blood pressure may rise in the emotions of fear and anger. The different distributions of blood in these emotional states are also apparent to the more casual observer. We are all familiar with the red face of anger and the pale face of fear. However, it is less obvious that the distribution of blood to the *stomach* also depends on the nature of the emotional state one is in.

Some of the most interesting observations on the involvement of the stomach in the diverse emotions come from the work of Wolf & Wolff on 'Tom'. Tom was a New Yorker of Irish origin who had suffered severe damage to the oesophagus from drinking hot clam chowder at the age of nine. It proved necessary to bring a portion of the lining of his stomach out through an opening in the abdominal wall, and it was through this opening that Tom had to feed himself by putting food directly into the stomach. Tom's misfortune was a fortunate accident for science, since it allowed direct observation of the changes wrought in the mucous lining of the stomach by the spontaneous or deliberately provoked occurrence of the different states of emotion.

It appears from these observations that the stomach has two basic, and opposing, patterns of reaction under different emotional conditions. One could be described as a pattern of *increased* function, with increased blood supply, engorgement of the mucosa, increased secretion of hydrochloric acid, and increased activity of the stomach muscles. The other is a pattern of *decreased* function, the mucosa going pale as blood is withdrawn, and both acid secretion and muscular activity undergoing diminution. There is a parallel between these reactions and those seen in the rest of the body: when the face goes red, as in rage, so does the stomach lining; and the pallor of fear occurs in the stomach as in the face.

We should not, however, over-emphasise the specificity of the physiological changes which have so far been correlated with the various emotional states. Even the limited physiological differences between fear and anger are perhaps better described as a more general distinction between states of activity and states of passivity. This view is supported by the fact that when psychiatric patients whose predominant emotion is that of depression are given the mecholyl test, their blood pressure responds in the same way as it does in people whose major emotion is fear; and Tom's stomach, too, was pale during states of depression just as it was at times of fear. Depression and fear are at first sight qualitatively quite different emotions, but they do share a behavioural passivity which contrasts sharply with the greater activity of anger. This inference is supported by a study carried out by Elmadjian, Hope & Lamson on hockey players, some of whom actually took part in the game while others simply watched. There were large increases in noradrenaline and smaller increases in adrenaline excretion during active play, but increased secretion of only adrenaline among those who just watched.

Failure of the James–Lange theory

The rather limited degree of correlation between particular emotions and events in the autonomic nervous system and the endocrine system implies that we should look elsewhere for the cues people use to identify their emotional state. One suggestion has come from the work of Stanley Schachter and his colleagues at Columbia University, New York.

Schachter's group observed the behaviour of people injected with adrenaline in a variety of different circumstances. They found that this hormone enhanced the intensity of emotion generally, but *which* emotion was experienced depended on the total experimental situation: if the subject was with someone acting angrily, he too experienced anger; in the company of someone playing the fool, he experienced a kind of euphoria; and if he was watching a slapstick film, he found it particularly amusing. Furthermore, the subjects' behaviour, as well as their judgements of their own emotional state, were quite different depending on whether they had been informed, not informed, or misinformed about the nature of the drug which they had been given. These results suggest then that physiological sensations are only part of the evidence on which we base a judgement of the mood we are in; also important is our knowledge of the circumstances leading up to and surrounding our sensations.

Unfortunately, recent efforts to replicate Schachter's findings have met with limited success. Marshall & Zimbardo repeated the essential features of Schachter's original experiments. But they found that physiological arousal induced by adrenaline was not affectively neutral, as proposed by Schachter. Rather, there was an increased likelihood that the subject would report *negative* emotions (anxiety, anger, etc.) even when he was in the company of someone playing the fool. Christina Maslach used an interesting alternative method to induce the symptoms of physiological arousal normally caused by adrenaline injection. She gave instructions under hypnosis that the subject of the experiment would experience these symptoms in response to a particular signal, but would not recall the instructions themselves. The signal was given when the subject (no longer hypnotised) was in the company of someone playing the fool or acting angrily, as in the original Schachter experiments. The hypnotic suggestions were successful in creating a state of physiological arousal in response to the signal (increased heart rate and changes in skin resistance characteristic of the response to adrenaline), and also in producing post-hypnotic amnesia for the instructions. As in the Marshall & Zimbardo study, the effect of this increased physiological arousal was to increase reports of *negative* emotions (e.g. sad, angry, apprehensive) rather than positive ones (happy, peaceful, confident), and this was so whether the subject was left in the company of an angry or euphoric partner. These findings rather strongly suggest that the kinds of physiological changes that are produced by adrenaline (increased heart rate, tremor, sweaty palms, etc.) are not affectively neutral, but are used as evidence that one's emotional state is an unpleasant one.

Evidently, the environmental situation in which physiological arousal occurs does not play the unfettered role in determining emotional experience that Schachter proposed. This role is still further constrained in that we interpret the situation at least partly in the light of the physiological arousal to which it gives rise (as proposed by James and Lange). An elegant experiment by Valins illustrates this point. He had young men listen to what they took to be the amplified sound of their own heart beat as they looked at pictures of attractive women. In fact, the information about heart beat had been falsified: while the subject inspected certain pictures, his heart rate was made to appear faster than it actually was. Later on the men were asked to rate the attractiveness of the women in each picture. They judged the pictures which had been accompanied by false heart rates as more attractive than the others. Thus they must have decided that those women were attractive *because* they (apparently) made their hearts race – just as the James–Lange theory would predict.

But this small success for the James–Lange theory is insufficient. It remains true that the physiological changes which accompany the different emotions have too little specificity to define them. Even in the experiments by Marshall & Zimbardo and by Maslach which ran counter to Schachter's predictions, the emotions that were induced by peripheral physiological arousal were diffusely negative – they were not precisely 'fear' or 'anger' or 'depression'. Given the available evidence, then, we must conclude that physiological arousal can heighten emotional experience, and that (if it is of the adrenaline/noradrenaline variety) it tends to make that experience an unpleasant one; but it does not uniquely specify the nature of the experience. Conversely, the situation in which the arousal occurs plays some role in determining the nature of the emotional experience (or, more accurately, one's report of the nature of the experience); but it too fails uniquely to specify it.

Peripheral physiological arousal, then, is not sufficient for a particular emotional experience. Nor is it necessary for one. Human beings deprived by injury to the spinal cord of the greater part of the sensory influx from the internal organs nevertheless report emotional experiences. And animals so deprived experimentally continue to manifest emotional behaviour. The important experiments of Wynne & Solomon have shown that dogs lacking the use of the sympathetic nervous system, or even the entire autonomic nervous system, can learn to avoid a shock by jumping to and fro in the shuttlebox described in Chapter 4 (Fig. 4.8); and, if surgery is carried out after the learning of this response is complete, they do not differ from normal animals in their persistence in performing it. Other experiments have used chemical means to destroy the sympathetic nervous system. This may be done by injecting young animals with an antiserum which prevents the growth of the sympathetic nerves (antiserum to nerve growth factor), or with a toxin (6-hydroxydopamine) which has a special affinity for these nerves. Mice or rats treated in these ways, like Wynne & Solomon's dogs, can nevertheless learn to avoid shock.

Given the results of these experiments, the temptation is to let the pendulum swing so far from the James–Lange theory as to deny that the autonomic nervous system has anything at all to do with fear and fear-motivated behaviour. There are many other experiments, however, which demonstrate that this conclusion would be too drastic: though the autonomic nervous system and the endocrine system are not indispensable for the experience of fear or the occurrence of fearful behaviour, they are certainly involved in them.

The degree of this involvement appears to vary with the complexity of the task. Here we must make some preliminary distinctions which are not fully dealt with until Chapters 9–11. First, we distinguish between escape and avoidance behaviour. In an escape task the animal cannot prevent the onset of a painful stimulus (usually an electric shock), but it can terminate the stimulus (e.g. by running). In an avoidance task, the animal can prevent the painful stimulus from occurring by making an appropriate response before it is due to occur. Second, we distinguish between active and passive avoidance. In passive avoidance tasks, the way to avoid shock is to refrain from making a response (which can often be achieved simply by remaining still). In active avoidance tasks, the animal must perform a specified response (e.g. running or pressing a bar) to avoid shock. Finally, we distinguish between two kinds of spatial avoidance tasks: one-way and two-way. In a one-way task, the animal avoids shock by running from one place in which the shock is presented to another which is always safe. The two-way task is exemplified by the shuttlebox: to avoid shock the animal must continually return to a place from which it has just run away (see Fig. 4.8).

Now, roughly speaking, escape and passive avoidance are simpler than active avoidance, and one-way active avoidance is simpler than two-way active avoidance. And, also roughly speaking, the effects of damage to the sympathetic nervous system are greater, the greater the complexity of the task. Wynne & Solomon, using the shuttlebox, found that although dogs with surgical destruction of the sympathetic nervous system were able to learn avoidance behaviour, they were slower than normal animals, showed less sign of emotional disturbance, and (if surgery was carried out before learning) were far less persistent in repeating the avoidance response over a long series of trials. King similarly observed a considerable impairment in shuttlebox avoidance in rats in which the sympathetic nervous system had been destroyed with 6-hydroxydopamine; but one-way active avoidance was impaired only marginally and escape performance not at all. Van-Toller used antiserum to nerve growth factor to destroy sympathetic nerves in mice; passive avoidance behaviour was normal in these animals.

The same pattern has emerged after removal of the adrenal medulla, the part of the endocrine system which is directly controlled by the sympathetic nervous system and which is responsible for the secretion of adrenaline: shuttlebox avoidance is impaired, but passive and one-way active avoidance usually are not. These results clearly indicate that the sympathetic nervous system and the adrenal medulla are involved in avoidance behaviour. But they are incompatible with a primary role in the determination of fear as such: loss of fear would be expected to reduce avoidance behaviour of all kinds, independently of particular experimental circumstances or the particular manner in which the animal must act to avoid a painful stimulus.

The sympathetic nervous system and the adrenal medulla are the executive organs for Cannon's emergency reaction; but Selye's general adaptation syndrome has also been shown to be involved in avoidance learning, as well as in other situations which can be expected to evoke fear. Increased activity of the adrenal cortex has been observed in human beings before surgical operations, in bomber crews during flight, and in monkeys working to avoid electric shock; in the latter case, moreover, the degree of adrenocortical activation was accurately graded in intensity according to the level of stress imposed on the animal. Adrenocortical activation may also be produced by frustrating an animal, that is, by not giving it a reward that has usually followed responding. This has been shown in experiments from Seymour Levine's laboratory. Rats were first trained to press levers for food reward. When the reward was discontinued, there was a substantial elevation in the secretion of corticosterone. Dantzer has shown the same phenomenon in pigs. In another of Levine's experiments, rats were fed at a regular time each day. There was a rise in corticosterone levels just before feeding time; and the levels could then be made to fall either by feeding the animal or simply by presenting stimuli usually associated with feeding. Thus the output of the adrenal cortex is highly sensitive to purely psychological variables, such as the expectation of reward or the disconfirmation of this expectation.

The secretion of adrenaline from the adrenal medulla, of noradrenaline from the adrenal medulla and the sympathetic nerves, and of glucocorticoids from the adrenal cortex all appear to have a generally beneficial effect upon behaviour. Frankenhaeuser, for example, has found that people who respond to stress by relatively large increases in the levels of circulating adrenaline, relative to those who show smaller increases, are faster, more accurate, and more persistent on a variety of simple laboratory tasks. The same pattern held also for noradrenaline secretion, though less clearly. In addition, there have been a number of experiments with animals in which injections of ACTH (the anterior pituitary hormone which causes the release of the adrenocortical hormones) have restored the capacity for efficient avoidance learning which had first been impaired by removal of the pituitary. Furthermore, in animals with the pituitary intact, injection of ACTH is able to increase the persistence of avoidance behaviour. Another hormone secreted by the pituitary (but from the posterior lobe) during stress is vasopressin (see Fig. 5.2). This hormone too increases the persistence of avoidance behaviour.

On the face of it these findings are at least consistent with the James–Lange theory, since one could argue that perception of the peripheral physiological changes caused by these various hormones gives rise to a central emotional state which then motivates (in a way we consider in Chapters 9–11) avoidance responding. It is indeed possible that this is part of the explanation of the effects of some hormones. Peter Salmon, for example, working in my laboratory, has shown that one may reduce the intensity of a rat's behavioural response to frustration by injecting a drug (practolol) which blocks responses to noradrenaline peripherally, but which does not enter the brain. This suggests a surprising degree of involvement of peripheral physiological changes in determining behavioural responses even to quite subtle psychological stressors.

But this may be an unnecessarily complex way in which to interpret the effects of hormones on behaviour. For an important series of experiments from De Wied's laboratory in Utrecht has demonstrated that hormones can effect behaviour by a *direct* action on the brain. As noted above, injections of ACTH increase the persistence of avoidance behaviour. But this effect is not produced by release of glucocorticoids from the adrenal cortex (the peripheral target organ for ACTH). On the contrary, the effect of these hormones is to *reduce* the persistence of avoidance behaviour. De Wied has been able to disentangle these two opposing effects of ACTH and glucocorticoids, respectively, by using small fragments of the full ACTH molecule which have no effect on the secretion of adrenocortical hormones, but which nonetheless have dramatic effects on the persistence of avoidance responding. The effect of these ACTH fragments is essentially the same in intact animals and in animals whose adrenal glands have been removed. But the effect is abolished by destruction of a small region of the brain, the parafascicular nucleus in the thalamus. Conversely, direct injection of ACTH into this region of the brain (though not only here) increases the persistence of avoidance behaviour in the same way as a peripheral injection.

The behavioural effects of vasopressin are also due to a direct action on the brain. Recent research has done much to elucidate this action. It turns out that some groups of neurons in the brain probably use vasopressin as their transmitter substance (see Fig. 5.6). Injected vasopressin, or vasopressin released naturally by the pituitary during stress, may therefore act upon receptors in the brain whose primary function is to respond to the vasopressin released by neighbouring neurons. Interestingly, a strain of rats (the Brattleboro strain) which suffers from a hereditary lack of vasopressin has been shown by De Wied's group to be particularly poor at learning to avoid shock. This is almost certainly due to the lack of vasopressin in the brain rather than in the periphery.

You may now see why, on the first page of this book, I used the term 'neuro-endocrine system' to capture the complex interrelations that bind the brain and the endocrine system. These relations are so close that any attempt to give priority to the one or the other is almost certainly doomed to failure. The James–Lange theory was one such attempt; and, although it undoubtedly contains a germ of truth, it seems unlikely that the germ will flower further.

You may also see why, at the beginning of this chapter, I described the belief that behaviour is the result of specific physiological events within the organism as a 'fundamental assumption' of experimental psychology rather than a particular hypothesis or theory. For, faced with the failure of the James–Lange theory of the emotions, we do not discard this assumption: instead we try to frame an alternative particular theory of the emotions which, like the James–Lange theory, is consistent with it. Having tried the endocrine and autonomic nervous systems and found them wanting (though by no means unimportant, as we shall have still further occasion to see), we do not have far to seek for our next repository of a physiological theory of the emotions. As Schachter's work has shown, we rely in our judgments of our own emotional state at least to some extent on the knowledge of what has led up to the sensations we are feeling, and of what is happening at the time we are feeling them. Psychologists call such bits of knowledge by the grandiose name of 'cognitions'. Where else could cognitions reside but in the brain?

6   An excursion into social biology: fear and sex

As a pattern of response to a wide variety of dangerous or painful circumstances, the general adaptation syndrome described in the previous chapter contains some intriguing features.

To begin with, there appears to be a quite fundamental opposition between stress-responses and sexual behaviour. This might seem to be natural enough: after all, a time of danger is not normally a good time for courtship. But then, on the other hand, a time of danger might just be the time that the species needs new recruits, if it is to survive. Just this need, no doubt, lies behind the curious fact that in time of war the birth rate goes up and (a superb example of the feats of adaptation achieved by natural selection) the ratio of male to female births swings towards an increased number of boys. So we perhaps ought to ask more questions of the suppression of sexual behaviour which occurs in the general adaptation syndrome before we accept it as a natural part of the scheme of things.

An even more puzzling feature of the general adaptation syndrome is the reduction in the organism's ability to close its wounds and to fight infection, both consequences of the release of glucocorticoids from the adrenal cortex. Would not animals be more likely to survive under stressful conditions (which might well include assault from predator or bacterium) if the glucocorticoids stayed at home in the adrenal cortex? Natural selection should be able to do better than this.

And perhaps it can. Perhaps we should be looking around for some kind of selection pressure which we have not yet thought of, one which would promote the preferential survival of species which have developed this antagonism between stress-responses, on the one hand, and sexual behaviour, the repair of damaged tissues, and the fight against infection, on the other.

The antagonism between fear and sex

Before starting on this search, let us be sure the phenomena we are seeking to explain really occur. The opposition between stress and sexual activities, at any rate, has appeared in a number of other areas of research.

Consider, for example, the extensive work of C. P. Richter on the effects of domestication on the behaviour and physiology of the laboratory rat. Domestication of the Norway rat for scientific purposes began about a hundred years (or 300 generations) ago. Among the many selection pressures which must have been operative in this process (ability to thrive on a laboratory diet, in confinement, and so on), selection for reduced emotionality has certainly played a part. Compared to the wild rat, the laboratory animal displays less fear of Man, less savageness to Man, and less aggressive behaviour towards members of its own species. Experimental evidence using the kind of tests we discussed in Chapter 4 (Open Field test, emergence tests, etc.) has also supported the proposition that the laboratory rat is less fearful than his wild cousin. Along with these behavioural changes, there has also been a profound alteration in the laboratory rat's endocrine system.

Richter's major observations can be summarised by saying that the importance and efficiency of the adrenal glands has declined considerably in the laboratory rat, and that many of their functions have been taken over by the sex glands. To begin with, the weight of the laboratory rat's adrenals, expressed as a proportion of body weight, was found to be only a half to a quarter of the weight in the wild rat. The lowered efficiency of the laboratory rat's adrenals manifested itself in a number of ways. The adrenals are of vital importance in the conservation by the body of salt. When placed on a low-salt diet, the health of the laboratory rats rapidly declined, while the wild rats were unaffected. When the two groups of animals were exposed to various kinds of stress, the effects on the adrenals (measured by changes in their chemistry) were much more pronounced in the laboratory rat. The greater dependence of the wild rats on adrenal functioning was shown by the fact that removal of the glands led to their rapid death, whereas the laboratory animals were able to survive. The death of the wild adrenalectomised rats was hastened when they were made to fight with one another, indicating the importance of the glands in the wild rat for this kind of stress.

Other experiments demonstrated the transfer of function from adrenals to gonads which appears to have taken place in the laboratory rat. Many years previously, some classic experiments of Wang and Richter had shown that if the female laboratory rat is given access to a running wheel, it is possible to measure characteristic activity changes over her estrous cycle, peak activity coinciding with the time of maximum sexual receptivity. Removal of the ovaries abolishes this activity cycle and greatly reduces the total activity recorded, and the latter effect is also obtained in the male by removing the testes. Removal of the sex glands in Richter's wild rats, by contrast, made no difference to the activity levels recorded. The estrous cycles, both in the running wheel and in more direct measures of sexual functioning, were in any case much less regular in the wild rat. In addition, the wild rats reached puberty later and, in the early part of life, had lighter gonads and accessory sex glands. Also suggestive of a true difference in gonadal function in the two kinds of rat is the fact that the laboratory specimen will mate at any time of the year, whereas the wild rat normally mates only in the spring and autumn. Richter's wild females, in fact, were less fertile at all times and all ages. It seems, then, that there has been a great swing during the domestication of the Norway rat from adrenal function to gonadal function, from coping with danger to copulation.

If we turn from the endocrinologist's study of stress-responses to the psychologist's study of fear, a similar picture emerges. The most thorough investigation of this kind was conducted by E. E. Anderson in 1938. He gave 55 male rats an extensive series of tests, not only of fear and sexual behaviour, but also of learning ability, general activity, and the hunger, thirst, and exploratory drives. Emotional defecation was measured in the Open Field and in two different kinds of maze. The sum of the scores from these three tests was then correlated with seven separate measures of sex drive, including number of copulations, speed of running to reach a receptive female, intensity of electric shock endured in getting to a receptive female, etc. All seven measures of sex drive correlated negatively with the defecation score, with values ranging from -0.2 to -0.5. The highest correlations were yielded by the most direct measures of sexual behaviour (number of copulations and number of inseminations). In a second experiment, Anderson used four tests of fear (two of defecation and two emergence tests) and a further 91 male rats. Number of copulations and number of inseminations both correlated negatively again with all four tests of fear (-0.34 to -0.58). Furthermore, when Anderson compared the scores from his two experiments, the animals from the former one had lower emotionality scores and higher sex drive. These experiments, then, constitute powerful evidence that in the male rat, the more fearful the animal, the less vigorous he is likely to be in his sexual behaviour.

There is more equivocal evidence for the same relationship in the female rat. It is possible to bring many female mammals into peak sexual receptivity, or 'estrus', artificially by injections of estrogen and progesterone. Another of Anderson's experiments showed that this treatment is able to reduce Open Field defecation. Together with S. Levine, I have been able to confirm this finding, and to show that the effective hormone is estrogen, the addition of progesterone having no further effect. Anderson also observed rats which had entered estrus spontaneously and found that they too had lower defecation scores. However, it has been pointed out by Robert Drewett that the reduced defecation at estrus could be due to the reduced food intake observed at that time, an effect which is also controlled by estrogen. Defecation in the Open Field is normally rather independent of purely metabolic defecation. For example, the Maudsley Reactive strain, which was selectively bred for high Open Field defecation (Chapter 4), actually defecates less in the home cage than its Nonreactive counterpart. Nonetheless, it is possible that in particular cases a lowered Open Field defecation score is due to metabolic rather than emotional factors. That this is so for the effects of estrus, whether spontaneous or artificially induced, has been conclusively demonstrated by Slater & Blizard.

The evidence from Open Field defecation, therefore, cannot be used to argue that estrus is a time of lowered fearfulness. There is, however, a small amount of other evidence in favour of this conclusion. Anderson found that rats which enter estrus naturally have faster emergence times. Burke & Broadhurst, in a similar study, found higher Open Field ambulation during spontaneous estrus, as Levine and I did with artificially induced estrus. But this observation too is equivocal, since Martin & Bättig report that at estrus rats will also increase their exploration of a familiar maze, a situation in which the influence of fear should be negligible. Perhaps most convincing is the observation by P. Gray that estrous mice show less avoidance of a box in which they have experienced electric shock than do non-estrous mice.

It will be evident that the relation between estrus and fearfulness remains an open question. Nevertheless, even without conclusive evidence from this source, there appears in general to exist a fundamental antagonism, as much behavioural as endocrinological, between fear (or, more generally, behaviour induced by stress) and sexual functions. Furthermore, this antagonism appears whether we consider the organisation of personality (as in Anderson's work on the male rat) or functional changes within one individual (as in the general adaptation syndrome). Why does this antagonism exist? This is the sort of question to which the most meaningful answer for a biologist is likely to be couched in terms of the theory of natural selection and the evolution of species. In other words, what pressure from the environment has been conferring additional chances of survival on species which display the fear–sex antagonism? We may find such a selection pressure in the theory of population homeostasis which has been developed by a number of biologists, notably V. C. Wynne-Edwards and (for mammals) J. J. Christian.

Population density

The problem of population homeostasis is posed by the fact that on the whole and except where there are sudden large changes in the environment, the density of animal populations (i.e. the number of animals per unit space) remains at a fairly stable level. This level is rarely set by the breeding potential of the population, which is in many cases astronomically high. There must therefore be certain controlling factors which brake the increase in population to which unlimited breeding could otherwise give rise. Furthermore, since the fluctuations around the stable level are often comparatively small, these controlling factors must act in some way homeostatically: that is, the effect of the brake must become stronger the more the population increases beyond its stable level and weaker the more the population falls below its stable level. In other words, there appears to exist some kind of a 'density-stat', akin to the thermostat which controls a heating system at a desired set-point (Fig. 6.1).

Thus far, most students of the subject are in agreement. Where they disagree is on the exact nature of the controlling factors. Some workers (e.g. the noted English ornithologist David Lack) have stressed the importance of factors *external* to the population, such as predation, disease, and starvation. It has been suggested that the incidence of mortality from at least some of these factors will vary with population density in such a way as to furnish homeostatic control. Others (especially Wynne-Edwards) have argued that these factors are incapable of producing the degree of stability which is commonly observed and that it is necessary to postulate in addition certain 'intrinsic' controlling factors. These would be built into the species' own physiology and behaviour as a result of natural selection. Their survival value would lie in the prevention of what Wynne-Edwards calls 'overfishing'. Imagine a species which continued to increase in size up to the limit imposed by the resources of its habitat; then the generation born after this limit is reached will find its food supplies seriously depleted. Such a species would swing wildly between very large and very small numbers and, if its minimum size happened to coincide with particularly harsh environmental conditions, it would

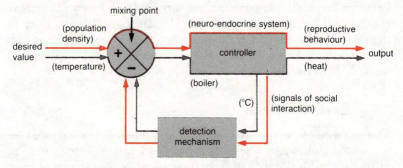

Fig. 6.1 The 'density-stat' for the control of population density. The homeostatic control of population density in mammalian species appears to involve the same kind of negative feedback on which a simple thermostat is based. When the 'desired value' is reached or exceeded, there is a shutdown on those activities of the neuro-endocrine system which are directed towards reproductive behaviour. The signals which convey to the 'mixing point' that the desired value has in fact been attained appear to be signals arising during social interaction with other members of the species. The actual desired value operating in a particular species (i.e. the population ceiling which is reached in the absence of any external constraints on continued population growth) is probably the result of evolutionary pressures acting through the gene pool.

run the danger of extinction. An intrinsic mechanism which prevented population density from rising too high, and thus obviated this danger, would have great survival value.

The difficulty for proponents of the hypothesis of intrinsic mechanisms of population control is to suggest ways in which such a mechanism might evolve. The problem can be simply stated. The hypothesis requires that certain individuals are restrained from breeding or suffer increased mortality at times when population density is high. If such a mechanism is to be acted upon by natural selection, these individuals must be chosen, not at hazard, but in virtue of their genotypes (that is, there genetic constitutions). But, *ex hypothesi*, the individuals who are most susceptible to the stress of population density are precisely those who must *fail to pass on* their genotypes, either because of restricted breeding or because of an early death. How, then, can such a mechanism evolve or, if it arises by chance, maintain its place in the gene-pool of the species? Various solutions have been proposed to deal with this dilemma, as we shall see later. None of them has been shown to work and all of them have met with objections. However, the evidence that intrinsic mechanisms of population control exist, at least in mammals, is so compelling that the challenge is no longer to the student of population homeostasis to demonstrate the new facts, but to the geneticist to cope with them.

The general adaptation syndrome as a density-stat

Wynne-Edwards has suggested that a great diversity of physiological, behavioural, and social mechanisms have evolved in different species to provide intrinsic homeostatic control of population density. These include territorial behaviour in birds, dominance hierarchies or 'peck-orders', and many other forms of social behaviour, cannibalism, and, most importantly for our present purposes, the general adaptation syndrome. The argument over these suggestions is likely to be long and heated. Here I intend only to consider the evidence that in mammals, the general adaptation syndrome (GAS) may act in the way required by Wynne-Edwards' theory. Note that if it does, we have an immediate explanation not only for the suppression of reproductive functions which is part of this syndrome, but also for the decreased capacity to cope with tissue-damage and infection which is an even more puzzling feature of it. For these changes are exactly what we would expect if the GAS has the function of restraining or reducing population density when this gets too high. On this view, it is not merely the case that high population density is one form of stress which activates the GAS; rather, the GAS has perhaps evolved precisely so as to act as a density-stat and has come in the course of time to be set into operation also by stimuli other than those which signal high population density.

Let us be clear about what must in fact be demonstrated if we are to say that the GAS acts as a density-stat. It will be necessary to show that: (1) an increase in population density activates the GAS and does so in proportion to the magnitude of the increase; (2) activation of the GAS by high density involves a sufficiently large increase in mortality or decrease in fertility for the increase in population density to be halted or reversed.

In addition, it would be valuable to obtain information on two further points. (3) What is the nature of the stimuli by which increases in population density are signalled to the members of the population? Wynne-Edwards has termed such signals, and the various forms of social behaviour which give rise to them, 'epideictic', meaning 'meant for display'. In the engineering of automatic control systems, they would be called the 'error signals' for the density-stat. (4) Which individuals are selected for increased mortality or decreased fertility in time of density stress? This cannot be a matter of chance if the density-stat is under genetic control.

Most of the facts we shall be discussing in support of these points concern rodents. The first thing we should be sure of, therefore, is that rodents do have a density-stat. The simplest way to demonstrate this is to place a group of individuals in a confined space, but with unlimited food and shelter within the space, and to allow them to breed freely. Calhoun carried out such an experiment with rats in 1952. His population never exceeded two hundred, though he estimated that the food, space, and shelter with which they were provided could support a number well over five thousand. Thus population density appeared to be severely limited by purely intrinsic factors. Similar demonstrations have been made with mice.

The next thing we must demonstrate is that high density activates the GAS. In view of the central role played by ACTH and the adrenocortical hormones in the stress syndrome, we would expect to find signs of their activity under conditions of high density. These signs include increased adrenal weight; depletion of lipids, ascorbic acid (Vitamin C), and cholesterol in the adrenal cortex; a decline in circulating eosinophils and in the weight of the thymus (both of which are the result of increased secretion of adrenal glucocorticoids); and increased excretion in the urine of metabolites of the adrenal hormones (especially 17-hydroxy-corticosteroids). They have been observed repeatedly, most often in experiments on mice, rats, and voles, but also in experiments which have used rhesus monkeys caged in groups. They have also been seen in human beings grouped together as bomber crews. Christian has shown, moreover, that the *degree* of adrenocortical activation is closely dependent on the intensity of the density stress imposed. Thus, when mice were caged in groups of different sizes, the increase in the weight of the adrenals and the decrease in the weight of the thymus were positively related to group size (Fig. 6.2). Another of Christian's experiments reversed the usual procedure. Instead of imposing an *increase* in population density, he artificially reduced the density of a population of wild rats by trapping a proportion of them. This caused a reduction in the weight of the adrenals by over one-third.

The evidence, then, that increased population density activates the pituitary–adrenocortical system is good. Does it also reduce fertility?

In the female, crowded conditions have been found to interfere with practically every link in the chain of reproductive physiology and behaviour. The estrous cycle is completely suppressed when mice are caged together in all-female groups. Christian & LeMunyan caged male and female mice together in groups of forty, with the result that no pregnancies occurred at all over a period of 6 weeks, though a few days is sufficient for conception to take place when a single male is caged with a single female. In this same study, when the crowding was reduced to ten animals of each sex, all the females became

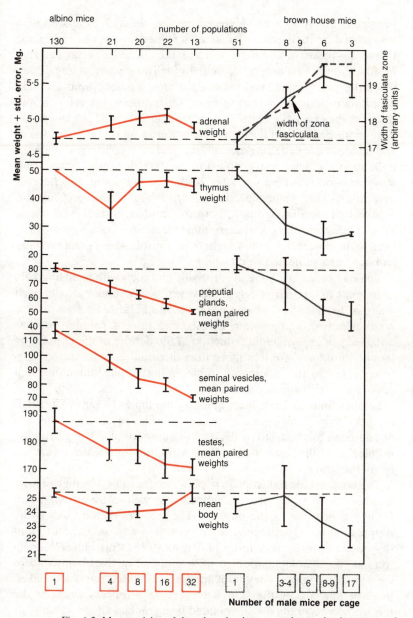

Fig. 6.2. Mean weights of the adrenals, thymus, and reproductive organs of male mice, albino on the left and a wild strain on the right, from populations of fixed size. Population size is plotted on the abscissa on a logarithmic scale.

pregnant, but the onset of pregnancy was delayed compared with uncrowded animals, the number of fertilised ova implanted in the uterus was reduced, and there was a marked increase in intra-uterine mortality, only seven females actually delivering any young. Even when the young conceived under such crowded conditions are delivered, their trials are not yet over, for the mother's milk is less ample than usual even if she is isolated before parturition. Thus the litters from mothers crowded during pregnancy grow up smaller than usual, and this decrease in size is especially marked in large litters. Decreased milk supply is also responsible for the fact that more pups from crowded mothers die soon after birth. There is even an effect of crowding during pregnancy on the *offspring's offspring*: Christian reports that even though pups born to crowded mothers were themselves allowed to mate and carry their young in isolation, the young born to them were lighter in weight than controls whose grandmothers had not been subjected to crowding.

These observations were mostly made in studies with fixed numbers of animals in laboratory cages. However, similar findings have been made in field studies. In one of Christian's field studies, an inverse relation between population density and rate of pregnancies was obtained; an experimental reduction of the density of the population by one-third gave rise to a more than threefold increase in the pregnancy rate. So the laboratory studies reflect a phenomenon which plays a very real role in nature.

Parallel findings have been made in the male. In mice, Christian reports that increasing population density leads to a fall in the secretion of androgen by the testis, a delay in the onset of puberty, a decrease in the size of the testes themselves, and inhibition of the formation of spermatozoa.

The homeostatic reduction in population size caused by high density is achieved by an increase in mortality, as well as by a decrease in fertility. Since one of the most marked features of the GAS is the suppression of anti-inflammatory processes, we would expect to find this occurring under conditions of density stress. Christian & Williamson examined inflammation in mice grouped in fives compared to animals caged singly. They implanted a pellet of cotton wool soaked in turpentine under the skin. This sets up a process whereby the foreign body is gradually surrounded by a granular tissue which seals it off from healthy tissue. When this granular tissue was removed at the end of a week, it was found to be 20% lighter in the grouped animals than in the segregated controls. An experiment by Davis & Read on the effects of crowding on resistance to infection came up with equally striking results. Mice were injected with larvae of the worm *Trichinella spiralis*. Each mouse lived in a separate cage, but some of them were placed for 3 hours a day in groups of five and six, while the other eleven were left undisturbed. The grouped animals had no food or water when they were together, thus reducing the possibility of transfer of the worms from one animal to another. On the fifteenth day after infection the animals were sacrificed and their intestines examined. Only three of the segregated mice were infected at all, with an average of nine worms apiece. Every one of the grouped mice, in contrast, was infected with an average of 32 worms. Other experiments, also with mice, have shown a depression of antibody formation and decreased time of survival after infection with tuberculosis.

Other forms of mortality have sometimes been reported as a result of increased population density. Sudden death in convulsions has been observed in mice and voles in the laboratory and in voles and snowshoe hares in the wild. In the field studies the deaths occurred after a build-up to the peak in the population cycle. There is evidence in support of Christian's suggestion that these deaths are due to a drastic fall in blood sugar as part of the 'exhaustion' stage of the GAS. Observations on wild populations of voles, for example, showed that the symptoms of the dying animals could be reproduced by injections of insulin (which lowers blood sugar level), that the deaths could be prevented by injections of glucose, and that the dying animals had abnormally low levels of sugar in the blood and liver. More minor effects of density stress have also been reported, including suppression of growth, weight loss, and reduced thyroid activity.

The signals which operate the density-stat

There seems little doubt, then, that the GAS can act as an efficient density-stat. What are the signals which set it into operation?

First, we should say what the error signal, or epideictic stimulus, is not. It is not living space *per se*. Christian found that the relation between population size and adrenal weight was not disturbed when he increased the floor space for his populations by a factor of 42; and Thiessen & Rodgers found the same relation even though they exactly equated living space per individual in their different-sized groups. Nor is it the novelty which is necessarily involved in those experiments in which hitherto segregated animals are grouped together. Southwick has shown that the sustained decrease in circulating eosinophils (a much-used sign of adrenocortical activation) which is produced by placing mice in fresh *groups* every day is not produced by placing them in fresh *cages* every day. Nor is it fighting *per se*, though this often occurs under crowded conditions, nor yet the wounding which may result from fighting. Christian found no correlation between the number of scars carried by grouped mice and the degree of adrenal weight increase which the grouping had produced. Furthermore, fighting is a mutual activity, yet the degree of adrenal weight increase which it produces depends on the social situation in which each of the two animals involved finds itself: when two voles fight, the one which is on its own territory shows far less adrenal hypertrophy than the interloper.

We are left with the conclusion that the stimuli which signal density stress are *those which are received in the course of social interaction*, and especially in the course of those competitive activities which underlie the acquisition of territory and the formation of status hierarchies. This thesis has been elaborated at great length by Wynne-Edwards, who goes so far as to suggest that these and many other forms of social behaviour have evolved precisely so that they may serve as part of the same kind of density-limiting mechanism as we have seen the general adaptation syndrome to be. Indeed, he defines a society as '*an organisation capable of providing conventional competition*' and claims that '*the social organisation is originally set up to provide the feed-back for the homeostatic machine*'.

The importance of social factors is very clear when we consider the characteristics of the individuals who are particularly susceptible to density stress. It is reasonably well established that these are individuals low in the dominance hierarchy which is a feature of the social organisation of many species, including rodents; they are also often the younger animals or lighter in weight, two things which tend to go together with low social status. Thus in mice, rats, and dogs, it has been reported that the activation of the adrenal cortex caused by grouping animals together is less, the more socially dominant the animal. A stranger on another animal's territory may be considered a subordinate, since it will usually adopt the submissive posture; thus it is consistent with this general pattern that a strange vole introduced into a cage containing a resident pair of voles shows more pronounced symptoms of stress than do the residents. A similar dependence of the effects of density stress on social status has been observed for suppression of breeding in cockerels, male and female mice, and voles; for suppression of growth in mice; and for density-dependent mortality in rabbits and voles.

Not only the individual's social status, but also the composition of the social group is important in determining the outcome of density stress. In the wild rat, Barnett observed that fighting between males only occurred if they were caged with females or if a stranger was introduced after a group had already been formed. (The females did not fight at all.)When they were caged with females, the males suffered a high mortality, though this was not directly due to the wounds received, these usually being superficial, Working with the white-footed mouse, Southwick found that increased population density affected adrenal weight (in subordinate animals) only if there was much fighting in the population; fighting could be induced in a peaceful population by the introduction of strangers, who themselves suffered a very high mortality (60% for males, 40% for females). Mason's observations on human bomber crews showed that the degree of adrenocortical activity was very similar in all members of a particular group, but the various groups differed markedly from one another, again indicating the importance of group composition.

In addition to social factors, genetic factors also help determine which individuals succumb to density stress. Bronson & Eleftheriou were able to induce an adrenocortical response by crowding the house mouse, but the deer mouse was unaffected by similar treatment. Within the species of house mouse (*Mus musculus*), Christian reports that the wild animal is more likely to die of density-induced shock than the laboratory animal. The greater freedom with which the crowded laboratory rodent, compared with its wild cousin, will breed is no doubt part of the same pattern.

The GAS, then, is ideally suited to act as a density-stat: (1) it is set into operation by density stress; (2) the degree to which it is activated is proportional to the intensity of the density stress; (3) it includes a very large reduction in fertility and an increase in mortality, able to act as a powerful brake on a rising population density; (4) the signals of high density which activate the GAS are those arising from social interaction, of the kind involved in competition between conspecifics and the establishment of social dominance hierarchies; (5) the individuals most affected by density stress are usually low down in the social hierarchy.

Population density and natural selection

The trouble with these conclusions, as I have already noted, is that they are very difficult to reconcile with the theory of natural selection.

Given the overwhelming success of natural selection as an explanatory concept in general biology, there can be no question of abandoning it when we come to the mechanisms that control population density. But equally the laboratory demonstrations of intrinsic controls over fertility and mortality that respond to population density are impressive; they cannot simply be ignored. Various suggestions have been made for ways out of this dilemma.

Wynne-Edwards suggested that selection might operate at the level of the group rather than (or, better, in additon to) the individual level. But attempts to put this suggestion onto a formal basis have so far failed. W. D. Hamilton and J. Maynard Smith each proposed a mechanism known as 'kin-selection', whereby, in a sufficiently isolated and inbreeding community (by no means uncommon in nature), the death or failure to reproduce of one individual may ensure the survival of a sufficient number of close relatives for the loss of his own genes to be more than compensated by the survival of the identical genes in his relatives. This mechanism can probably work under restricted conditions; but they are probably *too* restricted to account for the apparently widespread existence of density-dependent intrinsic controls among mammalian species.

A further possibility has been proposed by Dennis Chitty. It calls on mechanisms that are well described by E. O. Wilson in his book *Sociobiology* (p. 87). (An allele is one of several possible genes which may occupy a particular locus on a chromosome.)

Different genotypes can be subject to various density-dependent controls, with the result that the population fluctuates in size as one genotype replaces another. Suppose that when allele *a* predominates, the population equilibrates at a high level under the control of density-dependent effect *A*. However, selection favours allele *b* over *a* at this higher density. As *b* comes to predominate, the population shifts to a lower equilibrial density, mostly under the control of a new density-dependent effect, *B*. But at the level dictated by *B*, allele *a* is favoured by selection, and the stage is set for the move back up to the higher level. Thus genetic polymorphism and the corresponding differences in density-dependent control can be coupled in a reciprocally oscillating system to create a population cycle. (Wilson, 1975)

Population cycles of this kind have been described in several species of rodents. C. J. Krebs and his colleagues studied a population of wild voles that show a remarkably regular cycle, with peak density occurring about every 4 years. They measured the frequency over the cycle of two alleles (for which convenient biochemical markers existed) at each of two genetic loci. In agreement with Chitty's hypothesis, they found rapid changes in the frequency of the different alleles over the cycle, one increasing in frequency at times of population increase, the other increasing at times of population decline. This is particularly compelling evidence that genetic factors play a role in determining survival under density stress, although it should be noted that the functions, if any, of the genes Krebs measured in responding to density stress are unknown.

Chitty's hypothesis has some support, then, and it can take us some way towards reconciling intrinsic population controls with natural selection. But Krebs' work also revealed a further factor in population control – emigration or dispersal – whose importance had long been suspected but not previously so clearly demonstrated. In one relevant experiment Myers & Krebs fenced off a two-acre area in Indiana and compared the population dynamics in the enclosed population of voles with the dynamics observed in an unenclosed population. Their most dramatic observation was that population density within the enclosure rose to levels up to 20 times as great as normal. This apparently occurred because emigration from the enclosed area was prevented, showing that such emigration normally plays a powerful role in controlling population density in free-ranging voles. In subsequent experiments Krebs' group was able to identify the emigrating individuals in free-ranging populations. These differed genetically (on the same biochemical markers as before) from the non-emigrating individuals. Thus the changes in genotype observed over the population cycle in the free-ranging voles were probably in large measure due to the different genetic characteristics of emigrating and sedentary individuals, respectively.

In a recent experiment Butler has been able to bring the phenomenon of emigration into the laboratory. He enclosed mice in a circular area containing apertures through which the mice could exit, but only by swimming across water-filled trays. The mice soon established dominance hierarchies and territories. But about 20% of the animals chose rather to swim out of the arena. These émigrés were overwhelmingly (97%) those which had lost in the competition for social rank within the enclosure – precisely the kind of animal which, in Christian's laboratory experiments with fully enclosed populations, would be most likely to show density-dependent loss of fertility.

These observations on the role of emigration cast a new light upon the significance of Christian's experiments, and they do much to reduce the mystery of the genetic basis of intrinsic population controls. For we can now see that animals whose reproduction is restrained at times of high population density may *not* be giving up reproduction altogether. Rather, they may be delaying reproduction until they encounter better conditions in which to rear their young. Krebs' observations suggest that one way to do this is by emigrating; and many of the émigrés in his study were young females just reaching reproductive maturity – animals whose fertility would be particularly likely to be restrained in Christian's experiments, but which would be particularly well suited to reproduce rapidly in a freshly colonised habitat. Thus individual animals are not faced with the choice imagined by Wynne-Edwards – to breed for the survival of their own genes, or not to breed to ensure the survival of their group. Rather, they are faced with a choice between two strategies, each in different ways favouring the survival of their own genes: to obtain high social rank and breed in spite of high population density, or to emigrate to pastures new and breed there.

As noted above, Krebs' observations indicate that which of these choices is made is probably to some extent dependent on genotype; and he suggests, for example, that animals that refuse to quit their original habitat are genetically predisposed to display a high level of aggression. But we should not over-emphasise the role of genetic factors in determining an individual's choice between staying or emigrating. Butler's experiment strongly suggests that the determining factor is social rank: animals low in social rank emigrate. Many field studies are in agreement with this conclusion. But social rank frequently depends to an important degree on a variety of non-genetic factors – e.g. an animal's age, weight, or priority of entrance into a particular territory. Thus, if a population happens to attain excessively high density when an animal is young, still of low weight, and before it has established a territory, that animal may be forced to emigrate; but if high density is reached later, the same animal is more likely to be among those that remain in the original habitat and continue to breed there. In short, the same animal must in many cases come equipped with mechanisms promoting *either* survival strategy distinguished above; which strategy it adopts then depends upon the particular environmental circumstances it encounters as it matures. Laboratory studies of dominance confirm this inference. Social rank is not a fixed genetic trait – it varies with time, place, and the composition of the social group. If, for example, one takes animals that are subordinate in several groups, and puts them together in a fresh group, a new hierarchy quickly emerges; and the same thing occurs if one puts together animals that were all dominant in their original group.

A combination of the above arguments, then, can do much to reconcile intrinsic population controls with the theory of natural selection. These arguments hold that species which possess such intrinsic controls come equipped with mechanisms promoting two strategies for the survival of their own genes. One strategy is to attain high social rank and continue breeding in the original habitat in spite of high population density; the other is to emigrate and attempt to colonise a fresh habitat. Which strategy is adopted depends to some extent on genetic factors, as Krebs' observations show. But his observations also show that the full range of genotypes which he measured remain represented in both the sedentary and the emigrating populations, so that each can continue to respond to density stress in the same balanced manner as before. This flexibility in response to density stress may reflect the operation of coupled genetic oscillators, as proposed by Chitty (see the passage cited from E. O. Wilson, above). Or it may stem from the role played by environmental factors (such as youth or priority of arrival in a particular territory) in selecting the individuals which emigrate.

Though this account seems plausible as an explanation for the existence of intrinsic, density-dependent controls over fertility, density-dependent *mortality* from intrinsic mechanisms remains a mystery. Nonetheless, if we accept the role of the GAS as a density-stat, we can at least see some kind of rationale for the suppression of the processes of tissue-repair and resistance to infection which are otherwise such puzzling features of this syndrome. As for the antagonism between fear and sexual behaviour which appears not only in the GAS, but also (as we have seen) in other psychological and endocrinological data, we now have a reasonably good understanding of why this antagonism should occur. It appears to reflect the close relationship, traced in this chapter, which exists between submissive behaviour among individuals low in social rank, on the one hand, and the restraint of reproductive behaviour which, at least temporarily, these individuals need to display.

There is one further aspect of the relations between fear and sexual behaviour on which the general set of ideas developed in this chapter can throw some light: the differences in fearfulness between the two sexes. This topic is dealt with in the next chapter.

7 The route from gene to behaviour: sex differences and fear

According to Wynne-Edwards' general theory of population homeostatis, the male is usually the 'epideictic sex'. That is, it is he who is charged with the execution of the forms of behaviour which signal to the community the existing population density and also with the burden of reacting to the signals received from others. The reason for this sex difference, according to Wynne-Edwards, is that the female bears the main burden of reproduction, so that her time and energy are too precious to be readily spared for social competition. By taking this role on himself, the male ensures a more efficient division of labour between the sexes, and so a species that is as a whole better fitted for survival. Whether this is the correct explanation or not, it does seem generally to be the males who display many of the forms of social behaviour believed by Wynne-Edwards to serve an epideictic function, such as swarming or flocking, singing or displaying bright plumage, competing for rank in a social hierarchy or fighting in a ritual tournament, or striving for possession of a conventional prize or a territory. In many species females do not actively take part in the formation of a social hierarchy, taking on instead the social status of their male partner. In the light of Wynne-Edwards' theory, then, we would expect that males would be more likely than females to respond to density stress with the general adaptation syndrome.

Although little attention has been paid to differences between the sexes in the study of density stress, such evidence as there is is consistent with this hypothesis. Christian, for example, housed mice in all-male, all-female, or mixed-sex groups. The change in adrenal weight induced by grouping was much greater in males than females; also, when the females from the all-female and mixed-sex groups were compared, those from the mixed-sex groups showed a greater degree of adrenal hypertrophy. Thus, not only is the male more reactive himself to grouping, but also the competition between the males increases the stress to which the females are exposed. In another study, Christian found a 25% adrenal weight increase in a freely growing laboratory population of male rats compared with a 14% increase in the females.

Fearfulness in the two sexes

We might also expect that a lowered threshold for the occurrence of the general adaptation syndrome in response to density stress would entail a generally heightened level of emotionality. And, indeed, we find better evidence for a greater male sensitivity to stress if we turn to psychological studies of fear. The existence of this evidence is all the more impressive in that, by and large, experimental psychologists have been reluctant to study sex differences. It is as though the accepted social dogma that men and women are *equal* has led to the conclusion (totally illegitimate even outside the realm of strictly sexual behaviour) that they are not *different*. This feeling has been so strong that when differences between the sexes turn up in an experiment, they are treated as a nuisance, to be eliminated by 'refinement' of one's measuring devices or, more simply, by restricting future investigations to one sex only (usually the steady, dependable male). And this has been so, even when the investigation has concerned, not men and women, but male and female rats. Yet still the pattern in the data comes through.

In general, the differences observed between male and female rodents conform to this pattern: as Maudsley Reactive rat is to Maudsley Nonreactive (Chapter 4), so male is to female. Thus male rats have been reported to defecate more and ambulate less in the Open Field, to emerge into a novel environment more slowly and explore it less readily, and to freeze more in response to a novel sound. I have seen more grooming behaviour and less exploratory rearing in males compared to females when the experimental animals are placed in a novel cage, paralleling the similar observations made for the Maudsley strains. In the golden hamster, Swanson has observed less male ambulation in the Open Field and slower emergence times. Differences in avoidance behaviour between the sexes have also been noted. They take the form of female inferiority in passive avoidance (rather direct evidence of lessened fear; see Chapter 9) and female superiority in shuttlebox avoidance to match the same kinds of difference between the Nonreactive and Reactive strains. As we shall see later in the book, drugs (such as Valium or Librium) which reduce fear and anxiety in man change avoidance behaviour in the rat (both sexes) in the female direction: that is, they impair passive avoidance and improve shuttlebox avoidance. These drugs also improve performance when rats can avoid shock by pressing a bar. Assuming that this change is due to lessened fear (an assumption we shall justify in a later chapter), we must predict that females will be superior to males at this type of avoidance. This prediction has been confirmed by van Oyen and his colleagues in Amsterdam. Ulcer formation as the result of psychological conflict is also less severe in the female than the male rat. All these observations, then, are consistent with the hypothesis that the female rat is less fearful than the male.

If the male is generally more sensitive than the female to stress, we would expect this to show up particularly strongly in the inhibition of sexual behaviour itself, since this is such an important part of the operation of the density-stat. One of the leading authorities on sexual behaviour, F. A. Beach, has summarised the relevant evidence. Most of it is anecdotal or lacking in proper controls. However, a general picture emerges which is consistent with the hypothesis of greater male fearfulness. Male dogs, foxes, minks, rats, cats, porcupines, and cockerels appear to show a greater reluctance to copulate in a novel environment, in an environment previously associated with pain, and in the presence of other animals higher in the social hierarchy, than do their female counterparts. A striking example of the effect of the presence of social superiors on male sexual behaviour is the report by Guhl, Collias & Allee of 'psychological castration' in the cockerel. Four roosters were caged together with a number of hens and soon developed a rigid dominance hierarchy. The lowest ranking male was attacked by the others whenever he attempted to mate with one of the hens. As a result, his sexual behaviour was so suppressed that even after removal of the other three roosters he made no attempt to tread the females.

Now when we are dealing with laboratory animals, all exposed to the same environment, we can be reasonably certain that any differences between the sexes are of primarily genetic origin. So the existence of sex differences in fearfulness in laboratory animals may be taken as further evidence for the importance of genetic factors in the control of this type of behaviour. The importance of genetic factors is further brought out by the dependence of the sex differences themselves on the animal's genetic constitution, or 'genotype'. For example, the programme of selective inbreeding carried out by Broadhurst in order to create the Maudsley Reactive and Nonreactive strains of rats has resulted in a remarkable attenuation of the sex difference in Open Field defecation which was clearly present in the original population of animals used to start the breeding programme (Fig. 7.1). Similarly, a comparison between inbred and outbred mice carried out by Bruell found that the sex difference in emotionality among the former (greater male emotionality) was increased still further in the outbred mice.

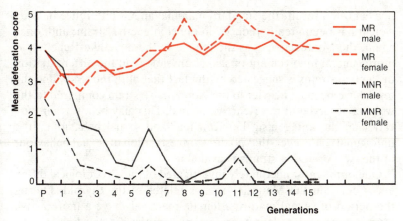

Fig. 7.1. Disappearance of the sex difference in defecation scores in the Open Field as a function of generation of selection of the Maudsley Reactive (MR) and Maudsley Nonreactive (MNR) strains of rats. The point marked 'P' on the abscissa represents the parental generation, in which, as is normal for unselected populations of rats, the males defecated significantly more than the females.

Sexual physiology and fearfulness

The influence which selective breeding has had on the sex differences originally present in the Maudsley strains may also be viewed in another light. Broadhurst's breeding programme managed to alter the frequency of the genes which determine high levels of fearfulness, increasing them in the Reactive strain and decreasing them in the Nonreactive strain. But genes cannot specify behaviour directly: they can only alter the activity of those physiological systems which in turn control behaviour. At some point, then, the behavioural scientist has to investigate the physiological route which links genes to behaviour. Is it possible that the physiological basis of sex differences lies in some way on the route from 'fear' genes to fearful behaviour? Is it possible, in fact, that the pattern, as Reactive to Nonreactive, so male to female, is more than a suggestive parallel and has a true causal basis? Some of the physiological changes wrought in the Maudsley strains by selective breeding tend to just this conclusion.

The first indication of the possible involvement of sexual physiology in the differentiation of the Maudsley strains concerns body weight. The Nonreactives are much lighter than the Reactives, this difference being largely confined to the males; crosses between the two strains closely resemble the Reactives in body weight, so it is Nonreactives which have become lighter rather than Reactives having become heavier. As a result, the standard mammalian sex difference in body weight has been greatly reduced in the Nonreactive strain; and, once again, the Nonreactive is as the female to the male-like Reactive. A more general involvement of sexual physiology in the differentiation of the two strains is suggested by the fact that both the ovary and the testis have become heavier in the Nonreactive strain compared to the weights observed in the Reactive animal. This may be a further indication of the antagonism between fear and sexual behaviour. It is unfortunate, however, that no direct investigation of sexual behaviour in the two Maudsley strains is available.

One of the most adequately investigated of the physiological differences between the Maudsley strains relates to the thyroid. During the period of sexual maturity, female mammals have a more active thyroid than males. It is therefore no surprise (in the light of our hypothesis) that the Nonreactive rat also has a more active thyroid than the Reactive. This difference between the strains lies not in the thyroid itself, but in the amount of stimulation it receives from the pituitary in the form of thyroid-stimulating hormone.

There is some indication of a functional link between these thyroid differences between the Maudsley strains and their emotionality differences. Feuer & Broadhurst were able to increase defecation and decrease ambulation by administering a drug which depresses the activity of the thyroid, and to decrease defecation and increase ambulation by administering thyroid-stimulating hormone, though only the first of these four findings reached an acceptable level of statistical significance. Also using a thyroid depressant, Watson reduced the Nonreactive rearing rate to that shown by Reactives. Most striking of all, Blizard & Chai tested two strains of mice that had been selectively bred for high and low levels of thyroid function. The relatively hypothyroid strain defecated more than the hyperthyroid strain in an Open Field test – exactly the same pattern of results (but obtained in exactly the opposite manner, starting from thyroid function rather than emotional behaviour) that has been observed in the Maudsley strains.

There are also indications that the difference in thyroid activity in the two strains has led to differences in the central nervous system. In his work on experimentally produced cretinism (a form of mental retardation which is due to an insufficiency of thyroxine in early life), Eayrs has shown that this disease can be recognised in the rat by a characteristic alteration of the electrical responses of the cerebral cortex to stimulation of certain subcortical structures. This same type of electrical response has been found in Maudsley Reactive rats by Eayrs, Glass & Broadhurst.

It seems, then, that the thyroid differences between the Maudsley strains may be a way-station of some importance on the route from gene to behaviour: they appear to be involved in some of the behavioural differences between the strains and to be related to quite substantial alterations in the brain. But there are some formidable problems of interpretation involved: is the direction of causation gene → endocrine gland → behaviour or gene → behaviour → endocrine gland? In other words, does the altered thyroid function in the Maudsley strains determine the differences in emotional behaviour; or does the difference in susceptibility to stress (which can depress thyroid activity) lead to the differences in thyroid functioning? (The same question can be asked with regard to the changed weight of the gonads, for we have seen that stress can reduce this too.) Of course, it is entirely possible – indeed probable – that both effects are at work in differentiating the Maudsley strains. It is not beyond the bounds of experimental ingenuity to distinguish between these various possibilities; but the experiments remain to be done.

A further important observation relating to changes in the central nervous system in the Maudsley strains concerns the substance called serotonin in the United States and 5-hydroxytryptamine in England. This is known to be a neurotransmitter (Fig. 5.6) in pathways that innervate the limbic system of the brain (see Fig. 13.10). This system, as we shall see in Chapter 13, plays a major role in the control of emotional behaviour, and especially in anxiety. It is therefore of considerable importance that Maas, working in Bethesda, Maryland, found a higher concentration of serotonin in the limbic system of the one of two inbred strains of mice which also had the higher Open Field defecation score. He then examined the brains of the Maudsley strains, which were specially established in the Bethesda colony for this purpose. The Reactive strain had a higher concentration of serotonin, again in the limbic portion of the brain. Furthermore, the same mysterious pattern appeared again: in the Maudsley strains, the males had higher concentrations of serotonin than the females. Once more, then, as Reactive to Nonreactive, so male to female.

Work in my own laboratory has also been concerned with differences in the brain between the sexes and between the Maudsley strains. This research deals with a part of the limbic system comprising the hippocampus and septal area (together making up the 'septo–hippocampal' system; see Figs. 13.9 and 13.10). We shall have much to say about this part of the brain in Chapter 13. To anticipate a little, the hippocampus displays a prominent, high-voltage slow electrical rhythm (similar to the rhythms recorded from the human scalp in the electroencephalogram, or EEG; Fig. 7.2). The frequency of this 'theta'

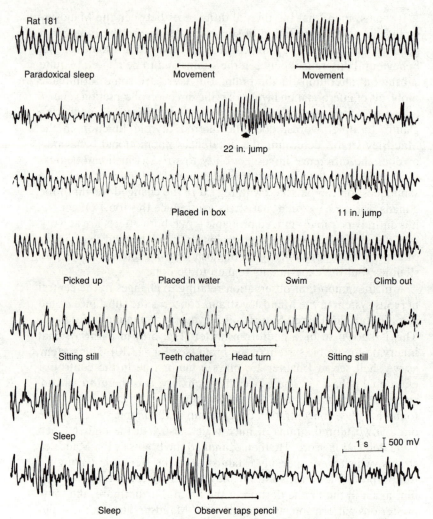

Fig. 7.2. Electrical activity at a single hippocampal site during sleep and various types of behaviour in the rat. Note the following: theta during paradoxical sleep, struggling when held in the hand, swimming and head movement; large-amplitude irregular activity during sitting still while alert and while chattering the teeth; irregular slow activity and 'spindling' during slow-wave sleep and small-amplitude irregular activity when the rat was awakened but did not move about. Note also the following: increased theta frequency and amplitude associated with twitching during paradoxical sleep and with jumping in avoidance tasks; different frequencies and amplitudes of theta associated with head movements, swimming, jumping 11 in, and jumping 22 in. Calibration: 1 sec, 0.5 mV; half-amplitude filters, 0.3 and 75 Hz. Electrode placement: CA 1, hippocampus. From Whishaw & Vanderwolf (1973).

rhythm in free-moving rats ranges from about 6 to 12 Hz (cycles per second). The rhythm is controlled by pacemaker cells located in the septal area. It is possible to implant electrodes chronically in the septal area. Then, by stimulating the septal area via these electrodes at a frequency within the naturally occurring range of theta frequencies, one can artificially elicit ('drive') the hippocampal theta rhythm (recorded from other implanted electrodes) at the stimulation frequency (Fig. 7.3).

In our experiments we measured the threshold current which, when delivered in this way to the septal area, was just capable of driving the hippocampal theta rhythm. In the male rat, the threshold varied reliably as a function of stimulation frequency, displaying a minimum precisely at a frequency of 7.7 Hz (Fig. 7.4). As we shall see in Chapter 13, the minimum threshold at 7.7 Hz in the male rat is abolished by a range of drugs which reduce fear and anxiety. Thus the 7.7-Hz minimum in the 'theta-driving curve' appears to be related in some way to this emotional state. This inference is supported by the fact

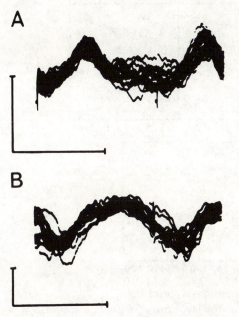

Fig. 7.3. Two examples of driven hippocampal theta waves in response to septal stimulation in the free-moving rat. Calibration: 0.5 mV and 100 msec. Each panel is the superimposition of about 10 successive sweeps on an oscilloscope. Stimulation parameters: 0.5 msec pulse width, 130 msec inter-pulse interval and intensities 100μA (A) and 80μA (B). (A) From James *et al.* (1977).

that, while the 7.7-Hz minimum is found in male rats of several different strains, including the Maudsley Reactive strain, it is absent in Nonreactive males. Furthermore, it is absent in the females of all the strains we have studied: as Reactive is to Nonreactive, so male is to female. These findings are summarised in Figs. 7.4 and 7.5.

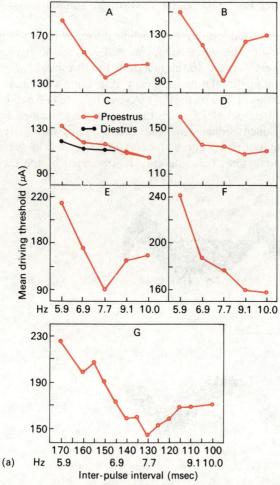

(a)

Fig. 7.4. (a) Thresholds for septal driving of hippocampal theta rhythm as a function of stimulation frequency. (A) Mean of five Wistar male rats; (B) single male rat; (C) mean of four females during proestrus and diestrus; (D) mean of six ovariectomized females; (E) mean of three male rats tested at 49, 59, and 69 days after castration, respectively; (F) mean of same three male rats as in (E), tested at 70, 72, and 79 days after castration, respectively; (G) results from single intact male tested with 5-msec increments of interpulse

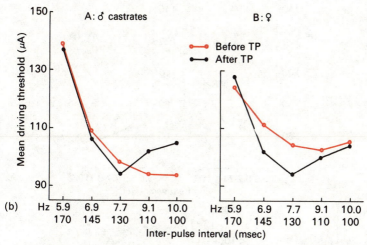

(b)

Caption for Fig. 7.4 (*cont.*)

interval (abscissa). Data from Drewett *et al.* (1977). (b) Effect of daily injection of testosterone propionate (TP) on theta-driving curves (i.e. threshold currents able to drive the hippocampal theta rhythm plotted as a function of the frequency of the septal stimulation current) in castrated male rats (A) and ovariectomized female rats (B). Data from Drewett *et al.* (1977).

In further research on the theta-driving curve, we have shown that the existence of a minimum threshold for theta-driving at 7.7 Hz in the male rate depends upon activity in a bundle of nerve-fibres that ascends from a group, or 'nucleus', of cells in the brain stem known as the 'locus coeruleus' (Fig. 13.2) and innervates much of the fore-brain, including the septo–hippocampal system. These cells use noradrenaline as their transmitter, so the fibre bundle is termed the 'dorsal ascending noradrenergic bundle' (see Fig. 13.3). Neil McNaughton, Peter Kelly and I destroyed this bundle by injecting it with a poison (6-hydroxydopamine) that selectively attacks cells containing noradrenaline. Male rats treated in this way lacked the 7.7-Hz minimum in the theta-driving curve (see Fig. 13.18). This and other evidence (discussed in Chapter 13) suggests that anti-anxiety drugs eliminate the 7.7-Hz minimum in the theta-driving curve by impairing conduction in the dorsal noradrenergic bundle. The difference between the sexes shown in Fig. 7.4 is so far less clearly understood. Drewett and I have shown that it depends upon testosterone, secreted by the testis, since castration of the male eliminates the 7.7-Hz minimum and injection of this hormone restores it (Fig. 7.4). Since cells in the locus coeruleus take up testosterone, it is possible that the hormone acts upon these cells so as to increase activity in the dorsal noradrenergic bundle.

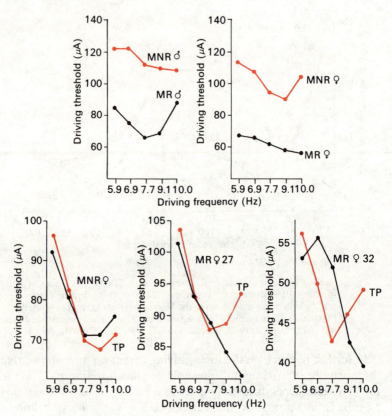

Fig. 7.5. Theta-driving curves (as in Fig. 7.4b) in untreated Maudsley Reactive (MR) and Nonreactive (MNR) males and females (above), and in a group of four MNR females and two individual MR females before and after daily injection of testosterone propionate (TP) (below). Data from Drewett *et al*. (1977).

The difference between the Maudsley strains in the theta-driving curve (Fig. 7.5), however, is not due to differences in circulating hormones. The data summarised above suggest, therefore, that it may be due to differences in the functioning of noradrenergic neurons in the brain. In particular, one might expect these neurons to function at a lower level in Nonreactive rats. David Blizard has indeed established that the two Maudsley strains differ in the level of activity of noradrenergic neurons, not only in the brain, but also in the periphery, where noradrenaline is a key transmitter in the sympathetic nervous system (Fig. 5.1). However, the direction of Blizard's findings is at first sight surprising: his results show that both the amount of noradrenaline and the rate at which the transmitter is released (a measure of activity in the relevant nerve-cells) are generally *higher* in Nonreactive relative to Reactive rats.

This observation poses a puzzle, not only for the chain of argument pursued above, but also in relation to Blizard's own findings in the peripheral nervous system. For noradrenaline release is known to increase heart rate (Chapter 5); yet – as one might expect, given the general evidence that Reactive rats are more emotional than Nonreactive rats – heart rate is higher in Reactives (the strain with the lower levels of noradrenaline), especially in response to mild stress, such as handling, bright light, or loud noise. Fortunately, further observations reported by Blizard's group have done much to resolve this puzzle. For it turns out that although blood levels of noradrenaline (originating from the terminals of noradrenergic neurons of the sympathetic nervous system) are higher in Nonreactive rats under resting conditions, the *increase* in noradrenaline (and adrenaline) release in response to mild stress is greater in Reactive rats. (Once again, by the way, there is a parallel here with sex differences – though we have to cross many species to find it: Marianne Frankenhaeuser has found that men respond to stress with a greater elevation in noradrenaline and adrenaline release than do women.) Another report from Blizard's group shows in addition that Reactive rats have a greater number of noradrenaline receptors in the heart (upon which the transmitter acts so as to increase cardiac output) than do Nonreactive rats.

It seems likely, therefore, that in the periphery (in spite of the lower resting levels of noradrenaline release observed in Reactive rats) the net functional output of the noradrenergic portion of the sympathetic nervous system, especially in response to stress, is higher in Reactive than in Nonreactive animals. It is not yet known whether a similar pattern of strain differences holds for noradrenergic functioning in the brain; but such a pattern would be consistent with our observations on the theta-driving curve (Fig. 7.5)

There are, then, a number of diverse strands of evidence which suggest that the physiological basis for the differentiation between the fearful Reactive rat and the fearless Nonreactive rat includes systems which are involved in determining differences between the sexes of a quite general kind. Indeed, it is only on this hypothesis that it is possible to see much pattern in the complex set of physiological changes which selective breeding (in a striking demonstration of the intimate relations which exist between psyche and soma) has produced in the Maudsley strains. But the conclusion that the physiology of sex differences is part of the physiology of fearfulness is a very important one indeed. For much more is known about the physiology of sexual behaviour and sex differences than about almost any other aspect of physiological psychology. The genetic basis of sexual differentiation is also comparatively well understood. And, furthermore, both the physiology and the genetics of sexual behaviour and sexual differentiation remain very similar among all the members of the mammalian family. Thus the involvement of the physiology of sex differences in the determination of fearfulness is likely to offer the experimental psychologist a relatively easy way in which to tackle the physiological basis of fearfulness, not only among animals, but also in Man.

The physiological basis of sex differences

Given these premises, then, it becomes both relevant and important to enquire what the physiological basis for sex differences in fearfulness might be.

One possibility is that these differences are due to the presence of the female sex hormone, estrogen, in the female and its absence in the male. This possibility can be excluded, however, since the sex differences in defecation, ambulation, passive avoidance, and shuttlebox avoidance all persist after the sex glands have been removed from both sexes. The same argument applies to the presence of androgen in the male and its absence in the female. The persistence of sex differences in emotional behaviour after gonadectomy also eliminates the possibility that these differences are simply due to the number of females which, on any given occasion of testing, are likely to be in estrus, which, as we have seen, may involve a reduction in emotionality. Thus it appears that we have to seek some more permanent feature of maleness or femaleness to account for our sex differences.

We may find this more permanent feature in the concept of brain sex. The female differs from the male both behaviourally and endocrinologically in that she is cyclic and he is not (Fig. 7.6). This cyclicity is apparent in the rat's estrous cycle and in the primate and human menstrual cycle. Hormonally, the cycle depends on the way in which the pituitary stimulates the gonads; but experiments (Fig. 7.7) have made it clear that in rodents and several other small mammalian species, the pituitary is in turn under the control of the brain, and probably of the hypothalamus. They have also shown that it is possible to alter the way in which the brain controls the pituitary by interference with the endocrine system at a sufficiently early age. In the rat the critical period is the first 1–5 days of life. If the testis is removed from the 1-day old rat, it grows up for most endocrinological and behavioural purposes as though it were a female (though it retains

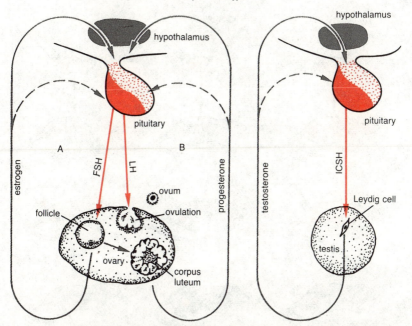

Fig. 7.6. Interplay of sex hormones differs in the female mammal (*left*) and the male (*right*). In the cyclic female system the pituitary initially (A) releases a follicle-stimulating hormone (FSH) that makes the ovary produce estrogen; the estrogen then acts on the hypothalamus to inhibit the further release of FSH by the pituitary and to stimulate the release of a luteinising hormone (LH) instead. This hormone (B) both triggers ovulation and makes the ovary produce a second hormone, progesterone. On reaching the hypothalamus the latter hormone inhibits further pituitary release of LH, thereby completing the cycle. In the noncyclic male system the pituitary continually releases an interstitial-cell-stimulating hormone (ICSH) that makes the testes produce testosterone; the latter hormone acts on the hypothalamus to stimulate further release of ICSH by the pituitary. Broken arrows represent the earlier theory that the sex hormones from ovaries and testes stimulated the pituitary directly.

male external genitalia). That is, if primed with appropriate hormones in adult life, it displays predominantly female sexual behaviour; and if implanted with an ovary, it causes that ovary to function in a normally cyclical fashion. Conversely, if an infant female rat is injected with a small quantity of the male sex hormone, testosterone, before it is 5 days old, it grows up endocrinologically and behaviourally as a male (again, apart from the external genitalia). That is, primed with the appropriate hormones in adulthood, it displays predominantly male sexual behaviour; and it fails to cause an implanted ovary to function in a normally cyclic fashion.

If we can experimentally alter the sex of the brain in this way, may we not be able to alter the sex differences in emotionality? Fig. 8.5

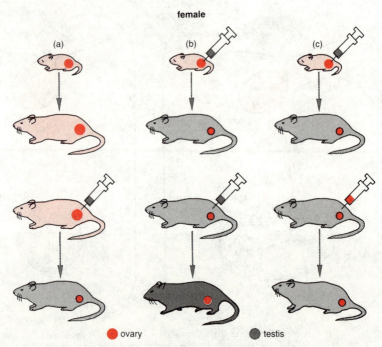

Fig. 7.7. Masculinised female rats were produced by injection of testosterone (black syringe) at birth. In column (a) a normal female (red) is injected with male hormones when mature; the animal exhibits some male sexual behaviour (grey). In column (b) the female is injected with male hormone in infancy; when reinjected at maturity, it exhibits full male sexual behaviour. In column (c), in spite of an injection of female hormone (red syringe) at maturity, the masculinised female fails to exhibit female sexual behaviour.

shows the results we obtained when we injected 5-day old female rats with testosterone and tested them in adulthood in the Open Field. There is a marked rise in the defecation score, abolishing the usual sex difference in one experiment and even reversing it in another. Ambulation, too, is changed in the male direction, being considerably decreased. Swanson has reported very similar findings with the same technique in the hamster, female ambulation scores being decreased and their time to emerge into a strange environment being increased until they were indistinguishable from those of the male. Conversely, castration of the neonatal male rat has been shown by Pfaff & Zigmond to decrease Open Field defecation and increase ambulation. The same pattern of change has been demonstrated by Beatty's group for shuttlebox avoidance: injection of testosterone to the neonatal female depressed performance (a move in the male direction), and castration of the neonatal male improved performance.

male

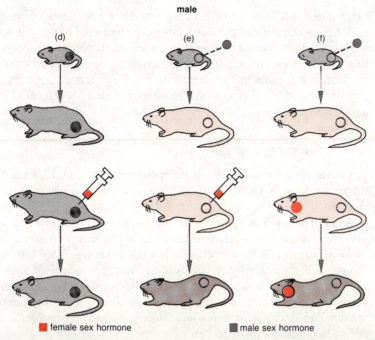

■ female sex hormone   ■ male sex hormone

Feminised male rats were produced by injection of estrogen and progesterone or by ovary implants only when the males had been castrated at birth and thereby deprived of testosterone during the critical first days of life. In column (d) a normal male (grey) is unaffected by the injection of female hormones (red syringe) at maturity. In column (e) a castrated male is similarly injected; it then assumes the female's permissive sexual posture. In column (f) the same behaviour is produced by implanting an ovary.

The main lines, then, of at least one of the routes in mammals from genes to fearful behaviour have begun to emerge. In the beginning, there are the genes which specify either a testis and male genitalia or an ovary and female genitalia. In early infancy the testis secretes a hormone which makes the hitherto undifferentiated brain male; the brain of an animal endowed with an ovary continues unaltered, i.e. female. The male brain guides the development of male physiology and, above all, endocrinology; the female brain does the same for the female body. The differentiation of the endocrine system which results from this includes not only the interplay between the pituitary and the gonads, but also the thyroid and other endocrine organs which display sex differences. Along with the differentiation of the endocrine system, there develop in the brain itself systems for the control either of male sexual behaviour or of female sexual behaviour. And here too the differences between the sexes extend beyond sexual behaviour itself: the brain specified as male by hormones from the infant testis determines the display of male emotional behaviour, with a relatively high level of fear and a male sensitivity to stress, especially the stress of social intercourse which activates the general adaptation syndrome; the brain left unaltered in infancy as female controls the organisation of female emotional behaviour, with a lower level of fear and less sensitivity to the stress of social life.

Sex and fear in Man

Sex is always with us; but it has been around rather a lot for the last three chapters. What do the conclusions we have reached in these chapters have to tell us about sex and fear in Man?

A major theme has been that of the antagonism between fear and sex. The prevalence of disturbances of sexual behaviour under conditions of stress and in the various human neuroses is well known, though it has usually led to entirely the wrong conclusions. This is most obvious in Freud's readinesss to attribute nearly all neurotic behaviour to aberrations of the sex drive, whereas, in the light of our mammalian heritage, it is far more likely that sexual problems are not causes but results of fear or neurosis.

A second important theme has been the stressful effects of high population density; and we have seen the way in which population density is signalled to the members of the population by the social interaction involved in the establishment of status and dominance hierarchies. The fact that (according to the statistics for New York State) mental disorder of all kinds is twice as great in the city as in rural areas is perhaps an example of the effects of crowding in Man. But there are so many differences between town and country that it is difficult to apportion cause and effect in the interpretation of such statistics. More compelling evidence has been gathered by Galle and his colleagues in a careful study of the effects of population density (measured in a number of different ways) in Chicago. These workers found that even when allowance was made for the correlation of crowded living quarters with class and race, two measures of crowding in the home predicted higher rates of mortality and of admission to mental hospitals. Laboratory studies of the effects of crowding on human subjects have produced only small and inconsistent effects. However, Freedman has reported an intriguing pattern of sex differences. When groups of young men were crowded together in a small space, there was a rise in competitiveness, in negative emotions, and in punitive feelings; but when the same experiment was done with young women, crowding produced exactly the opposite kind of change. This sex difference, of course, takes the direction that would be predicted by the arguments rehearsed in this chapter. Another possible example of the effects of population density in Man is the greater incidence of ulcers in the town than in the country. The great increase in ulcers this century is perhaps a response to increasing urbanisation; and it is significant that the increase in this typical stress disease has occurred mainly in young males – who would be the most sensitive to stress in most mammalian societies.

The third major theme has been the greater susceptibility of the male mammal, compared with the female, to stress and fear. This, too, as we have just seen, appears to find its expression in Man. In 1945 duodenal ulcer was 10 times more common in males than in females, though the incidence has recently been rising in women. The role of sexual physiology in this sex difference is suggested by the fact that it does not appear before puberty. Furthermore, the incidence of ulcers rises in women after the menopause, whether this occurs naturally or as the result of surgery, suggesting that female sexuality affords positive protection against this disease. Conversely, arthritis, which can be treated by the administration of glucocorticoids and is probably due to an insufficient secretion of these hormones, is far more common in women than in men. There are also sex differences in the activity of the sympathetic nervous system and the secretion of adreno-medullary hormones. Frankenhaeuser has found that men respond to a wide variety of psychological challenges with a greater elevation of the levels of circulating adrenaline and noradrenaline than do women; I have already commented on the parallel between this observation and the different ways in which the two Maudsley strains respond to stress. Where sexual behaviour is itself involved, the male susceptibility to stress is again seen: all kinds of sexual disturbance, including many which can be plausibly attributed to fear, are more common in men than women.

However, there are substantial reasons for caution in applying the generalisations we have drawn in this chapter, largely from studies of rodents and similar small mammals, to Man. Some of these reasons are endocrinological, some psychological.

Endocrinologically, as noted earlier, one key difference between the sexes is that the female is cyclic, the male not. In sub-primate species females display an estrous cycle, which in primates becomes the menstrual cycle (see Fig. 7.6). The menstrual cycle is more complicated, but based upon essentially the same hormonal events as the estrous cycle. In small mammals (rats, hamsters, and guinea-pigs – a group collectively given the mythical species name 'ramstergig' by Frank Beach) it has been demonstrated that the estrous cycle depends ultimately on controlling centres in the brain. A ramstergig female brain contains mechanisms which programme cyclicity, a male brain does not; and this sex difference in the brain depends in turn upon the action of testosterone at a critical time in early life upon brain sites which control the release of pituitary homones (see Fig. 7.7). However, during the last decade Knobil's research on the rhesus monkey has shown that the ramstergig model is not applicable to this primate species. In the rhesus monkey, the determining factor is not in the brain, nor even in the pituitary, but in the *gonad*: the hormones (estrogen and progesterone) released from an ovary give rise (by the signals they send to the brain) to a cyclic output of follicle-stimulating and luteinising hormones from the pituitary (Fig. 7.6); the hormone (testosterone) released from a testis does not have this capacity to induce cyclicity; and the type of brain (genetically male or female, exposed or not exposed to testosterone in early life) is unimportant. Furthermore, the available evidence indicates that the human menstrual cycle is controlled in the same way as the rhesus monkey's.

Knobil's experiments, then, show that one cannot take it for granted that the same route to sexual differentiation is followed in primates as in the ramstergig. It is not known, however, whether similar shifts from the brain to peripheral endocrine glands as the site of sexual differentiation have taken place for other sexually dimorphic primate characteristics, besides the cyclicity of release of sex hormones. Indeed, there is some evidence that in the case of aggressive behaviour, the site of sexual differentiation lies in the brain in primates as it does in rats and mice.

In both rodents and primates the male is usually the more aggressive sex. In rodents, this sex difference depends on the same principles that are operative for male sexual behaviour. High levels of aggression are typically seen when the levels of circulating testosterone in males are high. Castration of the adult male mouse causes a fall in aggressive behaviour, and replacement of testosterone by injection restores aggression to its normal level; injection of testosterone to an adult female mouse (unless it is extremely prolonged) fails to affect aggressive behaviour. In the rat a slightly more complex pattern was seen by van de Poll's group in Amsterdam: injection of testosterone to males (castrated in adulthood) *increased* aggression if the animal was allowed to *win* its fights but *decreased* aggression if it lost; neither of these effects was seen in female rats treated in the same way. In contrast to these findings with testosterone, ovarian hormones have little direct effect on aggressive behaviour. Thus the requirements for high levels of aggression are high levels of testosterone, a male brain, and success in agonistic encounters.

These inferences are supported by the results of experiments with the neonate. Castration of the neonatal male rat or mouse lowers adult levels of aggression and eliminates the response to testosterone in adulthood. Conversely, injection of testosterone to the neonatal female, combined with testosterone injections again in adulthood, produces a level of aggression typical of the normal male. Similar phenomena have been observed in the rhesus monkey by Young, Goy & Phoenix. In this species, the juvenile male's behaviour in social and play situations is very different from that of the female: more aggressive, more rough-and-tumble, and richer in threatening facial expressions. Injections of testosterone to pregnant rhesus mothers had the consequence that their female offspring displayed the aggressive play patterns of the male.

One obviously cannot perform Young's experiments with human beings. But occasionally either nature or physicians make mistakes which mimic those experiments. Money & Ehrhardt studied a group of girls who had been exposed *in utero* to abnormally high levels of androgens. In some cases this was the result of a genetic malfunction in the girls' own adrenal glands, which over-produced androgenic hormones. In others the mothers had been treated with hormones to reduce the probability of spontaneous abortion, but (unknown to the physicians at the time) these hormones had in addition androgenic properties. In both cases, the girls exposed prenatally in this way to high levels of androgens were reported to show unusually high levels of tomboyish behaviour during their childhood – perhaps an analogue of the increased rough-and-tumble play observed in Young's rhesus monkeys. Thus the propensity towards aggressive behaviour may depend to some degree on an action of androgens on the developing brain in our species as it does in both rodents and rhesus monkeys. But note that the same girls studied by Money & Ehrhardt showed no abnormal tendencies in their so-called 'gender identity' – they thought of themselves as girls, expected to have romantic attachments with and marry boys, and so on.

What should we expect, then, for the sexual differentiation of fearful behaviour in our own species? If we were to argue from Knobil's observations on the endocrine cycle or Money & Ehrhardt's on gender identity, we would not expect fearful behaviour to depend on the exposure of the developing brain to androgens. But if we were to argue from the data on aggressive behaviour, a role for early testosterone would not be surprising.

Evidence in support of such a role has been reported by Jacklin. She and her co-workers measured a number of hormones at birth (in the umbilical cord) and related them to timidity (as assessed by reactions to toys of various kinds) over the first 18 months of life. Timidity in boys (but not girls) was greater, the lower the levels of neonatal testosterone and progesterone, and the higher the levels of estradiol. Note that this relation between testosterone and timidity takes the opposite direction to the relation that holds in the rat. But (as we shall now see) this change in direction is consistent with other differences between rodent and primate species.

So far we have concentrated on the endocrinological reasons for caution in applying ramstergig data to Man. There are equally compelling psychological grounds for caution. As we have seen, the evidence is very good that female rats, on average, are less fearful than males. We also saw some evidence that the same pattern holds in our own species. But there is much stronger evidence that it does not. Almost all neurotic conditions, such as phobias, obsessive–compulsive neurosis, and neurotic depression (see Chapter 14), are much more common in women than men. Surveys of the normal population similarly reveal generally greater fear among women and girls than among men and boys. Conceivably, cultural factors are responsible for this pattern. Perhaps women are more willing than men to admit to their fears or to consult their doctors about them. This kind of hypothesis is plausible. But I know of no evidence in its support and one strong reason to doubt it: data from at least two other primate species (chimpanzees and bonnet macaques) fit the human, not the rodent pattern. Thus the shift to greater female fearfulness appears to be a primate rather than a specifically human invention.

Why should such an evolutionary shift have taken place? Tony Buffery and I have suggested an answer to this question. We proposed a hypothesis to explain the facts that: (1) the sex difference in aggression takes the same direction in rodents and primates (males more aggressive); and (2) the sex difference in fear takes different directions in rodents and primates (male rodents more fearful, female primates more fearful).

Our hypothesis turns on arguments like those deployed in the previous chapter when we discussed population density. That discussion assumed a close relationship between fearful and submissive behaviour, respectively, and similarly between aggressive and dominance behaviour. Now, in agreement with Wynne-Edwards' hypothesis that the male is usually the epideictic sex, dominance interactions in general play a much larger role in the lives of male than female mammals. This is as true for primates (including Man) as for rodents. But dominance interactions are not totally absent from the lives of females. It is at this point that an interesting difference emerges between rodent and primate species. Among rodents dominance interactions are almost exclusively a male affair. Males fight with males and establish a social hierarchy among themselves. They do not fight with females, nor do females fight much with each other; and females do not figure within the male dominance hierarchy (whether as dominant or as subordinate), nor do they form a strong hierarchy among themselves. Among most primate species the pattern is different. Males and females usually form a single dominance hierarchy in which, on average, males are dominant over females (though some females dominate some males). In forming this hierarchy, males interact socially with both males and females, and females interact with one another.

Consider now what these different social structures require of male and female rodents and primates, respectively. Among rodents, males must come equipped to display both dominant and subordinate behaviour, depending on the likelihood that their opponent of the moment is weaker or stronger than themselves (as demonstrated in van de Poll's experiment described above). Female rodents have much less need for either kind of behaviour, since they do not often engage in dominance interactions. Equating dominance and submission to aggressive and fearful behaviour, respectively, it follows that females should be both less aggressive and less fearful than males, as observed among rodents. Among primates, since both sexes engage in dominance interactions, both sexes must come equipped to display both dominant and subordinate behaviour. But the male is on average dominant over the female. It follows that the male on average should be better than the female at displaying dominant behaviour and worse at displaying subordinate behaviour, i.e. the male should be more aggressive and less fearful than the female, as observed among primates. The negative relation between neonatal testosterone levels and timidity observed in young boys by Jacklin (see above) suggests that though human males show less (not more) fear than females, the route by which this sex difference is achieved still depends (as does the converse sex difference in the rat) on the action of male hormones on the developing brain.

Constructed as it is *post hoc* to account for the known facts, the Gray & Buffery hypothesis is difficult to test. There has, however, been one attempt to do so. J. M. Warren argued from the hypothesis that there should be no sex differences in fearfulness in a species in which the formation of hierarchies plays little part in social life. He accordingly tested cats, which lead essentially solitary lives. As predicted, male and female cats did not differ in fearfulness (tested by the animals' reactions to novel stimuli).

In these different ways, then, our human mode of responding to stress appears to reflect pressures which have been at work at different stages of our evolution as a species. A good analogy is that of the onion. The systems that control behaviour come in layers wrapped around each other. The oldest layers, in the heart of the onion, go back to the earliest stages of mammalian, even vertebrate, evolution; others, more recent, stem from our heritage as a primate; and still others, of course (but these are in some ways the hardest to objectify and bring under laboratory control), lie at the surface of the onion and are unique to our own species. When new layers are added, the old ones remain. Sometimes the new layers simply supply a more advanced means to reach an ancient end. But sometimes the effects of old and new layers are to some degree in conflict with each other. Something like this appears to have happened in regard to the sex differences in fearfulness which have occupied our attention in this chapter. In some ways, as we have seen, the older pattern of greater male fearfulness and susceptibility to stress has persisted in our own species; but in other ways the pattern has gone into reverse. Disentangling these different effects will not be easy; but it is an essential task if we are properly to understand our own nature.

8 The early environment and fearfulness

In our attempt to trace the route from 'fear' genes to fearful behaviour in the last chapter, we needed to pass through the earliest period of mammalian infancy. But by this time the environment, too, has begun to exercise a powerful influence on the eventual behaviour of the adult animal. In fact, the new-born babe has already had a great deal of experience whose effects will last throughout his life, if we may judge from results obtained in the psychological laboratory. The eventual personality of the rat may be affected by quite slight influences applied to the animal in the first few days of life; by the social environment with which he is provided by mother and siblings; by events which have occurred during his stay in the womb; and even by events experienced by his mother long before his conception (see Fig. 4.3).

The most extensive programme of research on the effects of the early environment on adult emotionality is associated with the names of two American workers, S. Levine and V. Denenberg. They and their collaborators have carried out a large number of experiments on the effects of exposing laboratory rodents to a variety of stimuli, often of a mildly noxious kind, during the first 1 or 2 weeks of life. (To place this time-scale in context, it should be said that the laboratory rat is normally weaned at about 3–4 weeks of age, that it becomes sexually mature at about 2 months, is a young adult by about 3 months, and has a normal life span of about 2 years.) Early on it was expected that these treatments would have adverse effects, if any, on the animal's development, but it soon became clear that, on the contrary, they were usually beneficial. It also transpired that, by and large, the type of stimulus used was of minor importance: very similar results have been obtained with treatments as diverse as electric shock, exposure to cold, shaking about, or simply picking the animal up and placing it in a small compartment outside the home cage (a treatment known as 'handling').

All of these treatments have been found most frequently (though with important exceptions) to reduce fearfulness when this is measured later in the adult. Open Field defecation is reduced and ambulation is increased; freezing in the Open Field is also reduced; emergence from the home cage into a novel environment is speeded up; resumption of drinking after it is interrupted by the administration of an electric shock occurs more quickly; susceptibility to conflict-produced ulcers is reduced; and the speed and efficiency of avoidance learning in the shuttlebox (Fig. 4.8) is increased. Furthermore, effects of this kind have been reported to last up to 1½ years, by which time the rat is well on the way to old age.

The exceptional cases in which infant stimulation experiments have resulted in *increased* adult fearfulness have usually involved relatively high-intensity stimulation; or they have employed the mouse, and, for this smaller animal, the treatments applied may represent an effectively higher intensity of stimulation than for the rat. In consequence, the effects of infant stimulation on the adult animal can often be described by the kind of curvilinear relationship depicted in Fig. 8.1: both too little and too much stimulation during infancy have relatively harmful effects on the adult animal, which shows an increased degree of fear, as measured by the defecation response and poorer learning of avoidance behaviour, as well as decreased body weight, when compared to animals which have been exposed during infancy to an intermediate degree of infant stimulation. In general, however, too little stimulation during infancy appears to have more undesirable consequences than too much.

The early social environment

In the infant treatment experiments, the experimental animals are subjected to stimulation over and above the normal stimulation to which the laboratory routine exposes the controls. An alternative approach is to *deprive* animals of some of the stimulation to which they would normally be exposed. This technique has been used in the study of the effects of *social isolation* (which of necessity involves sensory restriction as well), especially with higher species. In general, the results of the various isolation experiments agree well with the picture emerging from the early handling studies, namely that if stimulation is to deviate from the optimum level, too little is worse than too much.

As we might expect, the effects of social isolation are most dramatic in the highly social primates. Harlow raised rhesus monkeys in total social isolation (and with a great deal of sensory restriction), without their mothers or other monkey companions. When they were allowed to join other animals at 6 months, there were gross disturbances in the way they played with one another, even compared to other animals which had been reared in the semi-isolation of bare wire cages from

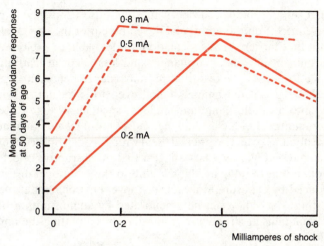

Fig. 8.1. Avoidance conditioning performance in the shuttlebox in mice at 50 days of age, at three intensities of shock, as a function of intensity of shock (abscissa) to which the animals were exposed in infancy.

which they could see and hear, but not touch one another. Disturbances in their emotional behaviour were apparent in a number of ways. They sat for long periods clutching themselves or, when caged in pairs, each other, apparently in great fear. When playing with the semi-isolated animals, the totally isolated monkeys never threatened their companions, even though they were frequently subjected to abuse and aggression. However, their aggressive/defensive behaviour was submerged rather than eliminated. When they were put together with other monkeys which had been subjected to the even more severe deprivation of 12 months' total isolation, the 6-months-deprived animals 'attacked their more helpless playmates in a violent and uncontrolled fashion'.

R. E. Miller used his co-operative conditioning paradigm (Chapter 3) to study Harlow's socially deprived monkeys further. In this design the receiver monkey is trained to perform a response that will avoid shock by picking up signs of fear in the facial expression of the sender monkey (the only one of the pair to perceive the warning of impending shock). Compared to normally reared monkeys the social isolates were impaired both in responding to the sender's facial expression and in sending appropriate cues themselves. Thus social isolation during infancy grossly disturbs the normal development of both the expression of emotion and the capacity to decode emotional expression in others.

Combined sensory and social deprivation has also been shown to have adverse effects on the emotional development of the dog. Melzack reports that terriers bred in isolation are less able to explore a novel environment in a directed manner and less adept at learning an avoidance response. The disturbance in avoidance behaviour was so severe that the isolated animals failed even to learn not to put their noses into a candle flame and, indeed, spent *more* time close to the experimenter after having had their noses burnt by him. Similar results were obtained when he pricked them with a needle. Subsequent results obtained by Fuller & Clark suggest that the disruptive effects of early isolation in the dog are largely due to the emotional disturbance produced by exposure to a radically novel environment at the time of testing, since they can be ameliorated by administration of the major tranquillising drug chlorpromazine, or by petting the animals just before the session.

The lowly rat, too, is susceptible to the effects of alterations in his early social environment. A series of studies from Ader's laboratory, for example, has shown that in agreement with the findings on dog and monkey, total social isolation applied after the rat is weaned increases his level of adult fearfulness. Early weaning is another variable which, in Ader's hands, appears to result in increased emotionality in the rat: weaning at 15 instead of 21 days of age led to greater adult susceptibility to conflict-induced ulcers, though in males only. As well as bringing a strange kind of comfort to those psycho-analytic theories which see the source of all male ills in unsatisfied love of the mother, this result reminds us of the ubiquity of sex differences in emotional behaviour, the male as usual being the more delicate sex.

In general, then, early social isolation has adverse effects on emotional behaviour. However, in view of the effects of crowding on the general adaptation syndrome (Chapter 6), it is obviously not the case that increases in social stimulation in infancy are beneficial without limit. As in the case of simple sensory stimulation, there appears to be a curvilinear relation between social stimulation and adult emotionality, both too little and too much stimulation producing a high level of fearfulness. As an example of the effects of too much stimulation, Seitz grouped rat pups into sixes or twelves and fostered the litters so created; in adulthood it was the animals reared in the large litters which had a higher Open Field defecation score, a lower ambulation score, slower emergence time, and a greater tendency to startle at unexpected stimuli. Taken together with Ader's work on the effects of total isolation in the rat, this allows us to conclude that for this species, *either* to live singly *or* to live in twelves is to live dangerously.

In the squirrel monkey, too, Gonzalez *et al*. have shown that circulating corticosteroids (Chapter 5) are higher in animals grouped in sixes or caged singly than in animals caged in pairs.

In recent years the effects of social isolation have been subjected to considerable analysis in experiments with small rodents. These experiments have revealed two rather different effects. Dorothy Einon, a prominent investigator in the field, has summarised the relevant data as follows. In agreement with Ader's early results, one effect of isolation is to increase later timidity, as measured for example by reluctance to emerge into a novel environment or to eat there. This effect is seen in rats, mice, guinea pigs, hamsters, and gerbils; it occurs after isolation at any age; and it is reversed by subsequent periods of social housing. But, in addition, in the rat a quite different effect is seen if isolation is imposed between about 19 and 45 days of age (but not at other ages). This second effect consists of hyperactivity, slow extinction of rewarded behaviour, poor learning in spatial mazes, and slower learning of reversal problems (in which a previously rewarded response is no longer rewarded, while a response previously treated as incorrect now leads to reward). This syndrome bears many similarities to the syndrome seen after injections of anti-anxiety drugs (or lesions to the septo-hippocampal system of the brain; see Table 12.2.), implying that it may reflect *decreased* fearfulness, the opposite of the first effect described above. The second effect of isolation, unlike the first, is not reversed by later periods of social housing. Einon has suggested that it is due to the loss of opportunities for rough-and-tumble play during a critical period of social development: the isolated animal fails to develop the normal capacity to inhibit behaviour that is punished by stronger or more dominant peers. (As we shall see in later chapters, many features of the syndrome seen after anti-anxiety drugs or septo-hippocampal damage appear to be due to the loss of just such a capacity for behavioural inhibition.) In agreement with this hypothesis, Einon has shown that the second effect of social isolation is absent in mice, guinea pigs, gerbils, or hamsters (none of which engage in rough-and-tumble play fighting), but present at least to some extent in ferrets (which do).

As implied by Einon's play hypothesis, the young animal is sensitive to more subtle aspects of social interaction than just the sheer amount of social stimulation received. Evidence exists that the *kind* of animal encountered by the infant may also have permanent effects on adult personality. One way to demonstrate this aspect of social interaction is to alter the mother in some way and look for changes in the behaviour of her pups when they reach adult life. We may do this, for example, by applying some treatment to the mother between the birth of her litter and the time the litter is weaned. Control litters are, of course, left undisturbed. In one of Ader's experiments using this approach, the rat mother was simply removed from the home cage for a time on each of Days 1–10 or Days 11–20 of the pups' lives. This had very similar effects on the pups' adult emotional behaviour to those achieved with the low-intensity infant stimulation treatments: they were quicker to emerge into a novel environment and less suscep-tible to conflict-induced ulcers than pups reared by undisturbed mothers – i.e. there was an apparent reduction in fearfulness.

The fact that the pup's adult behaviour may be altered by treatments given to the mother raises the possibility that changes in maternal behaviour consequent upon the handling of the pup may mediate the effects of infant stimulation upon adult behaviour described at the start of this chapter. This hypothesis was proposed by Martin Richards in 1966 and has received some support. Observational studies in Levine's laboratory have shown, for example, that rat mothers do indeed treat handled and non-handled pups differently. Other exper-iments have demonstrated that the effects of early handling may vary depending upon the strain of the mother to which the infants have been fostered (to eliminate confounding between the neonatal mater-nal environment and genotypic or prenatal environmental effects). This finding shows that the effects of early handling may at least be altered by the behaviour of the mother, though it does not necessarily follow that they are uniquely mediated by maternal behaviour.

The prenatal environment

We do not have to wait until the pups are born before we try to alter the mother. The environment exists not from the time of birth but from the time of conception; and, although interest in the psycholog-ical effects of the intra-uterine environment has grown more slowly than in those of the early postnatal environment, there are now avail-able quite a few experiments which demonstrate beyond doubt that there is truth in the old wives' tale that it is possible to affect an individual's personality by frightening its mother while it is in the womb. The technique here is to apply some comparatively mild form of stress to the mother while she is pregnant and to compare the adult behaviour of her pups with the behaviour shown by the offspring of an undisturbed mother. Now this kind of procedure may also affect the mother's *postnatal* behaviour towards the pups and thus alter their behaviour by this route. If we wish to demonstrate a true effect of the *prenatal* environment on adult behaviour, then it is necessary to control for postnatal effects by the same kind of cross-fostering technique which, as we have seen, is essential for establishing the reality of a genetic effect on behaviour (Fig. 4.5); but now the cross-fostering is between litters born to treated and untreated mothers rather than between those born to mothers of different strains.

In an effort to make the stress applied in this kind of experiment as purely psychological as possible, W. R. Thompson devised the technique of training the prospective mothers *before mating* to avoid an electric shock on presentation of a warning signal. After the animals have become pregnant they are simply returned to the apparatus (a shuttlebox), exposed to the warning signal (but *no* shock), and prevented from performing the avoidance response: a procedure which is calculated to make most rats pretty frightened. Working with the Maudsley strains in Broadhurst's laboratory, Justin Joffe used Thompson's technique in an extensive experiment which not only demonstrates the effects of the prenatal maternal environment on adult fearfulness, but also highlights some of the other complexities inherent in mammalian development.

The design of Joffe's experiment is illustrated in Fig. 8.2. He used the two Maudsley strains, mated in all four possible combinations (a 'diallel cross'). This was in order to allow an evaluation of any interactions which might occur between the environmental treatments and the two genotypes involved, that of the mother and that of the foetus. Mothers were treated in one of three ways: the 'experimental' group was treated according to Thompson's procedure; the 'premating control' group was exposed to avoidance conditioning prior to mating, but not touched during pregnancy; and the 'normal' group was exposed neither to avoidance conditioning nor to any treatment during pregnancy. *All* the offspring were then fostered to untreated Nonreactive mothers, thus ensuring that any changes in offspring behaviour could be attributed to *pre*natal influences. The offspring of these Nonreactive mothers were in turn fostered to the three kinds of experimentally created mother, so that postnatally transmitted effects of the experimental treatments could also be evaluated. The tests used in adulthood were the Open Field and the same avoidance conditioning task.

Let us first attend to those features of Joffe's results which concern the prenatal environment. On the avoidance conditioning task, the offspring of the experimental group were significantly better than the offspring of the normal group, while the offspring of the premating control group did not differ from these. Thus the pregnancy stress

Red – Maudsley Reactive (MR)
White = Maudsley Nonreactive (MNR)
Pink = Cross between MR and MNR

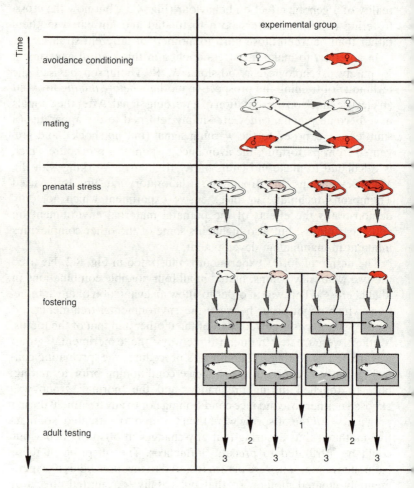

1 = prenatal and genetic effects   2 = postnatal effects

Fig. 8.2. The design of Joffe's experiment on prenatal stress. For a full explanation, see text. His design allowed the separate evaluation of prenatal environmental effects (by comparison across columns between the groups labelled 1 in the 'adult testing' row); of postnatal environmental effects (by comparison across columns between the groups labelled 2); and of the effects of the offspring on the behaviour of their foster-mothers (by comparison across columns between the groups labelled 3). Comparisons between the premating controls and the normal group disclosed the effects on offspring behaviour of the environment to which their parents were exposed before

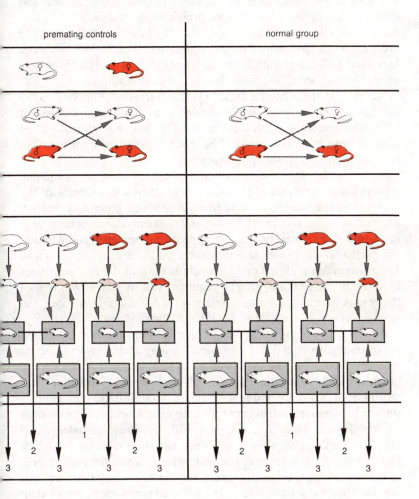

premating controls   normal group

offspring to mother effects

Caption for Fig. 8.2 (*cont.*).

their conception; comparisons between the experimental group and the premating controls disclosed the effects on offspring behaviour of the stress
applied to the mothers during the stay of the offspring in the womb. The
use of a 'diallel cross' (shown diagrammatically in the 'mating' row) also
allowed the evaluation of genetic effects, and of interactions between the
two genotypes involved (maternal and offspring) and the various kinds of
environmental effect.

applied in Thompson's procedure *had* affected the young then in the womb. Secondly, there were effects on the ambulation score in the Open Field test which differed according to genetic constitution and occurred only in the offspring of the *premating* control group: offspring from this group with Nonreactive *fathers* had their ambulation score reduced, while those with Reactive fathers had their ambulation score increased. Offspring of mothers which had received the full Thompson procedure did not differ from the normal controls.

Notice that these results mean: (1) that a treatment applied to the mother *before her pups are conceived* can affect them by way of influences transmitted prenatally; and (2) that this influence must actually have been *counteracted* in Joffe's experiment by the further treatment applied during pregnancy.

Notice further that the dependence of the effect of the premating treatment on the strain of the offspring's father is an example of the common phenomenon of an interaction between genes and environment; that is the *way* in which the environment affects behaviour is in its turn determined by the genotype of the organism subjected to the particular environmental influence. Furthermore, the design of Joffe's experiment allows one to conclude that, of the two genotypes involved in the prenatal effects (maternal and foetal), it was that of the foetus which was actually effective in his conditions, since the mother's genotype is of course not affected by the strain of the father.

A particularly interesting effect of prenatal stress has been described by Ingeborg Ward and her colleagues among male offspring of rat mothers exposed during the last week of gestation to any of a variety of stressors (restraint coupled with high-intensity light; crowding; malnutrition; or a conditioned fear stimulus). In adulthood such animals show a pronounced impairment in male sexual behaviour, including an inability to ejaculate, together with enhanced female sexual reflexes. Similar effects have also been shown in mice by Politch & Herrenkohl. If the prenatally stressed males are cross-fostered at birth to normal mothers, these changes in sexual behaviour persist, indicating that they are indeed due to the prenatal stress itself, rather than to any alteration in postnatal mothering. The effects of prenatal stress may be exacerbated, however, by social isolation after weaning. Conversely, group housing after weaning, injections of testosterone over a prolonged period in adulthood, or cohabitation with female rats are all able to ameliorate the stressed rats' unfortunate condition.

Ward has demonstrated the probable hormonal basis of these effects. As shown in Fig. 8.3, plasma testosterone levels in rat foetuses have a characteristic pattern, a marked rise occurring in males on

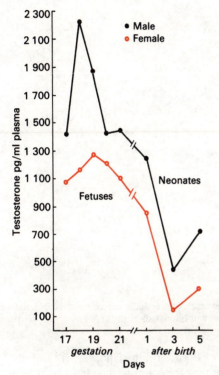

Fig. 8.3. Plasma testosterone levels of male and female rats from Days 17–21 of gestation and Days 1–5 postpartum. From Weisz & Ward (1980).

Days 18 and 19 of gestation. Prenatal stress obliterates this rise in testosterone levels (Fig. 8.4). Given the well-known role of testosterone in the masculinisation of sexual behaviour (see Fig. 7.7), this stress-induced change in testosterone levels is almost certainly responsible for the altered sexual behaviour displayed by the stressed animals when they reached adulthood. Support for this hypothesis comes from Dörner's group in East Germany, who were able to block the demasculinising effect of prenatal stress on adult sexual behaviour by administering testosterone to the stressed mothers (and so to the pups in their womb) and to the pups themselves just after birth. No changes are seen in the adult sexual behaviour of prenatally stressed females; nor would any be expected, since normal female sexual behaviour does not depend upon the action of hormones upon the developing brain as does normal male sexual behaviour.

Further analysis of the causal chain leading to these effects has implicated the endogenous opiates. It is known that stress induces

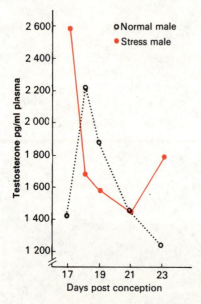

Fig. 8.4. Plasma testosterone levels of prenatally stressed and control males from Days 17–21 of gestation and the day of birth (Day 23). From Ward & Weisz (1980).

the release from the pituitary of β-endorphin (Chapter 5), and also the small doses of opiates suppress testosterone levels, probably by way of a reduction in the release of luteinising hormone (Fig. 7.6). It is possible, therefore, that prenatal stress causes the release of maternal endorphins, which then act upon the foetus to reduce testosterone levels. In support of this inference, Ward has shown that injections of morphine to pregnant rats give rise among male offspring to the same loss of masculine sexual behaviour, and increase in feminine sexual behaviour, as does prenatal stress. In addition, she has presented evidence indicating that the opiate antagonist naltrexone given to pregnant rats concurrently with prenatal stress (restraint plus bright light) is able to prevent the reduction in foetal testosterone levels that is otherwise caused by stress. A second possible causal pathway (not necessarily incompatible with the first) would involve the adreno-pituitary axis (Fig. 5.4). In line with this possibility, Politch & Herrenkohl have shown in mice that injections to pregnant mothers of either ACTH or corticosterone (both, of course, normally released under conditions of stress) impaired adult sexual behaviour in their male offspring.

Ward's experiments naturally invite speculation. Might these effects constitute yet one more aspect of the population density controls considered in Chapter 6? It is consistent with this view that crowding is one of the prenatal stressors shown to impair male sexual behaviour; it is also consistent that it is exclusively the male (Wynne-Edwards' epideictic sex) that is affected in this way. A second intriguing possibility is that the effects described by Ward may underlie some of the phenomena of human male homosexuality. Support for this possibility comes from Dörner's report that among homosexual men registered for medical treatment in a number of districts in East Germany, significantly more were born during the stressful period of the Second World War and the early post-war period than in the years before or after. However, there are many artefacts that can affect the results of this kind of social survey; more data are needed, therefore, to evaluate this hypothesis adequately.

The social psychology of the rat family

So far, we have been concerned with the effects we may have on the pups by altering the mother, whether we apply our experimental treatment to her after the pups are born, while they are in the womb, or even before they are conceived. But we must not forget the possibility that altering the *pups* may in turn alter the *mother.*

The Nonreactive females to which the offspring of Joffe's three kinds of experimental mothers (Fig. 8.2) had been fostered had all been tested in the Open Field before the experiment began. They were re-tested after their foster litters had been weaned. The normal decline in ambulation on a second occasion of testing appeared as usual in the females which had reared offspring of the untreated biological mothers. But it failed to appear in the animals which had given foster mothering to the offspring of the biological mothers exposed either to pregnancy stress or to the premating treatment only. We are forced to conclude that both experimental treatments had prenatally affected the pups and that the *different kinds of pup had been able to affect the behaviour of the foster mothers.*

Furthermore, besides these effects of pup on postnatal mother, Joffe was able to detect effects of postnatal mother on pup. The offspring of the Nonreactive foster mothers were themselves fostered to the three kinds of experimental mother. In adulthood, those which had been foster mothered by animals exposed either to the full Thompson procedure or to the premating treatment alone were significantly poorer at avoidance learning than those which had been fostered to untreated mothers.

Thus we are forced out of the tidy world of classical science, where there is an 'independent' variable which we can confidently manipulate and a 'dependent' variable which shows the effects of our manipulations. Instead we must enter the confused world of dialectics, in which everything we look at is both cause and effect and the final outcome is due to a dynamic interplay of forces exerted by *all* variables: the mother affects her offspring, the changed offspring in turn affect their mother, the changed mother again alters the behaviour of her offspring, and so on until some kind of equilibrium is reached.

The dialectics of the rat family contain at least one further dimension: the interactions between the pups themselves. We stumbled across this phenomenon (though we ought to have suspected its existence) in the course of our experiments on the physiological pathways which specify the sex differences in fearfulness (Chapter 7). Some puzzling inconsistencies in the results we were getting when we tested adult animals in the Open Field after castrating them or injecting them with hormone in infancy made us suspect that the treatments applied to the males in a litter were altering the behaviour of their sisters, and that the treatments applied to the females in a litter were affecting the behaviour of their brothers. We have carried out a number of experiments now in which these 'sibling effects' have appeared. The clearest instance is shown in Fig. 8.5.

In this experiment both males and females were injected either with a quantity of testosterone dissolved in peanut oil (a physiologically inactive substance) or the same amount of oil alone, as a 'placebo' control, in the four possible combinations: male placebo, female placebo; male placebo, female testosterone; male testosterone, female placebo; and male testosterone, female testosterone. In adulthood both defecation and ambulation in the Open Field depended not only on whether the animal had itself been injected with the hormone, but also on the treatment given to its brothers (or sisters). As we have already seen in Chapter 7, the masculinising effect of testosterone given in infancy to females is accompanied by a shift in their Open Field defecation score to the higher male norm; but this shift is greatly reduced if their brothers are simultaneously injected with the hormone. The direct effect of testosterone on the male is to decrease defecation; and this effect too is lessened if their siblings of the opposite sex are also injected with testosterone. The direct effect of infant testosterone is to decrease ambulation in both sexes. And, on this measure, the sibling effect works, not in opposition to the direct

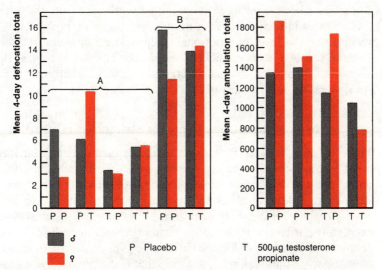

Fig. 8.5. The effects of injections of male sex hormone to infant rats (at 5 days of age) on defecation and on ambulation in the Open Field test in adulthood. On both measures the difference between the sexes which exist in the groups given placebo injection of oil only (the 'PP' groups) is abolished or reversed by the masculinising effect of the hormone on the females (the groups labelled 'PT' or 'TT'). The effects of the hormone are different depending on whether the siblings of the opposite sex within the litter are themselves treated with placebo or with the hormone. The results shown for defecation are from two separate experiments: experiment B was performed with Maudsley Reactive rats and with rats bred by crossing the Maudsley Reactive and Nonreactive strains; experiment A was performed with an unselected population of Wistar rats, which are the original parental strain from which the Maudsley strains were bred. The pattern of results for the PP and TT groups is identical in the two experiments.

effects of testosterone (as it does in the case of the defecation measure), but in harness with them: it is the animals coming from litters where both sexes were injected with the hormone which show most strongly the depression in ambulation scores due to testosterone.

We may bring some order into this puzzling pattern of results if we realise that the sibling effect *reduces* the direct effects of the hormone in the case of that measure (defecation) for which the direct effects are *opposite* in sign in the two sexes; and that it *enhances* the direct effects in the case of that measure (ambulation) for which the direct effects go in the *same* direction in the two sexes. Thus, in both cases, there is a tendency for the sexes to move closer when both have been treated with testosterone. Notice, however, that the animals injected with placebo do not display a sibling effect at all. It is as though the disturbance in normal development occasioned by the hormonal injection renders both sexes more susceptible to influence from the other sex; and this influence, when it occurs, is such as to impose on the whole litter the level of emotionality derived as an average of the levels of the two sexes considered separately. Thus we see again the same kind of possible mutual interaction between siblings within a litter which we saw to exist between mother and pups.

The physiological route from early environment to behaviour

There is, then, a great diversity of ways in which the early environment, social as well as purely sensory, may have permanent effects on the adult animal's susceptibility to fear. But the early environment is no more able than genes to produce the behaviour of the adult animal directly: it can only alter those physiological systems which eventually participate in the control and organisation of the observed behaviour. Just as we did in the case of the genetic control of fearfulness, then, we must ask the question: what is the physiological route whereby the early environment exerts its effects?

We have already seen the outline of one answer to this type of question, in the account of Ward's studies of the effects of prenatal stress on adult male sexual behaviour. That account emphasised the action of hormones (in particular, testosterone) upon the developing brain. It is possible that a similar account applies to the results of the infant stimulation experiments conducted by the Levine and Denenberg groups.

It is clear that some kind of basic physiological change is set into motion by the infant treatments. As well as affecting adult body weight, they cause an earlier appearance of body hair, earlier opening of the eyes, earlier locomotion, and earlier puberty. In addition, there are complex changes in the brain, including an earlier start to the process whereby neurons become covered by a myelin sheath (aiding conduction of the nerve impulse), increases in the weight of certain parts of the brain, decreased activity of cholinesterase, an enzyme which is important in the transmission of the nerve impulse at cholinergic synapses, and an increased susceptibility to electro-convulsive seizures. Levine has even reported that infantile stimulation can affect adult resistance to leukaemia, while, in the dog, earlier maturation of the electroencephalogram has been seen.

It is indeed remarkable that such widespread effects, both behavioural and physiological, can be produced by the simple treatment of picking up the pup for a few minutes a day during the first week or so of life. An important clue to the route by which they are produced is provided by the work of Levine and his collaborators. They have shown that one of the effects of the infant treatments is to hasten the maturation of the pituitary–adrenocortical system, which we know to be critical in the mammalian reaction to stress. For example, the adrenocortical response to cold stress (measured as a drop in the content of ascorbic acid in the gland) does not occur until 16 days of age in the unhandled animal, but was seen as early as 12 days in rats handled in the first few days of life. Similarly (Fig. 8.6), in handled animals as young as 3 days of age electric shock caused an increase in circulating adrenocortical hormones, whereas the unhandled animal shows no response at this age. The adrenocortical response to injected adrenocorticotrophic hormone is also greater in the handled animal as early as 6–9 days after birth.

It seems possible, then, that the physiological chain by which early experience affects adult emotionality may start with some modification in the sensitivity of the adrenocortical system. Levine's first guess as to the nature of this modification was to suppose that a kind of 'emotional immunisation' takes place, with the consequence that this system is made less responsive to subsequent stress. However, it soon became clear that this hypothesis is too simple; in some cases the handled animal does show a smaller adrenocortical response to stress in adulthood than the unhandled animal, but in other cases exactly the reverse is true.

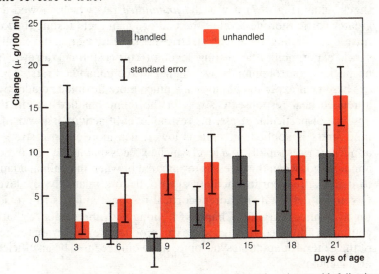

Fig. 8.6. Mean change in plasma concentrations of corticosteroids following electric shock (0.1 mA) in previously handled and nonhandled infant rats, at various ages.

The critical variables which determine whether it is the infant-treated animal or the untreated control which displays the greater adult adrenocortical reactivity appear to be: (1) the time which is allowed to elapse between application of the stress and measurement of the adrenocortical response: and (2) the intensity of the stress which is applied. The experimental results suggest that, with respect to (1), the infant-treated animal responds more rapidly when stress is applied, but then shows a more rapid return to the resting level when it terminates; and that, with respect to (2), the infant-treated animal adjusts the magnitude of the adrenocortical response more finely to the intensity of the applied stress.

An illustration of the first point can be drawn from an experiment by Haltmeyer, Denenberg & Zarrow, in which rats were shocked in adulthood after half had been handled throughout the first 20 days of life and half had been left undisturbed. The amount of corticosterone in the blood was assayed at various times after the termination of the shock. It was found (Fig. 8.7) that the handled rats had the higher corticosterone level at the time of shock termination but a lower rate of release of the homone over the next 15 minutes. Together with an experiment by Levine, Haltmeyer, Denenberg & Karas, this study also illustrates the second point. The Levine *et al.* experiment was identical to the one just described except that in adulthood the animals were exposed to a simple Open Field test instead of electric shock. Now (Fig. 8.7) it was the *unhandled* rats which showed the greater corticosterone level at the end of Open Field testing, in contrast to the finding made with electric shock. Moreover, in the electric shock experiment, the resting level of corticosterone, assayed in unshocked control animals, was higher in the unhandled rats.

These neat experiments suggest a much more flexible and adaptive adrenocortical response to stress in the infant-handled rat. In the absence of particular stress, his resting level of activity is lower; if a mild stress is applied, it remains lower; if a more intense stress is applied, it rises rapidly to a level which exceeds that in the unhandled animal; and, when this stress is terminated or when the animal adapts to it, his adrenocortical activity again falls to a relatively low level. Given the help that the adrenocortical response appears to provide for avoidance learning (Chapter 5), along with the evidence that a continued adrenocortical response can produce the deleterious effects of the general adaptation syndrome, this would seem to be an efficient arrangement.

Similar changes have been described by Pfeifer & Davis in a study of tyrosine hydroxylase in the adrenal medulla of infant-handled rats.

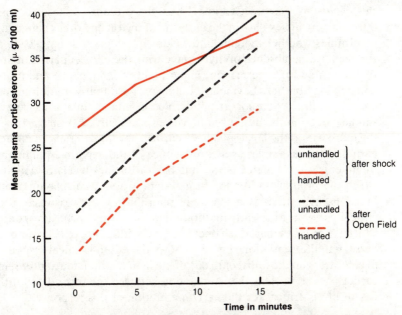

Fig. 8.7. Plasma corticosteroid levels at different intervals after either removal from the Open Field or exposure to electric shock in rats which were either handled or not handled in infancy. The rats handled in infancy respond to the *mild* stress of the Open Field *less* than do the nonhandled rats; they respond to the *stronger* stress of electric shock with an initially *greater* rise in the level of corticosteroids, but, compared to the nonhandled rats, their response gets less extreme with the passage of time after the termination of the stress.

Tyrosine hydroxylase is the rate-limiting enzyme in the synthesis of both noradrenaline and adrenaline (Fig. 5.5) and therefore plays a vital role in orchestrating both the alarm and resistance stages of the general adaptation syndrome (see Figs. 5.4 and 5.7). Pfeifer & Davis showed that early handling had complex effects upon the activity of this enzyme: basal, unstressed levels were lower; the increase in tyrosine hydroxylase activity caused by the first exposure to a novel environment or to foot-shock was enhanced; and there was a more rapid return of the activity of the enzyme to baseline levels after the end of 4 days of foot-shock. The parallel with Haltmeyer's findings in the experiments on corticosterone will be clear.

The experiments described above provide suggestive evidence that the adrenocortical system lies somewhere on the physiological route whereby the infant treatments produce their effects. Levine & Mullins have proposed an intriguing hypothesis as to the mechanism by which this occurs. We saw in Chapter 7 how the presence of testosterone during a critical few days in the early life of the rat and other mammals determines whether the animal develops into a male or female. In a similar way, the absence of thyroxine during the earliest period of life is responsible for a permanent impairment of intelligence and problem-solving abilities. Levine & Mullins propose that variation in the concentration of adrenocortical steroids in infancy may also have a similar effect on the infant brain. More specifically, they suggest that the more varied the levels of adrenocortical hormones released during early life (for example, as a consequence of the experimental infant treatments), the greater is the number of distinct levels of activity which the pituitary–adrenocortical system is able to display in adult life. This hypothesis is made more plausible by the existence of a critical period for the early maturation of the adrenocortical response which is similar to that which operates for the influence of testosterone on sexual differentiation. Fig. 8.8 shows the adrenocortical response to cold stress on the fourteenth day of life in rats stimulated at different earlier ages. It is clear that stimulation between the second and fifth day of life is critical for the early maturation of this response.

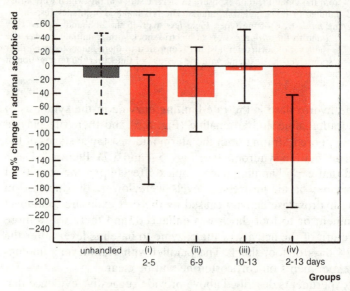

Fig. 8.8. Depletion of adrenal ascorbic acid in response to cold stress applied on the fourteenth day of life in rats handled for different periods (as indicated on the abscissa) in still earlier infancy. The bars represent the mean depletion; the lines, the range of scores within each group.

Recent data have put the Levine & Mullins hypothesis on a sounder footing. The release of ACTH from the pituitary (and so the release of corticosteroids, stimulated by ACTH, from the adrenal cortex) is under the control of the hypothalamus, which is itself influenced in this task by a number of higher brain structures (Fig. 5.4). A critical role in this concatenation of controls is played by glucocorticoid receptors in the hippocampus (Fig. 13.9). Glucocorticoid hormones, such as corticosterone, are released from the adrenal cortex, reach the brain, and bind to these receptors. This then sets in motion a chain of neural and hormonal events that leads to inhibition of pituitary release of ACTH, so providing a negative feedback loop to restrict further release of corticosterone. It is likely that this feedback loop is active in the return of corticosterone levels to baseline values upon the termination of a stressor, as in the kind of experiment conducted by Haltmeyer *et al.* (described above). The existence of these receptors was not known at the time Levine & Mullins proposed their hypothesis. But, with hindsight, we can see that they might provide a simple mechanism for the kind of flexible control over corticosterone levels which these workers attributed to the early-handled animal.

In line with this argument, Meaney *et al.* have recently reported that early handling gives rise to a substantial increase in the number of glucocorticoid receptors present in the adult rat brain. The increased density of receptors was found only in the hippocampus and the frontal cortex, two regions which (as we shall see in Chapters 13 and 14) have been crucially implicated, on independent grounds, in the experience of anxiety. It seems possible, therefore, that the chief way in which early handling affects adult levels of fear is by acting upon the developing brain (more particularly, upon the developing hippocampus and frontal cortex) so as to equip it with a more sensitive feedback control over corticosteroid release in response to varied conditions of stress.

The complexity of the interactions between genes and environment which determine just one aspect – fearfulness – of the adult personality of the rat is sufficiently daunting for the psychologist to perhaps lose all hope of ever attaining his ultimate goal: to unravel the full chain of events which results in the mature personality of an individual human being. We have seen effects on rat personality of simple treatments in early infancy, such as 'handling'; of the postnatal environment provided by the animal's mother and siblings; of events which have occurred during the animal's stay in the womb; and even of events which occurred before his conception. The extent or the direction of these effects may depend on the genotype of the infant or of its mother; and often two sorts of environmental effect can interact or even cancel each other out. Changes in offspring or siblings produced by these influences may in turn affect the mother or the other siblings, which may again be the source of further changes in a continuing dialectic.

Yet if the Levine–Mullins hypothesis concerning the physiological route of the early handling effects were to prove correct, we would begin to see an encouraging unity in our understanding of how both genes and early environment act to determine fearfulness in the adult animal. It seems possible that in both cases a critical role is played by the action of hormones in the earliest period of life on the developing nervous system. This is already known to be an important node in the process of sexual differentiation and in the development of normal intelligence, so the principles at work may be of still further generality. At any rate, it is on the discovery of such general principles in experimental work on lower animals that our chief hopes of progress in the understanding of Man's nature must rest.

9 Punishment and conflict

Our major concern so far in this book has been with the *origins* of fear, both phylogenetic (i.e. arising in the course of the evolution of species) and ontogenetic (arising during the lifetime of the individual animal). We have asked how particular stimuli acquire the capacity to evoke fear, and how species acquire their particular modes of expression of fear; we have considered the evidence that the degree of susceptibility to fear in the adult animal is in part due to the operation of hereditary factors carried in the genes, and also the evidence for the effects of the early environment on this aspect of the personality; we have tried to trace physiological routes whereby these kinds of influence may come to affect the behaviour of the adult animal; and we have seen how fundamental selection pressures have acted over the millenia to create organisms which possess a mechanism for the homeostatic control of population density (of the kind represented by the general adaptation syndrome in mammals), along with the profound antagonism between fear and sexual behaviour which such a mechanism entails.

For the remainder of the book, however, the centre of our interest will shift from the origins of fearful behaviour to its *organisation*. We shall ask questions about the kind of behaviour an animal displays when it is fearful; about the way in which it may learn new forms of behaviour under the prompting of fear; about the effects of fear on behaviour which is itself prompted by other motives; about the way in which the brain controls fearful behaviour; and about the kinds of disturbances in behaviour, termed 'neurotic' in Man, which may result from fear.

In considering the organisation of fearful behaviour, and the interactions between this and other forms of behaviour, it is as well to draw a distinction at the outset between the two kinds of fear-motivated behaviour termed by the psychologist 'passive avoidance' and 'active avoidance'. This distinction turns on the way in which the individual comes to be exposed to the feared stimulus, and what he must do to get away from it.

In passive avoidance situations, there is something the animal wants to do, but this brings him into contact with the feared stimulus; the only way to avoid the feared stimulus is then to give up the desired behaviour. In a typical example, the animal is hungry and in sight of food; but, if he touches it, he will suffer an electric shock. Not unnaturally, the shock in such a situation is termed a 'punishment'. (Technically speaking, a punishment is defined as a stimulus which if presented to the animal immediately after it makes some response, will decrease the probability of recurrence of that response. An electric shock of moderate intensity is calculated to have this effect on the members of most animal species.) In short, in passive avoidance situations, the animal is told '*Don't* do this, or else . . .'

In active avoidance situations, by contrast, the animal usually wants just to be left alone. But someone else (a circus animal-trainer, an experimental psychologist) is anxious that he acquire some new form of behaviour. The trainer, therefore, brings the animal into contact with the feared stimulus, and removes this stimulus only if the animal displays the desired behaviour. In other words, the animal is punished for making *any response other than one* – the one specified by trainer or experimenter. A typical example from the psychological laboratory is the shuttlebox task which we have met several times already (Fig. 4.8). Here, the only form of behaviour which will enable the animal to avoid the punishing shock is to jump from side to side of the apparatus when the appropriate signal is presented. The animal is told, in effect, '*Do this*, or else . . . '.

In this chapter and the following one, we shall be concerned with various forms of passive avoidance behaviour, postponing consideration of active avoidance until later.

Approach–avoidance conflict

In a passive avoidance situation, then, the animal (or man) is simultaneously impelled by the signals it receives from its environment towards two incompatible forms of action: one is to approach a particular object or stimulus and the other is to keep away from it. For obvious reasons, this sort of situation is also spoken of as an 'approach–avoidance conflict'. It will surprise no one to be told that conflict of this kind is extremely common in the lives of both animals and men. But it is even commoner than one might have thought. For example, Moynihan has used ethological techniques to analyse the displays used by the black-headed gull in courtship and fighting into a balance between attack and withdrawal (Fig. 9.1). For our own species, a group of social psychologists at Oxford University have been working under the direction of Michael Argyle to show how a similar kind of analysis into approach and avoidance tendencies can be applied to the factors which determine the distance at which two people choose

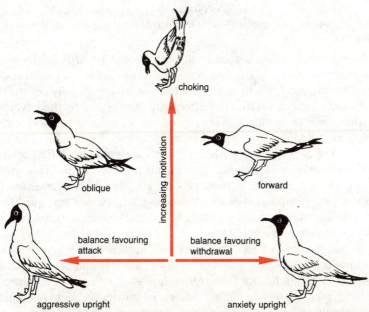

Fig. 9.1. The displays of the black-headed gull used in courtship and fighting are thought to involve degrees of attack and withdrawal. In the 'oblique' and the 'aggressive upright' displays, aggression predominates; but the oblique display is relatively more intense. In the 'forward' and 'anxiety upright' displays, the balance favours withdrawal. The 'choking' display seems to represent a balance of attack and withdrawal elements, with very strong motivation.

to stand apart from one another in various social situations, the bodily stance they adopt, and the degree to which they look at each other.

The ubiquity of approach–avoidance conflicts in animal behaviour is nowhere more apparent than in the effects of a novel stimulus. Novelty is both an effective stimulus for fear and the chief stimulus for exploratory behaviour; thus, of its very nature, novelty is the cause of an approach–avoidance conflict. Any mother who has watched her child play peek-a-boo with a stranger will have seen the behaviour which occurs during such a conflict. Lorenz gives a nice description of the same kind of behaviour in the raven:

A young raven, confronted with a new object, which may be a camera, an old bottle, a stuffed polecat, or anything else, first reacts with escape responses. He will fly up to an elevated perch and, from this point of vantage, stare at the object literally for hours. After this, he will begin to approach the object very gradually, maintaining all the while a maximum of caution and the expressive attitude of intense fear. He will cover the last distance from the object hopping sideways with half-raised wings, in the utmost readiness to flee. At last, he will deliver a single fearful blow with his powerful beak at the object and forthwith fly back to his safe perch. If nothing happens, he will repeat the same procedure in much quicker sequence and with more confidence. (Lorenz, 1956)

In general, animals appear to be attracted towards mildly novel stimuli and to avoid extremely novel stimuli. Since each exposure to a stimulus decreases its novelty, stimuli which are initially beyond the limit at which their attractive effects are outweighed by their aversive effects will (provided the novelty of the stimulus is not so great at first that it drives the animal completely away) gradually pass into the range at which the attractive effects are stronger. A fine balance of this kind provides an excellent means by which animals are drawn to explore a novel environment – an activity which is essential for their future survival – without exposing themselves too quickly to whatever dangers it might contain. Lorenz's raven appears to have been behaving in exactly this way.

The attractive effects of novelty may appear even when the novel stimulus is one which would normally excite fear and act as a punishment. Hebb, in his book *The Organisation of Behaviour*, emphasises the degree to which human beings in particular seek out mildly frightening situations, such as horror films or roller-coaster rides, for amusement. The same sort of thing can be experimentally demonstrated in the rat. Halliday, for example, showed that rats would explore the striped arm of a Y-shaped maze (whose other arms were painted grey) *more* if they had previously been given an electric shock in a striped box.

The results of Halliday's experiment run counter to what we would expect on a common-sense basis, showing how difficult it may be to predict the *outcome* of an approach–avoidance conflict. Yet such predictions may be of considerable importance in many practical situations. Parents may wish to use the most effective ways of eliminating unwanted forms of behaviour in their children or, conversely, they may wish to train them to persist in the face of adversity. Psychiatrists may need to do exactly the same for their patients. As we shall see, quite small difference in the techniques they use may have the outcome of increasing the persistence of the undesired behaviour, or eliminating the behaviour it is wished to encourage.

Miller's equilibrium model

Both parents and psychiatrists, however, may take heart from the existence of a large corpus of experimental findings concerning approach–avoidance conflicts; one, moreover, which permits a more definitive summary than is yet common in experimental psychology. Much of our knowledge in this area springs from the elegant experiments conducted by Neal Miller and his colleagues. On the basis of these experiments Miller has constructed an important theory of conflict behaviour. This theory follows a classic pattern which has served many other sciences well: it is an equilibrium model in which an animal in an approach–avoidance conflict will come to rest at that point in his environment at which the forces favouring approach to, and avoidance of, the simultaneously feared and desired goal are equal to each other.

Miller's basic experimental situation is one in which a rat is first trained to run down an alley for a reward (e.g. food) and is then given an electric shock each time it reaches the goal. Animals trained in this way show obvious signs of conflict. Under appropriate conditions they will start off towards the goalbox, stop, and then oscillate around this stopping point, slow approach behaviour being interrupted by sudden withdrawals.

If we suppose that this behaviour results from conflicts between separate approach and avoidance tendencies, the first step is to specify the conditions which affect the strength of each. Fig. 9.2 shows some of these conditions in the form of a block diagram. Not surprisingly, we expect the approach tendency to be stronger if, for example, the animal is hungrier and the food reward is greater; similarly, the avoidance tendency will be stronger, the more intense the electric shock.

But Miller's critical postulate concerns distance from the goal. We would obviously expect both the approach and avoidance tendencies to get stronger, the closer the animal gets to the goal. But the strength of the approach and avoidance tendencies cannot be affected in an identical manner by distance from the goal. If they were, whichever was stronger at the furthest point from the goal would also be stronger at the nearest point to the goal and, in that case, the animal would either stop as far away as possible from it or approach it completely: the *part*-approach to the goal which is such a conspicuous feature of conflict behaviour simply could not occur. The only way in which the behaviour of part–approach, followed by hesitations and oscillations, can occur is if the strength of avoidance increases more rapidly with nearness to the goal than does that of approach; and this is the postulate which lies at the heart of Miller's conflict model.

The resulting approach and avoidance functions are shown in Fig. 9.3. It can be seen that the model now predicts that if the animal is placed further away from the goal than the point at which the two theoretical functions cross, he should move towards the goal until he

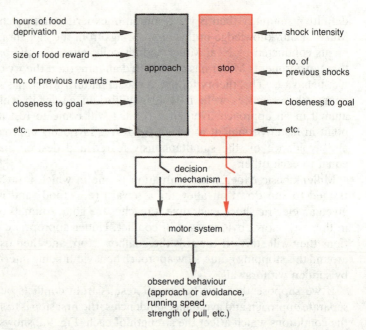

Fig. 9.2. Miller's analysis of the factors which affect the strength of the approach and passive avoidance tendencies in a conflict situation drawn as a block diagram. The 'decision mechanism' closes *either* the switch which allows the 'motor system' to be controlled by the 'approach' mechanism *or* the switch which allows it to be controlled by the 'stop' or 'passive avoidance' mechanism, but not both. Which switch is closed depends on the strength of the inputs to the decision mechanism from the approach and stop mechanisms, and these in turn depend on the factors listed to the left and right of the figure.

reaches that point; if, on the other hand, he is placed nearer the goal than the equilibrium point, he should retreat towards it. Given both physical and psychological inertia in this system, we should observe the oscillation around the equilibrium point which is in fact typically observed. The effects of those conditions which affect either the approach or the avoidance tendency, but not both, are conceived as raising or lowering the entire function to which they relate without affecting its slope. Thus, increasing hunger or decreasing shock intensity should have the result that the animal will approach more closely to the goal, perhaps even reaching it; and the reverse operations should have the reverse effect.

These deductions have been confirmed in a series of experiments from Miller's laboratory, especially those conducted by J. S. Brown, who introduced a neat technique by which it is possible to measure, quite literally, the 'strength' of the approach and avoidance tendencies. The rat is trained to run wearing a special harness which can be used to stop him at any chosen point in the runway. Since it is attached to a spring, it is possible to measure the strength with which the animal pulls against the harness at that point. Using this device (with animals trained only on the one habit, approach or avoidance), Brown was able to verify directly the crucial postulate that the strength of the avoidance tendency falls off more steeply with distance from the goal than does the strength of the approach tendency (Fig. 9.3.D).

Stimulus generalisation and sexual deviation

A further series of experiments showed that stimulus similarity affects approach and avoidance functions in the same way as distance from the goal. According to the principle of 'stimulus generalisation' (Fig. 2.5), the magnitude of a conditioned response falls off as the similarity of a test stimulus to the original conditioned stimulus becomes less. Again, then, we would expect both the approach and avoidance tendencies to be decreased if the original training situation is changed. Using runways of varying degrees of similarity, Miller has shown that this is indeed so, and that the decrease in the strength of the avoidance tendency is greater than the decrease in approach. Moreover, the effects of stimulus similarity summate with those of distance from the goal, so that, for example, an animal will approach closer to the goal when the stimulus characteristics of the alley are changed in some way.

These effects of stimulus similarity are very important. They are probably involved in the causation of patterns of behaviour in which a response is directed towards an apparently inappropriate stimulus. Consider a man who is, for whatever reason, frightened of women, but sexually attracted to them. Such a man might be unable to direct sexual behaviour towards women, but be able to do so with individuals of a moderate degree of similarity to women. The outcome might well be that he was able to behave sexually only with young girls or young boys. Alternatively, he might still direct his sexual behaviour towards mature females, but find that there was a point at which the avoidance tendency overcame the approach tendency. The outcome now might be a 'peeping Tom' syndrome, or a fetishism directed towards articles of female clothing. Support for the latter possibility comes from an experiment by La Torre, in which simulated rejection by a preferred female companion caused male undergraduates to rate pictures of women as less attractive than pictures of women's underwear.

As we have seen, the critical feature of Miller's analysis is the assumption of a steeper fall-off in avoidance tendencies with change or distance from the goal. It would obviously be more satisfactory if

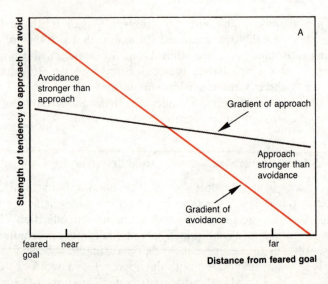

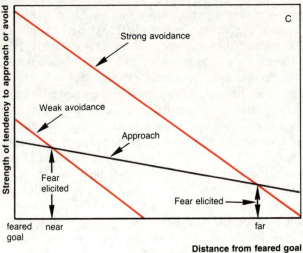

Fig. 9.3. Miller's equilibrium model for approach–avoidance conflict. The strength of both the approach and the avoidance tendencies falls off with distance from the goal, but the strength of the avoidance tendency falls off more rapidly (A). Factors which raise or lower the strength of the approach tendency other than distance from the goal are thought to raise or lower the height of the approach gradient without affecting its slope (B). Similarly, factors which raise or lower the strength of the avoidance tendency raise or lower the height of the avoidance gradient (C). In every case, the animal should approach towards the goal until the two gradients cross and then oscillate about this point. The fear elicited depends on the height of the

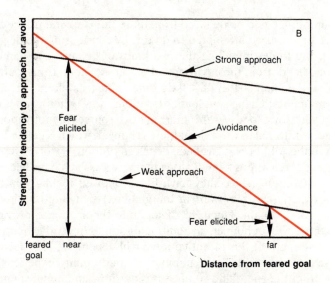

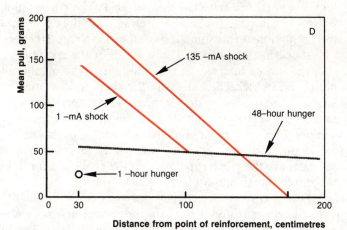

Caption for Fig. 9.3 (*cont.*).

avoidance gradient at this equilibrium point. Thus the animal may experience more effective fear with a strong approach tendency than with a weak approach tendency (B), or, still more paradoxically, more fear with a weak avoidance tendency than with a strong avoidance tendency (C). Empirical data (D), obtained from rats wearing a special harness which made it possible to measure the force of the approach and avoidance tendencies separately, when they were trained to run in a runway with either only food or only shock, give strong support to Miller's model.

this assumption could be deduced from more fundamental principles. Miller & Brown have attempted to do this by considering the degree to which the forces impelling the two kinds of behaviour originate, on the one hand, from the environment or, on the other, from sources internal to the animal. It is only the former which should be affected by environmental change or spatial location. Now, when a rat runs down an alley for a food reward, there are two sources of motivational excitement. Internal to the animal there is hunger, which is dependent on the number of hours it has been deprived of food. External to the animal there are various stimuli along the runway which have become *conditioned* stimuli (since they have been followed regularly in the animal's experience by reward in the goalbox) signalling the imminence of food. In the language of learning theory, these provide, besides the 'drive' of hunger, an additional form of 'incentive' motivation. The strength of this incentive force will depend on such factors as the size of the reward, the number of conditioning experiences, distance from the food, and so on. Consider now an animal running down a runway towards an expected electric shock. The same stimuli along the runway will have become, as well as conditioned sources of incentive, conditioned fear stimuli, since they have regularly been followed by shock in the goalbox. The strength of the conditioned fear will depend on the intensity of the punishment, the number of times it has been experienced, and so on. But there is no continuing *internal* source of motivation to do for the avoidance tendency what hunger does for the approach tendency. Thus the avoidance tendency should be more purely dependent on environmental change and distance from the goal than the approach tendency. In this way, then, it is possible to deduce the greater steepness of the avoidance function than the approach function in Miller's conflict model.

If this analysis is correct, the greater steepness of the avoidance function should be eliminated if we train the animal in such a way that the controlling cues for avoidance behaviour are independent of the environment. One way to do this is to use a 'Sidman avoidance schedule' (Fig. 9.4). In this, the animal is shocked at regular intervals unless he performs a designated avoidance response: this response postpones the occurrence of the next shock by some length of time. If the animal keeps up a high enough rate of response he avoids most of the shocks; any pauses in his responding, however, are followed by renewed shock. Thus the warning signal is effectively, 'a long time since I last responded', and this, of course, is dependent on cues which are internal to the animal. Using this schedule with both monkeys and rats, Hearst obtained avoidance behaviour which showed very little decrease when the environment was changed; and,

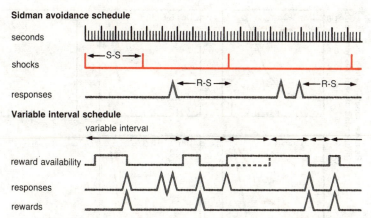

Fig. 9.4. A diagrammatic representation of the Sidman avoidance schedule and the variable interval schedule. On the Sidman avoidance schedule, the animal is given electric shocks at regular intervals (the 'shock–shock' or 'S–S' interval), and each response (e.g. pressing a lever) postpones the next shock by a specified amount of time (the 'response–shock' or 'R–S' interval). By responding sufficiently frequently, the animal can avoid shock altogether. On the variable interval schedule, rewards become available (but only if the animal responds) at intervals which are randomly distributed around some mean value (which specifies the particular variable interval schedule used). Responses which occur before the next reward availability has been reached are not rewarded. If two reward availabilities are reached without the animal having responded during the first (as indicated by the dashed line), the next response produces only one reward. Thus, if the rate of response drops too low, the animal will not get all the rewards it could.

in the same experiment, behaviour rewarded with food in a way which was *less* dependent on the animal's own behaviour was found, in fact, to show a *steeper* fall-off with environmental change (Fig. 9.5).

It would seem, then, that correctly to apply Miller's equilibrium model to conflict situations, we need to know which tendency, approach or avoidance, is more dependent on environmental sources of motivational excitement. However, if this is known or can be estimated, the model can offer considerable help in devising techniques of modification of the behaviour resulting from conflict situations. And indeed, as we shall see in our final chapter, it is being used in the treatment of many forms of neurotic behaviour in Man.

The effects of punishment

Prediction of the outcome of conflicts can also be helped by a consideration of the particular variables which favour the approach or avoidance tendencies. Let us suppose that our aim is to eliminate by

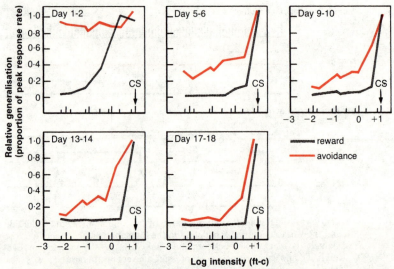

Fig. 9.5. Generalisation gradients for concurrent approach and avoidance during successive stages of discrimination training for one monkey. The avoidance behaviour was reinforced on a Sidman avoidance schedule and the approach behaviour was rewarded on a variable interval schedule (see Fig. 9.4), in both cases only in the presence of the brightest light intensity (indicated as 'CS' in the figures). When response rates were measured in the presence of lights of lower intensity, they showed an orderly decline as the difference between the CS and the particular light intensity increased; but this decline was greater for the approach behaviour than for the avoidance behaviour.

punishment some form of existing rewarded behaviour. Perhaps we want to stop little Albert (now a little older) from grabbing the chocolate biscuits before they are offered. What factors should we bear in mind when we deliver our rebuke?

One obvious variable is the intensity of the punishment we employ. In experiments in which rats press levers or pigeons peck keys for a food reward and are simultaneously punished for doing so by the delivery of an electric shock, it is found that the degree of suppression of the punished behaviour is closely dependent on the intensity of shock used. With a relatively low intensity, the suppression is only temporary, and the response rate eventually returns to the pre-punishment level, even though punishment continues to be delivered; at an intermediate shock intensity, partial suppression is obtained, the subject always showing some lasting reduction in response rate; with shocks of very high intensity, complete and permanent suppression is obtained. Delay of punishment after the emission of the punished response lessens its effectiveness, as does irregularity of punishment.

Thus a swift, certain punishment can be as effective as a more intense punishment applied in a dilatory or inconsistent manner.

If the aim is, in fact, complete suppression of a response, the most effective technique is to provide the subject with an alternative response which will lead to satisfaction of the drive (hunger, thirst, sexual desire, etc.) on which the approach tendency is based. If this is done, quite a low intensity of punishment can achieve a lasting suppressive effect. Whiting & Mowrer made this point clearly in an experiment in which rats were first rewarded with food for taking one route through a maze and then punished for doing so. When the rats abandoned this route, they rewarded them for taking a different route to the same goal. The old route was not used again. In a similar experiment, Richard L. Solomon trained puppies by swatting them on the nose with a rolled-up newspaper for eating horsemeat, while allowing them to eat food pellets. This method of training was so effective that they were ready to starve when given the opportunity later to eat only horsemeat, though they eagerly ate food pellets when these were offered.

The strong and long-lasting response suppression obtained by Solomon in this experiment was probably due to another circumstance as well. This is the fact that the punishment was delivered during the 'consummatory' act of eating as distinct from the 'instrumental' behaviour leading to the acquisition of food. (By a 'consummatory' act is meant the innate behaviour patterns which are directed towards stimuli which act as rewards for the species concerned. Thus, eating, drinking, copulation are consummatory acts, whereas the learnt forms of behaviour which bring the animal into contact with food, water, or a sexual partner are 'instrumental' responses.) Solomon has emphasised the particularly strong effects of punishment applied during consummatory behaviour:

One would think that consummatory acts, often being of biological significance for the survival of the individual and the species, would be highly resistant to suppression by punishment. The *contrary* appears to be so. Male sexual behaviour may be seriously suppressed by weak punishment. . . . Eating in dogs and cats can be permanently suppressed by a moderate shock delivered through the feet or through the food dish itself. . . . A toy snake presented to a spider monkey while he is eating can result in self-starvation. (Solomon, 1964)

The difference between punishment applied to the consummatory act and punishment applied to the instrumental behaviour preceding the consummatory act is probably an instance of a more general factor of *temporal order*. Suppose we run a food–shock conflict experiment of the kind used by Neal Miller. Let us hold everything in the experiment constant (degree of hunger, amount of food, number of trials, intensity of shock, etc.) except that in one case we allow the shock slightly to precede the delivery of the food, and in the other we have the delivery of the food slightly precede the occurrence of the shock. The difference in the result of the experiment which can be brought about by this reversal of the temporal order of reward and punishment can often be dramatic. If the shock comes *first*, the animal soon adapts to it, until he treats it, not as a punishment, but as a signal that food is available, and starts to approach the food-cup and eat immediately the shock occurs. If the punishment comes *second*, while the animal is eating, he shows, in contrast, signs of intense fear and completely gives up eating in the experimental situation. A recent experiment by Rodriguez & Logan illustrates this principle. These workers offered thirsty rats a choice between one alley in which they could run to obtain water which was then followed by an electric shock, and another alley in which the shock occurred before water delivery. The rats preferred the alley in which shock was delivered before water; in the other alley, they ran slower and often refused to run at all.

It seems likely that this great dependence of the outcome of a conflict on the temporal order obtaining between reward and punishment reflects something very fundamental about the functioning of the brain. This something is probably connected with classical conditioning. As we know, if two stimuli regularly follow each other in time, classical conditioning is the phenomenon that the earlier of the two stimuli comes to evoke a reaction appropriate to the later one. 'Backward conditioning', in which the second stimulus would elicit a response appropriate to the earlier stimulus, occurs only under special conditions if it occurs at all. Thus, if shock is followed by food, it may become a conditioned stimulus for a salivary response. But if food is followed by shock, it comes to elicit a conditioned defensive reaction. So the nervous system makes a fundamental distinction between the first and second of two stimuli which repeatedly follow each other in time.

It can be seen, then, that the greater effectiveness of punishment applied to consummatory acts than to instrumental behaviour can be derived from the variable of temporal order. For, in the former case, punishment is applied after reward whereas, when the instrumental behaviour is punished, provided the animal is not totally stopped in his tracks by the first few punishments, reward is experienced after punishment.

It has been suggested by Solomon that the difference between punishments applied during consummatory and instrumental behaviour may be related to the development of the 'conscience'. Solomon breaks this notion down into two components: resistance to temptation beforehand and guilt feelings if the forbidden act is accomplished. As H.J. Eysenck put it, in his defence of the view that conscience may be considered as nothing more than a cluster of conditioned responses, 'we might perhaps anticipate that punishment which is administered before a given type of activity is indulged in would lead to a pronounced reluctance later on to indulge in that particular activity, whereas punishment which is administered during the activity, i.e. after it has begun, would perhaps lead to later guilt feelings.'

The results of Solomon's experiment with dogs trained to eat dry food pellets instead of the horsemeat they would normally prefer were in agreement with this analysis. Dogs swatted on the nose while *approaching* the forbidden horsemeat took far longer than those swatted on the nose while *eating* it to resume eating the meat when this was the only food available – i.e. they showed greater resistance to temptation. But, when they did succumb to the temptation of eating meat, they did so with tails wagging and no subsequent signs of distress. The dogs which were punished while eating the horsemeat, in contrast, showed a great deal of emotional upset both during and following the 'crime'. Although the presence of the experimenter was not necessary for this display of emotion, it was intensified when he subsequently entered the room. Thus it bore many of the hallmarks of the human emotion of guilt.

Resistance to punishment: counter-conditioning

It may perhaps have occurred to the reader that proper use of the temporal order of punishment and reward might help us train someone to be particularly *resistant* to punishment. This problem has been studied in several experiments from Miller's laboratory. These show that, under some conditions (but, as we shall see, there are others), simply exposing an animal to punishment a number of times does nothing to reduce the subsequent effectiveness of the punishment; that is, 'adaptation' to the punishment does not take place. For resistance to punishment to be increased it is necessary that the punishment be followed by a reward, a process termed 'counter-conditioning'.

In one experiment, for example, a group of rats was trained to run down a runway and also exposed to a shock of gradually increasing intensity in the goalbox. A second group was trained on a number of trials with reward only and then exposed immediately to the shock of highest intensity reached by the first group. A third group was treated the same as the first group, except that instead of receiving shock in the presence of food in the goalbox, they were shocked outside the runway altogether. The disruptive effects of the high-intensity shock were much greater in the second group than in the first, but the third group was affected to exactly the same extent as the second. Thus, accustoming the animals to the gradually increasing shock intensity was only effective if the shock was delivered before reward.

Neil McNaughton and I have shown that by pairing an electric shock with reward in this way, it is possible to make the shock less capable of disrupting behaviour even at times when it is not paired with reward. Our experiment made use of the Estes & Skinner conditioned suppression technique (Fig. 4.7). Rats were first trained to press a bar for food reward. Next, a tone was paired as a CS with foot-shock as a UCS, both stimuli being presented to the animal while it was pressing the bar. The tone came to suppress bar-pressing, in the manner shown in Fig. 4.7. At this point we began to interrupt the experimental sessions with 'intrusion periods', each lasting 1 minute. During an intrusion period the bar was withdrawn from the Skinner box and there occurred one delivery of food and one foot-shock. The animals were now divided into two groups. For one group, the foot-shock (now as CS) immediately preceded the food delivery (as UCS) in every intrusion period. The other group served as controls: they received exactly the same number of food deliveries and shocks as the first group, but in a random temporal relationship to each other. Note that the two groups continued to be treated identically outside the intrusion periods, that is, when the bar was present in the Skinner box, bar-pressing was rewarded with food, and tone CSs were followed by shock UCSs. The results of the experiment are shown in Fig. 9.6: the group for which shock was paired with food *in the intrusion periods* came to display less suppression of bar-pressing (relative to the controls) to the tone paired with the same shock *outside* the intrusion periods. Thus the classical conditioning (counter-conditioning) of the shock (as CS) to the food during the intrusion periods had weakened the capacity of the shock itself to serve as a UCS at other times. Dearing & Dickinson have similarly shown, using rabbits, that pairing shock with water robs the shock of some of its capacity subsequently to act as a punishment and suppress water-rewarded bar-pressing.

These experiments demonstrate 'Pavlovian counter-conditioning', i.e. counter-conditioning that clearly depends on the pairing, in a classical conditioning paradigm, of an aversive event (as CS) with an appetitive event. Now, in both McNaughton's and Dearing & Dickin-

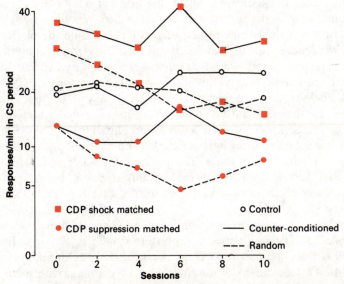

Fig. 9.6. Failure of chlordiazepoxide (CDP; 5 mg/kg, injected intraperitoneally) to reduce Pavlovian counter-conditioning. Within each treatment group there are two behavioural subgroups: for 'counter-conditioned' groups a single shock was presented within each intrusion (i.e. a 1-minute period when the lever was withdrawn from the testing box) and was immediately followed by a single food pellet; for 'random' groups there was no relationship between the delivery of shock and food. The data plotted are response rates (square-root transformed) on the lever during presentation of the conditioned stimulus (CS) which warned of shock delivery, after covariance to allow for differences in the rate of responding before delivery of the CS ('preCS' rates). The statistics (analysis of covariance) demonstrate strong linear trend differences between the counter-conditioned and random groups in the CS rates ($p < 0.001$) but not the preCS rates. CDP had no effect on these trend differences, whether shock intensities were the same in the drugged and undrugged groups ('shock matched'), or whether shock intensities were increased in the CDP groups so as to produce the same levels of suppression of lever-pressing as in undrugged animals ('suppression matched'). From McNaughton & Gray (1983).

son's experiments, the aversive event was itself an *unconditioned* punishment (electric shock). It is also possible to conduct a Pavlovian counter-conditioning experiment in which the aversive event is a *conditioned* punishment, i.e. a stimulus which has acquired its aversive properties through previous classical conditioning with an unconditioned punishment. Naively, one might suppose that counter-conditioning of such a conditioned fear stimulus would be easier to achieve than the counter-conditioning of unconditioned punishment. However, the data do not clearly support this assumption.

Some attempts to counter-condition conditioned fear stimuli have been successful. For example, Wilson & Dinsmoor trained rats to remain on a small platform to avoid shock on a grid floor below. They measured fear by the time it took the animal to step down from the platform (with shock now turned off). Animals fed on the platform descended more quickly than unfed controls. The classical conditioning component of this procedure is, however, not very explicit. In a more tightly controlled study, Scavio established a tone as CS for shocks delivered to the region of the eye ('paraorbital shock') in rabbits. He then extinguished the eyeblink response under three conditions: presentations of the CS alone; presentations of the CS and water, but without pairing between the CS and water; or presentations of the CS paired with water. The presence of water at all during extinction hastened the loss of the eyeblink response, but a significantly greater effect was obtained when the CS actually predicted water delivery.

However, other experiments have failed to confirm these findings. Lovibond & Dickinson used a procedure similar to Scavio's. They established a clicker as the CS for paraorbital shock in rabbits and then tried to counter-condition the clicker by pairing it with sucrose, but to no avail. They also did the reverse experiment, first establishing the clicker as CS for sucrose and then trying to counter-condition it using shock. This experiment was successful: the conditioned jaw-movement response to the clicker was eliminated more rapidly when sucrose was omitted *and* shock was paired with the clicker. This is a general pattern of results to which we return in Chapter 12 (in the discussion of 'retardation' experiments): it appears to be easier to transform conditioned appetitive stimuli into conditioned aversive stimuli than vice versa.

On balance, then, the evidence for counter-conditioning of unconditioned punishment is stronger than that for counter-conditioning of conditioned fear stimuli. Now, given that it is possible, by pairing unconditioned punishment with a following reward, to rob the punishment of much of its effect, it is not surprising that a normally punishing stimulus can come to act as a cue or signal for the availability of reward. In this way, it is possible to get an *increased* probability of response by 'punishing' the response in question. This effect was demonstrated 35 years ago in a series of runway studies by Muenzinger and his collaborators in which rats were shocked for making the 'right' response (i.e. the choice that led to food) in a T-shaped maze. They learned *better* to go where the food was than non-shocked controls, and as well as rats shocked for making the wrong choice (i.e. the one not leading to food).

A study by Holz & Azrin shows that by pairing punishment with a following reward, it is even possible to turn the normally punishing stimulus into a 'secondary' reward. This is a stimulus which is not innately rewarding for the species concerned, but becomes rewarding as a result of pairing with an innate reward. In the Holz & Azrin experiment, pigeons were rewarded for pecking at a key if and only if the pecks also produced an electric shock. Under these conditions, when food reward was discontinued entirely, pecking continued at a higher rate if the response *still produced the shocks*.

Such behaviour in human beings is usually termed 'masochistic', masochism being a condition in which the individual seeks out painful stimuli which are shunned by the majority of his fellows. Masochism has been the subject of a number of more or less extravagant explanations among psychiatrists of one persuasion or another. Yet some of the seeds of masochism may lie in just this kind of temporal relation between reward and punishment. Given a number of conditioning experiences in childhood in which pain has been followed by powerful rewards (and it is easy to imagine this occurring in the normal course of events), the development of masochistic forms of behaviour is by no means an improbable outcome.

A final determinant of the outcome of an approach–avoidance conflict, and one which has surprisingly powerful effects, is the nature of the response actually elicited by the punishment. If this is compatible with the approach response, an animal can be trained to continue in its approach behaviour in spite of quite severe punishment. This is demonstrated in another experiment from Neal Miller's laboratory.

Fowler & Miller trained rats to run down a runway for a food reward and exposed them to shock in the goalbox under two conditions. In one group the shock was delivered to the *forepaws*, causing the animals to lurch *backwards*; the other group was shocked on the *hindpaws*, causing them to lurch *forwards*. In addition, shock intensity was varied in different groups and there were non-shocked controls. The usual conflict result was achieved in the groups given shock to the forepaws: they ran slower than the non-shocked controls, and the more intense the shock the slower they ran. But the groups shocked on the hindpaws actually ran *faster* than the non-shocked controls, and this effect too was greater, the more intense the shock (Fig. 9.7). Furthermore, when running speed was analysed as a function of distance along the runway, it was found that the speeding-up effect of hindpaw shock was greater, the nearer the animal was to the point of shock.

To account for these remarkable results, we may say that owing to the fact that the animal's initial response to hindpaw shock is that of running forward, he gets into a sequence in which running forward is followed by *two* rewards: food and the termination of shock. If this

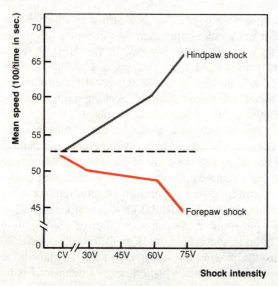

Fig. 9.7. Speed of running towards a goalbox in which both food and shock were delivered, as a function of shock intensity in volts and whether the shock was delivered to the hindpaws or the forepaws.

sequence is able to survive the first few trials after shock is first introduced, it becomes sufficiently fixed in the animal's behaviour to overcome the tendency to stop and freeze which is more usually initiated by classically conditioned fear stimuli. Whether or not this is the correct explanation of these results, they serve to remind us of the importance of seemingly minor aspects of an experimental procedure. By shifting shock from forepaw to hindpaw, Fowler & Miller were able completely to reverse the usual outcome of an approach–avoidance conflict. There is no reason to doubt that such minutiae may play an equally important role in human life; and here too we may see more of the seeds of masochism.

Resistance to punishment: non-associative effects

The creation of resistance to punishment by the process of counter-conditioning (discussed in the previous section) depends upon the formation of an association between punishment and reward. But it is also possible to increase resistance to punishment by non-associative means. Under some conditions the simple repetition of a painful stimulus, such as an electric shock, gives rise to a process of adaptation to the shock, shown by a lessening in its capacity to disrupt behaviour. Neal Miller, in a characteristically gritty phrase, has called this process 'toughening up'.

As an example of toughening up, consider some experiments by Jay Weiss and Howard Glazer, working in Miller's laboratory. A group of rats was given a single session in which they were exposed to inescapable electric shocks (*escapable* shocks of identical intensity and duration have different effects – a topic we shall take up later). Each rat was then trained in a shuttlebox in which it could escape or avoid shock by running from one side to the other of the apparatus in the manner depicted in Fig. 4.8. Compared to animals that had no experience of shock before being put in the shuttlebox (or to rats that had escapable shocks) the rats that had the inescapable shock were slow to learn the shuttling response. This general phenomenon was first described in dogs by Seligman & Maier, who called it 'learned helplessness'. Helplessness after inescapable shock has since been reported many times in many laboratories. The Weiss & Glazer experiments, however, added an important new twist to the story. These workers showed that if training in the shuttlebox was preceded, not by a single session of inescapable shocks, but by 15 such sessions at the rate of one a day, then the disruptive effect of the final session of inescapable shock on shuttlebox responding was completely eliminated. This is the kind of effect covered by Miller's phrase 'toughening up'. Note that during the repeated sessions of inescapable shock that give rise to toughening up, there are no rewards with which the shock might enter into an association. So toughening up appears to be a purely non-associative process: the shock loses its capacity to disrupt behaviour in virtue simply of repeated presentation.

When toughening up occurs, its effects spread far beyond the particular stimulus that is used to produce it. Weiss & Glazer's experiments, for example, showed that repeated exposure to inescapable shock produced tolerance for the disruptive effects of not only shock, but also forced swimming in cold water. The latter treatment, given just before shuttlebox avoidance learning, disrupts learning in the same way as does a single session of inescapable shock. But the disruptive effect of a cold swim was prevented if the rat was first exposed to 14 sessions of inescapable shock. Weiss & Glazer also demonstrated the converse phenomenon: the disruptive effects of a single session of either cold swimming or inescapable shock on shuttlebox avoidance were blocked by 14 prior cold swims. So repeated shock produced tolerance for shock and 'cross-tolerance' for cold swim, and vice versa. Even broader effects have been reported. Chen & Amsel demonstrated increased 'resistance to extinction' (see Chapter 10) in rats given repeated foot-shock at about 20 days of age; when tested as much as 7 weeks later the shocked animals were more persistent than unshocked controls in running to an empty goalbox that had once contained a food reward. Still more dramatically, Anisman, working with mice, has shown that inescapable foot-shock can produce increased resistance to disease. In his experiments tumours implanted in animals previously exposed to shock grew more slowly than in controls.

Miller's toughening up, then, is a process of such broad effect that it is more natural to speak of it as giving rise to a general 'tolerance for stress' rather than merely 'tolerance for punishment'. But toughening up does not always occur after exposure to punishing or other aversive stimulation. As we have already seen, Miller was able to demonstrate counter-conditioning by giving shock followed by reward in the goalbox of an alley under conditions in which shock delivered outside the alley failed to produce any non-associative increase in resistance to punishment. Thus one must ask what conditions favour associative and non-associative mechanisms in the production of tolerance for punishment. Little attention has so far been paid to this question. But scrutiny of the published reports of successful counter-conditioning and toughening up, respectively, reveals an apparently clear pattern: the intensity, duration, and number of presentations of shock that are necessary to produce toughening up seem to be much greater than those needed for associative counter-conditioning.

Displacement activities

We have seen something of the factors which determine the outcome of an approach–avoidance conflict. But conflict may have other effects which cannot be described simply as the dominance of the approach or avoidance tendency.

To begin with, it is frequently observed that an animal in a conflict situation engages in forms of behaviour which are appropriate neither to the approach tendency, nor to the avoidance tendency. These forms of behaviour have been studied in most detail by the ethologists, who term them 'displacement activities'.

There is considerable controversy over the mechanism by which displacement activities occur; but certain facts about the nature of these activities appear to be clear. The 'irrelevant' behaviour which is often seen during conflict is usually very common in the animal's repertoire in any case, consisting of such activities as grooming and preening, feeding and sleeping. The most likely time for the occurrence of displacement activities is when the two conflicting tendencies are about evenly balanced. For example, Tinbergen's studies of territorial fighting in the stickleback (see Chapter 2) showed that displacement 'nest-digging' is most common at the boundary of a particular fish's territory, where attack and flight are about equally likely. Finally, when displacement activities do occur, it is now clear that they are controlled by the same stimuli and other causal factors which control them when they occur in non-conflict situations.

It is natural to ask why an animal torn between, say, fight and flight should pause to clean its feathers or dig a hole in the sand. Some of the earlier explanations of displacement activities were highly dramatic; but it now seems that the explanation is comparatively simple. The two behaviour tendencies in conflict effectively prevent each other from controlling the animal's behaviour; control passes instead to the form of behaviour for which most eliciting stimuli or other causal factors are present, and this will tend to consist of activities which are frequently displayed by the animal anyway. There are probably special factors favouring the choice of grooming or preening as a displacement activity: the high level of excitation in the autonomic nervous system which occurs during emotional arousal will give rise to a number of bodily changes (e.g. raising of the hair, changes in the distribution of the blood) which will provide a host of stimuli for grooming. Whether this account of displacement activities in animals is also able to account for the apparently similar behaviour seen in Man in frightening or embarrassing situations (scratching, hair-smoothing, tie-straightening, and so on) is a moot question; but it does not seem unlikely.

Experimental neurosis

It has been clear, however, from the earliest days of the experimental study of behaviour that conflict can give rise to far more serious disorders than 'displacement' tie-straightening. Pavlov's experiments on 'experimental neurosis' involved only simple classical conditioning techniques; nevertheless, the element of conflict contained in them was sometimes strong enough totally to disrupt the animal's behaviour.

In the earliest of these experiments (conducted by Yerofeyeva in 1912) an electric shock was used as a conditioned stimulus signalling food. So long as the shock was applied to one part of the dog's body only, the original reaction to it was eliminated completely and replaced by a conditioned salivary response. But when the shock was applied to parts of the dog's body other than those at which the initial training had taken place, not only was there no transfer of the salivary response, but the established conditioned reflexes also disappeared and took a long time to return.

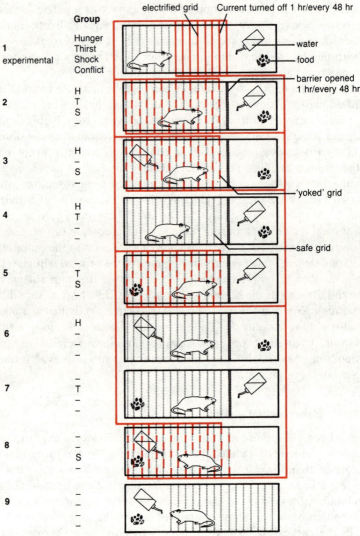

Fig. 9.8. The design of Sawrey, Conger & Turrell's experiment on the production of ulcers by conflict. The experimental group (group 1) was deprived of food and water, but could obtain both by crossing an electrified grid, which was made safe only for 1 hour in every 48 hours. The various other groups were exposed to the same degree of hunger or thirst in the combinations indicated by the letters to the left of the figure, but in their case food and/or water was cut off, not by an electrified grid but by a barrier which they could not cross. In addition, groups 2, 3, 5, and 8 were in boxes whose grid floors were wired in series with the electrified grid of the group 1 box. Thus, whenever a group 1 rat ventured onto the grid in search of food or water, and closed the shock circuit, the rats in these 'yoked control' groups

But it was a later experiment, reported by Shenger-Krestovnika in 1921, which really stimulated Pavlov's interest in experimental neurosis. In this famous experiment, a dog was given food after being shown a circle but not after being shown an ellipse; in consequence, it came to salivate in response to the circle but not to the ellipse. The ellipse was now progressively made more like a circle. When the ratio of the axes of the ellipse was reduced to 9:8, the dog could discriminate it from a circle only with great difficulty. He showed some sign of success on this problem for about 3 weeks, but then his behaviour was totally disrupted. He was quite unable to respond correctly not only on this difficult task, but also when presented with perfectly obvious ellipses and circles which had given him no trouble in the earlier part of the experiment. Furthermore, instead of coming to stand quietly in the apparatus, as in the past, the animal now showed extreme excitement, struggling and howling. Note that this remarkable result, which has recently been confirmed by Thomas & DeWold, was obtained without the use of any painful stimuli: merely a succession of some stimuli followed by food and others not followed by food. This is the first hint of a theme to which we shall return in the next chapter: the parallel between the effects of punishment and those of non-reward.

Conflict and ulcers

Now a conflict situation necessarily involves a sequence of rewards and punishments (or non-rewards acting like punishments). Thus the harmful effects seen might be in no way due to the *conflict*, but simply to the punishments received. To demonstrate harmful effects of conflict *per se* requires special experimental designs.

An experiment by Sawrey, Conger & Turrell meets this demand admirably. They were studying the conditions which give rise to gastric ulcers. The design of their experiment is set out in Fig. 9.8. There is one 'experimental group' which is exposed to conflict; and an elaborate, but necessary, series of eight 'control' groups to evaluate the effects of each of the separate factors in the conflict. The experimental group could get at food and water only by crossing an electrified grid;

Caption for Fig. 9.8 (*cont.*).

also received a shock. In this way, the various control groups were exposed, alone or in every combination, to all of the conditions to which the experimental group was also exposed, with the single exception of the element of conflict about whether or not to cross the grid in search of sustenance.

it thus experienced food and water deprivation, shock, and conflict about whether to take the food or avoid the shock. The various control groups experienced each individual condition alone or in every possible combination. As a further control, those animals which were in control groups getting shock were each 'yoked' to a corresponding animal in the experimental group, getting shock at exactly the same time as the latter, but independently of their own behaviour (so that no conflict was involved). At the end of the experiment, the animals' stomachs were examined. It was found that hunger and shock (but not thirst) could cause ulcers on their own, but that conflict *per se* significantly increased the degree of ulceration.

There are two alternative ways of interpreting this experiment. The conflict animals were the only ones to experience conflict, but they were also the only ones to exercise any control over the situation they were in: they determined how often and when they received shocks and how hungry they went, whereas the yoked controls merely suffered shocks and hunger to the same extent as the conflict animals. We could say, therefore, either that the critical feature causing ulcers in the conflict group was *worry* – shall I go forward, or shall I stay still? – or that it was *taking decisions*.

The harmful effects of decision-making have been emphasised on independent grounds by Brady and his colleagues. They trained two monkeys in a Sidman avoidance situation (Fig. 9.4) and used two others as yoked controls, i.e. they could do nothing except take the shocks that the experimental animals failed to avoid. Now, the element of conflict is minimal in the Sidman avoidance situation, since there is no competing drive to interfere with the animal's primary aim of avoiding shock; but there are, of course, decisions to be made – when to press the lever which will avoid shock. In the Brady experiment it was only the experimental animals (not the yoked controls) which developed ulcers. This would seem to indicate that it is decision-making which gives rise to ulcers; and, in an explicit analogy. Brady refers to the experimental animals as 'executive monkeys'.

There is rather strong evidence, however, that this interpretation is wrong, and that it is conflict, or worry, which is the important factor leading to the development of ulcers.

To begin with, the executive monkey experiment suffers from a flaw pointed out by Jaw Weiss. The monkeys were not assigned to the experimental conditions at random; instead, in each pair the animal which learnt the avoidance response more quickly was chosen to be the Sidman animal. Now we have learnt enough about individual differences in emotionality to recognise this as an illegitimate procedure: it is as likely that the results of Brady's experiment were due to the monkeys' pre-experimental personality as to the decisions made by the executive monkeys. These doubts are reinforced by the results Weiss obtained himself in an experiment with rats. This experiment repeated the same basic design used by Brady's group with monkeys, except that animals were assigned to the experimental conditions at random. The experimental group of rats received shock which they could escape or avoid by jumping on to a platform (in one experiment) or by pushing a panel (in a second one). The yoked controls again received shocks which were determined by the performance of the experimental animal to which they were each matched. But this time the effects of these treatments were much greater in the *yoked controls*: they defecated more, their rate of growth was slowed, they showed greater reluctance subsequently to drink in the experimental situation when thirsty (presumably indicating greater fear), and they developed more and more severe ulcers. Thus, the exercise of control over the amount of shock received offers protection to the 'executive rat' from the ulcers which shock can cause.

In interpreting this conclusion we must bear in mind that (not unnaturally) what matters for the executive rat is not the control which, theoretically, it *might* exercise over a stressor, but the control that it actually does exercise. This depends upon the difficulty of the task that the animal is set. Tsuda and co-workers, for example, compared two conditions under which executive rats could escape or avoid shock: in one, the effective response consisted of two consecutive pulls upon a disc (a schedule known as 'fixed ratio, or FR, 2'); in the other, the schedule was 'variable ratio, or VR, 5', i.e. the effective response varied unpredictably from trial to trial around an average value of five consecutive pulls upon the disc. Only in the former case did the executive rats develop fewer ulcers than the yoked controls; on the VR 5 schedule the executives (who avoided fewer shocks than executives reinforced on the FR 2 schedule) actually developed more ulcers than the controls. Similar relationships have been demonstrated between the levels of corticosterone in the blood and the difficulty of an escape–avoidance task. Thus Berger and co-workers compared two bar-pressing tasks: in one the rat had a 60-second warning and was able to avoid 70% of the programmed shocks; in the other, there was only a 10-second warning and the level of avoidance dropped to 16%. In the former case, the executive rat had lower levels of circulating corticosterone than yoked controls, but in the latter case corticosterone levels were equally high in executives and controls. A further important factor is the predictability of shock: among animals that have no control over shock the provision of a clear signal of the time at which each shock will occur reliably reduces ulcers.

Taking all these various findings together, the following conclusions seem warranted. First, it is worry, not control through the taking of decisions, which is the ulcerous element in conflict. In the Weiss experiment, where no conflict is involved, only decision-making, the animal with control over the situation is less disturbed by it. But in the Sawrey experiment (Fig. 9.8), the animals in control over the situation were also the only ones to suffer the conflict between approach (to the food and water) and avoidance (of the shock). Since it was these animals which developed the most severe ulcers, we must conclude that the effect of conflict *per se* may be so severe as to outweigh any advantage accruing from control of the situation. Second (perhaps an obvious conclusion), control provides protection against ulcers only to the extent that it is effective control and within the animal's level of capability. And, finally, if there is nothing the animal can do about the stressors to which it is exposed, it still helps to know when they will occur. We may relate this conclusion back again to the theme of worry: if you don't know when a shock may occur, you need to worry all the time; if you do, you can at least relax at times when shock is unlikely to occur. This is a theme to which we shall return in Chapter 11.

There is evidence at the human level too that conflict and helplessness are powerful causal factors in the production of ulcers. Wolf & Wolff's observations of Tom's stomach (Chapter 5) are important in this connection. It will be recalled that these workers described a pattern of increased function in the stomach in certain states of emotion. This pattern includes an increased blood supply, swelling of the mucous lining of the stomach, increased secretion of hydrochloric acid, and an increase in the amount of stomach movement. It was particularly prominent at times when Tom experienced emotions of resentment, but was unable to attack the object of his resentment. An example described by Wolf concerns a coal shortage in 1946. Tom was unable to get coal and blamed this on the politicians ('a bunch of crooks') and on the dealer who had promised to get him some but had not done so ('what kills me is not keeping a promise'). He felt angry and frustrated, but there was nothing he could do about it. During this period, Tom's stomach became especially red and turgid, its acidity became higher than it had been for a year, and bleeding occurred in several places. These, of course, are conditions which could easily lead to the formation of ulcers. There are remarkable parallels between Wolf's analysis of the conditions that affected Tom's stomach and experimental findings in the rat. As we have seen, shocking an animal can cause ulcers. But the incidence of ulcers can be reduced if the rat is provided with another rat to attack when it is shocked. The adrenocortical response to shock (which is often closely related to the development of ulcers) is similarly reduced if an animal is allowed to fight. Whether Tom would also have been better off if he had had greater opportunity to express his aggression, history does not relate.

Helplessness

As we have seen, it is clear from experiments on psychogenic ulcers that the opportunity to perform an effective 'coping' response that controls exposure to a stressor is able to reduce the deleterious effects of the stressor even when (by the use of yoked controls) actual exposure is held constant. The condition in which animals are denied such coping responses has recently come under increasingly detailed scrutiny. The key observation that has fuelled this interest was reported by Martin Seligman and two of his associates (Bruce Overmier and Stephen Maier) in 1967. These workers showed that dogs, exposed in a first phase of the experiment to shocks that they could neither escape nor avoid, later failed to escape shock in a different situation in which escape was now possible. Subsequently the same effect has been demonstrated in cats, fish, mice, and rats. It is important to note that the deficit in escape behaviour is not seen if, in the first phase of the experiment, animals receive an equivalent experience of shock that they can control (i.e. escape) by making an appropriate instrumental response. To demonstrate this, use is made of the same design, comparing executive animals and yoked controls, that we encountered in the section on ulcers. Seligman termed the impairment in escape behaviour (and comparable effects when other kinds of instrumental response are measured) 'learned helplessness'. Since much of the controversy that has surrounded the phenomenon turns on what, if anything, is learned, this term begs many important questions. Here I shall preserve the term 'helplessness', but strip it of 'learned', using it to refer in a purely descriptive manner to that class of phenomena in which exposure to uncontrollable aversive events gives rise to subsequent deficits in instrumental responding.

Helplessness is a striking phenomenon in its own right; however, interest in it has been all the more intense since Seligman proposed that it constitutes a valid model of human depression. This is a claim to which we shall return (see Chapters 13 and 14). Here, the issue which concerns us is this: what is the psychological nature of helplessness as this emerges from experiments with animals?

Essentially three answers have been given to this question. Seligman's answer is that the helpless animal fails to perceive the relationship between responses and outcomes even when these once again exist (this is the 'learned helplessness' hypothesis). According to this hypothesis, the animal learns in the first phase of an experiment on helplessness that his responses have no effect on the stressors to which he is exposed; then in the second phase of the experiment, when his responses *can* control the stressors, he is slow to detect and learn about the new response–outcome contingencies. The other two answers are much simpler. Several workers, including Weiss and Anisman, have independently proposed that exposure to inescapable stress impairs the animal's ability to initiate and/or sustain movement. In the second phase of the experiment, according to this hypothesis, the animal can learn a coping response but cannot perform it. The third answer, proposed by Jackson, Maier & Coon, is that the helpless animal feels less pain. There is now a wealth of evidence that pain and stimuli associated with pain (see Chapter 2) can give rise to an analgesic response – or, rather, to several different kinds of analgesic response, some mediated by endogenous opiates, some by non-opioid systems. According to Jackson, therefore, exposure to uncontrollable shock gives rise to such an analgesia – the animal then fails to escape shock in the second phase of the experiment because it does not hurt so much. Each of these hypotheses has been proposed as a full explanation of all the effects produced by exposure to uncontrollable stress, needing no help from the others. It is now clear that none of them is in fact capable of providing such a complete account. But each is probably correct in its major claim: exposure to uncontrollable stress apparently gives rise to difficulties in instrumental learning *and* to deficits in movement *and* to analgesia.

Some of the evidence that seemed to speak most strongly in favour of the learned helplessness hypothesis came from the so-called 'immunization' effect. This is the demonstration that training with escapable shocks before exposure to inescapable shocks blocks the development of helplessness. This was interpreted by Seligman & Maier as indicating that prior experience with controllable shock proactively interferes with learning that shock is uncontrollable (such proactive interference is a well-known phenomenon in, for example, human learning of verbal material). The force of this argument is, however, much weakened by evidence from Anisman's laboratory that deficits in escape learning caused by certain drugs can also be blocked by prior experience with escapable shock; and by evidence from Maier's laboratory that shock-induced analgesia is similarly so blocked. It is difficult to see how reactions to a drug can be altered by prior learning that stress is controllable, or how loss of analgesia could result directly from such learning. The immunization effect nonetheless remains a phenomenon of considerable interest, as it suggests that experience with a successful coping response has long-lasting and remarkably general effects, not only on subsequent behaviour, but even on subsequent reactions to physical treatments such as drugs.

The case for learned helplessness as a general account of the phenomena observed in helplessness experiments has also been weakened by the evidence from Weiss's group (considered earlier in this chapter) that exposure to multiple sessions of inescapable shock gives rise to toughening up – that is to say, the deficit in escape behaviour seen after one session of inescapable shock is no longer seen after 14 sessions. If a rat learns in one session that shock is something he can do nothing about, it is difficult to see why he should change his mind after 14 identical sessions.

If we turn to the other two hypotheses, there are clear demonstrations of both analgesia and activity deficits after inescapable (and not after escapable) shock. It is possible that both these effects contribute to the escape deficit that is the core of helplessness. However, there are also clear instances of deficits in escape and other coping responses that cannot be due to either analgesia or impairments in movement.

The strongest demonstration of a dissociation between analgesia and the escape deficit comes from Maier's laboratory. One source of endorphins (endogenous opiates) is in the pituitary (Chapter 5). It was reasoned, therefore, that removal of the pituitary might eliminate shock-induced analgesia, and so it did: yet the escape deficit also induced by shock persisted after this operation. Similar results were obtained by injection of a synthetic glucocorticoid (Chapter 5), dexamethasone, which blocks release from the pituitary of both ACTH and endorphins. Thus opioid analgesia, while undoubtedly a consequence of inescapable shock and sharing many of the same features (time-course, etc.) as the escape deficit, does not itself cause this deficit. As to 'non-opioid' analgesia, Anisman's group has reported that the experimental parameters that affect this and the escape deficit differ considerably.

Other experiments from Maier's laboratory have similarly shown that inescapable shock gives rise to learning deficits that are consistent with the learned helplessness hypothesis and cannot be secondary to an activity deficit. In one of these, inescapable, but not equivalent escapable shocks, impaired the animal's subsequent ability to choose the arm of a Y-maze in which it could escape shock. Such a choice measure is, of course, unaffected by the speed or vigour of the animal's movements. A second experiment is even more convincing. It depends upon the difference between the *conditioned suppression* of bar-pressing and *punishment*, in a strict sense, of bar-pressing. We must first consider, therefore, the nature of this difference.

Punishment and conditioned suppression

In a conditioned suppression experiment (Fig. 4.7) an initially neutral stimulus (e.g. a tone or light) is paired as a Pavlovian CS with an aversive UCS (typically, foot-shock) and comes in consequence to be able, when presented while the animal is performing an instrumental response (e.g. bar-pressing for food), to suppress performance of that response. Since essentially the same results are obtained whether pairing of the CS and UCS is carried out 'on the baseline' (that is, while the animal is performing the instrumental response) or 'off the baseline' (that is, under conditions in which the response cannot be performed, due for example to removal of the bar or to use of a different chamber), conditioned suppression appears to be a straightforward case of classical conditioning (Chapter 2). The CS acquires its changed properties in virtue of its association with the UCS, and the instrumental response (bar-pressing for food reward) which it comes to suppress plays no part in this learning process. In a punishment experiment, in contrast, the contingency established by the experimenter concerns not two stimuli, but a response and a stimulus, to wit, pressing the bar and shock. This relationship, barpress → shock, may be arranged to occur either in the presence of a specific discriminative stimulus (tone, light, etc.) ('discriminated punishment') or without any such special signal (but general apparatus cues are always present to play the role of implicit discriminative stimuli). The effect of punishment is, of course, to suppress bar-pressing – indeed, this capacity to suppress responses upon which it is made contingent is the defining characteristic of any 'punishment'. But now we have a problem: if it is sufficient to pair stimuli (e.g., apparatus cues) with shock for these stimuli to suppress bar-pressing in their presence, how can we tell that the specific punishment contingency (i.e. the relationship between bar-pressing and shocks) has some further suppressant effect upon bar-pressing over and above that caused by apparatus cues?

Connoisseurs of learning theory will recognise this problem as a special case of a more general issue, namely, how to demonstrate that in addition to the process of classical (stimulus–stimulus) conditioning, behaviour is also determined by instrumental (response–stimulus) conditioning. This is a central and difficult issue in the theory of learning. However, there is now wide agreement that, in principle, neither of these two processes of learning can be reduced to the other, although they may both reflect some more general learning mechanism, and that behaviour is determined in part by both. Unfortunately, such agreement on general principles does not relieve us of the burden of demonstrating a role for instrumental (over and above classical) conditioning in particular cases; and the particular case of punishment has proved to be one of the most difficult for which to make such a demonstration. Nonetheless, there are now a number of lines of evidence that show fairly conclusively that the response–shock contingency that is established when an experimenter punishes bar–pressing (or other such instrumental responses) indeed contributes to the consequent suppression of bar-pressing.

One way to approach this problem is to ask the animal whether *he* can tell the difference between response-contingent and response-independent shock. Rawlins, Feldon and I performed such an experiment with rats. During sessions in which they were bar-pressing for food they were presented from time to time with one of two distinctive stimuli. One signalled the delivery at random intervals of time of foot-shock that was independent of their behaviour; the other signalled the delivery of shock contingent upon bar-pressing. The shocks were of identical strength and programmed according to the same schedule. The question we asked was whether the animal would show greater suppression of bar-pressing in the presence of the signal for response-contingent shock (it was already known that shocks of equal strength produce greater response suppression when used in a punishment paradigm compared to one of conditioned suppression; see e.g. Fig. 9.9). The rats indeed learned to make this discrimination (though with difficulty). Furthermore, response rates differed significantly even during the initial segment of each signal presentation, before the first shock was delivered. Thus the rats were not reacting to the last shock received; they had genuinely learnt that the two stimuli signalled different modes of shock delivery.

A second experimental design that has contributed to the conclusion that punishment differs from conditioned suppression involves training the rat to perform two different food-rewarded responses in the same apparatus during the same session and then punishing only one of them. If response suppression were entirely a matter of stimulus–stimulus associations, one would expect both responses to be suppressed, and perhaps equally so, rather than just the one that is explicitly punished. In an experiment of this kind conducted by Russell Church, rats were trained both to press a lever and to pull a chain for food reward on a concurrent variable-interval schedule (Fig. 9.4): that is, each of these responses was rewarded on a variable-interval schedule independently of the other, ensuring that the animal kept switching back and forth between the two. Church then compared two conditions for the delivery of shock. In the punishment condition, during a warning signal shocks (programmed at random intervals) were delivered contingent upon performance of only one of the two food-rewarded responses; in the conditioned suppression condition the delivery of shock was independent of both responses. The results of this experiment are shown in Fig. 9.9. It can be seen that the degree of suppression of the unpunished response in the punishment condition is less than that of the punished response and equal to the suppression of both responses in the conditioned suppression condition.

These results rather strongly suggest that punishment gives rise to response suppression in part because of the stimulus–stimulus associations (in this case, between the warning signal and shock) that form a necessary part of the punishment contingency, but also in part because of an additional specific association between some aspect of the response and shock. However, even the latter could perhaps be regarded as a stimulus–stimulus association: in Church's experiment, for example, the rat might associate sight of and contact with the lever (if lever-pressing is the punished response) rather than the chain, and it may withdraw from the lever in consequence. Just such classically conditioned withdrawal from stimuli associated with shock has

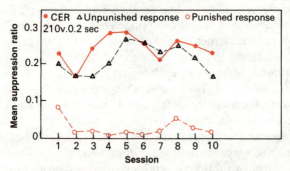

Fig. 9.9. Comparison of suppression produced by response-contingent and by response-noncontingent shock. Mean suppression ratio (0 = total suppression, 0.5 = no suppression) when one of two responses (bar-pressing or chain-pulling), but not the other, was punished during the warning signal, or when shock was not contingent on either response ('CER'). From Church, Wooten & Matthews (1970).

been reported by Karpicke & Dout. This problem of interpretation was overcome in an experiment by Bolles and his co-workers. They trained rats to earn food by performing two different responses on one specially-designed lever: pressing down on it or pulling it out from the wall. Then only one of these responses was punished. Initially, this punishment contingency gave rise to suppression of both responses; but the unpunished response soon recovered and came to be performed to the exclusion of the other. Mere association of the lever with shock could not have produced this result.

These experiments demonstrate, then, that punishment involves a true instrumental (response–shock) contingency over and above the purely stimulus–stimulus associations that give rise to conditioned suppression. Armed with this conclusion we may now return to the issue of helplessness. For Seligman's learned helplessness hypothesis predicts that exposure to inescapable shock should make the animal poorer at learning specifically *instrumental* contingencies, not classically conditioned stimulus–stimulus associations. In a test of this prediction Jackson, Maier & Rapoport compared the effects of inescapable shock upon subsequent *punishment* and *conditioned suppression* of bar-pressing. There was no effect of the prior inescapable shock upon the degree of conditioned suppression; but the suppression caused by punishment was reduced. This experiment is important from two points of view. First, it adds to the evidence that punishment involves something beyond the stimulus–stimulus associations that give rise to conditioned suppression (for otherwise the two forms of suppression should have been equally affected by prior inescapable shock). Second, it demonstrates a deficit in instrumental responding that cannot be attributed to an impairment in the initiation of movements (for the reduced effect of punishment was manifest as an *increase* in the rate of bar-pressing).

One may regard the Jackson, Maier & Rapoport experiment as showing that inescapable shock teaches the animal (in agreement with Seligman's hypothesis) that shock is independent of its instrumental responses but not necessarily independent of environmental stimuli. Other experiments show that the rat is capable of the converse learning – that stimuli and shock rather than responses and shock are independent of one another. Baker, for example, first exposed rats to presentations of noise and shock in which the two events were in random relation to each other. Subsequently the noise was used as a CS for shock in a conditioned suppression procedure. Compared to controls that had not had the prior experience of a random relationship between noise and shock, the pre-exposed rats were slower to develop conditioned suppression of bar-pressing. Exactly the same result was obtained if, in the second phase of the experiment, bar-pressing was punished in the presence of the noise. This is as one would expect, given the evidence from Church's work (Fig. 9.9) that there is a large element of conditioned suppression in the reduced bar-pressing caused by a punishment contingency. But Baker also looked at a further pre-exposure condition, in which noise and shock were presented in random relation to each other while the animal was bar-pressing. This procedure, of course, allows the animal to learn not only that the occurrence of shock is independent of the noise, but also that it is independent of bar-pressing. When the noise was subsequently used as either the CS in a conditioned suppression procedure or the discriminative stimulus for punishment of bar-pressing, there was again a retardation of learning to suppress bar-pressing in the pre-exposed animals. However, there was now an additional degree of retardation (compared to the effects of pre-treatment with random noise and shock outside the bar-pressing situation) which affected only the punishment condition. Thus animals can learn both *both* that stimuli and shock are unrelated *and* that responses and shock are unrelated, and adjust their later behaviour accordingly.

Everyone was right, then, but no one was completely right: uncontrollable aversive events appear to give rise to several *independent* consequences, including difficulty in forming new response–outcome associations, analgesia, and lowered activity levels. (The reduction in activity in response to shock seen after prior inescapable shock may, however, be secondary to opiate-mediated analgesia: Dugan & Maier have shown that, like shock-induced analgesia, the activity deficit is blocked both by the opiate antagonist naltrexone and the glucocorticoid dexamethasone.)

Indeed, the consequences of inescapable stress are even wider. Besides the ulcers considered earlier in this chapter, it leads to loss of appetite, weight loss, reduced aggressiveness and competitiveness, loss of grooming and play behaviour, and sleep disturbances. Furthermore, the difficulty in forming new response–outcome associations (Seligman's *learned* helplessness) extends beyond the escape learning paradigm that has most often been used to measure it. Rosellini has even shown (confirming one of Seligman's specific predictions) that exposure to inescapable (but not escapable) shock causes deficits in subsequent rewarded behaviour. As in the case of the escape deficit, this effect is not secondary to reduced motor activity, since Rosellini used a choice response (poking the nose in one of two holes for food reward). All these findings lend weight to the claim that uncontrollable stress causes in animals a state that is similar to depression in Man (see Chapters 13 and 14 for a further examination of this claim).

Conclusion

When an animal such as a rat is exposed to a régime of uncontrollable shock, we must suppose that (after an initial period during which it attempts to struggle, flee, and so on) it experiences no conflict. There is nothing the animal can do about its unfortunate situation, and the experiments we have reviewed make it plain that the rat is perfectly capable of appreciating this fact. So both conflict and no conflict, in their different ways, can be bad for you – or good for you, for we have also seen that repeated exposure to aversive events (whether coping is allowed or not) can give rise to an improved capacity to cope with later stress (see the sections on resistance to punishment).

During this discussion of both conflict and helplessness I have commented on a number of forms of human behaviour that might plausibly be attributed to the same kinds of process that have come under scrutiny in laboratory animals. These include various kinds of sexual deviation, such as paedophilia, fetishism, and the 'peeping Tom' syndrome; masochistic behaviour; the degree and mode of development of the conscience and feelings of guilt; the apparently inappropriate forms of behaviour decribed by the ethologists as 'displacement activities'; depression; and the development of ulcers. Less dramatically, but no less importantly, it is to an analysis of conflict behaviour that we would turn if we wished to train a criminal to mend his ways, or to teach our children perseverance in the face of adversity.

But adversity does not consist only of blows. The perseverance we wish to encourage in our children is more often perseverance in the face of disappointment than perseverance in the face of pain. The conditions which give rise to neurotic behaviour in the patient seen in the psychiatric clinic, or to aggression in the criminal seen in gaol, are more likely to be the frustration of their expectations than physical attack. To the extent, then, that an analysis of conflict behaviour is restricted to avoidance tendencies established on the basis of punishment, it is likely to be of limited use. We shall see, however, in the next chapter that there is a considerable body of evidence to suggest that the disappointment engendered by the absence of an expected reward acts in much the same way as the pain produced by punishment. This parallel between the effects of punishment and those of non-reward enormously extends the value of the kind of conflict analysis which we have examined in this chapter.

10   Fear and frustration

In the previous chapter we hinted at a theme which will dominate this one: the parallel between the effects of punishment and those of the omission of anticipated reward. So striking is this parallel that it has given rise to the hypothesis that 'frustration', as psychologists term the state of the organism which is set up by omission of reward, is functionally and physiologically very similar to – and perhaps identical with – the state of fear.

This hypothesis – the 'fear = frustration hypothesis' – may sound extremely implausible. After all, we are all able to say when we are feeling 'afraid' and when we are feeling 'disappointed' or 'frustrated'; how could we do this if the states of fear and frustration were identical? However, it will be recalled that in discussing Schachter's work on the human ability to recognise one's own emotional state (Chapter 5), we concluded that this ability rests in good measure upon a knowledge of the events leading up to or accompanying the physiological state we are in at the time. Thus, to be able to apply the words 'fear' or 'frustration' correctly to our own emotional reactions (or to those displayed by others), it is sufficient: (1) to recognise that we are in some kind of unpleasant emotional turmoil; and (2) to know whether this turmoil was evoked by the sight of a snake or by a broken date.

This kind of distinction has been made more formally by the influential Polish scientist Jerzy Konorski in his general theory of classical conditioning. Konorski distinguishes between 'preparatory' and 'consummatory' conditioning. Preparatory CRs are diffuse expressions of a general emotional state, involving such reactions as increased activity and approach if the UCS is appetitive (food, water, etc.); or changes in heart rate and other autonomic indices (Chapter 5), suppression of ongoing activity, and withdrawal if it is aversive. Consummatory CRs, in contrast, are precise, discrete reflexes, such as pecking or licking if the UCS is appetitive, blinking or leg flexion if it is aversive. If a shock is used as UCS, it will result in similar preparatory CRs (conditioned withdrawal, conditioned suppression, etc) no matter how it is delivered. But a shock delivered to the paw and to the cheek will result in quite different consummatory CRs – paw flexion and an eyeblink, respectively. Konorski suggests therefore that preparatory conditioning depends upon an association betwee the CS and

174

the general motivational and emotional properties of the UCS, whereas consummatory conditioning reflects an association between the CS and the specific sensory properties of the UCS. Under most circumstances both kinds of conditioning will take place in parallel, although certain special manipulations allow one to vary the one but not the other.

In the light of Konorski's work the fear = frustration hypothesis may be taken as asserting that the general motivational and emotional properties of the central states elicited by punishment and non-reward, respectively, are identical; and similarly that the preparatory, but not necessarily consummatory, CRs (understood again as central motivational/emotional states) set up by classical conditioning with punishment or non-reward respectively as the UCS are again identical. Note, however, that the assertion that the states of fear and frustration are identical in *kind* does not preclude the possibility of variation in *intensity*. I should think it very likely that the intensity of the reaction aroused by 'frustrative non-reward' (the technical term for the omission of an anticipated reward) is nearly always less than the intensity of the reaction aroused by a punishment. But it is possible that a very intense disappointment may set up a reaction which is as strong as, or even stronger than, the reaction caused by a mild punishment. At any rate, the fear = frustration hypothesis, in its strongest form, admits only of such quantitative variation, and not variation in kind, in the functional and physiological state aroused by punishment and frustrative non-reward.

In spite of its counter-intuitive nature, the fear = frustration hypothesis must in the end be judged, like any other scientific hypothesis, in the light of the experimental evidence. And the reader is asked to suspend judgment until he has had the opportunity to consider this evidence.

If we are to take this hypothesis at all seriously, it must first be shown that frustrative non-reward shares the most obvious properties of punishment. Now, to subject an animal to frustrative non-reward, the experimenter first rewards it for performing some response on a number of occasions, and then allows it to perform this response again but this time (or times) without rewarding it. As we know, the defining characteristic of a punishment is that it is aversive, that is, the organism will work to terminate or avoid it. The very least we must show, then, is that an animal will work to terminate or avoid frustrative non-reward. Unfortunately, it is logically impossible to terminate non-reward without a reward; so any behaviour which might be described as directed towards the termination of non-reward might also be described, and more plausibly, as being directed to the attainment of a reward. What we can do, however, is demonstrate the aversive properties of stimuli *associated* with frustrative non-reward.

The best demonstration of this effect is in an experimental situation devised by Adelman & Maatsch. They trained rats to run down the familiar straight alley for a food reward. They then removed the food from the goalbox, but carried on placing the animals in the startbox as before. Technically, this is described as 'extinction' and the number of trials for which the animal continues to run for zero reward is a measure of 'resistance to extinction'. Now, usually what is done in extinction is to allow the animal to run over and over again until eventually it gives up, stopping either in the stem of the runway or in the startbox itself. This, of course, does not allow you to decide whether the animal is finding the environment in which he experiences frustrative non-reward (i.e. the goalbox) aversive or whether he simply no longer bothers to get to an empty goalbox. However, Adelman & Maatsch introduced one important modification into the usual extinction scheme: they provided the rat with an alternative mode of exit from the goalbox by putting a ledge round it on to which the rat could jump. A control group of hungry rats was rewarded with food on the ledge for jumping up to it. A second control group was never given the food, either in the goalbox or on the ledge, but merely exposed to the goalbox the same number of times as the experimental group and allowed to jump out. The remarkable finding made by Adelman & Maatsch was that the experimental group, which was escaping from an environment in which it was exposed to frustrative non-reward, learnt to jump out more quickly even than the food-rewarded controls, while both did much better than the control group which was neither rewarded nor frustrated.

A second demonstration of the aversive properties of frustrative non-reward is due to Wagner. He also trained rats to run down an alley, but rewarded them on only a proportion of the trials, a procedure known as a 'partial reinforcement schedule'. On the trials when they were not rewarded, a distinctive stimulus (noise and light) was presented to them as they entered the goalbox. They were then tested in a shuttlebox, and it was found that they would jump across the barrier in order to turn this stimulus off significantly more often than control rats which had also experienced the same stimulus in the runway, but *not* in association with the non-rewarded trials.

An extension of Wagner's experiment has been reported by Helen Daly. She trained four groups of rats to run in an alley, all with 15 pellets of food as reward during initial acquisition. After the running response was fully established, the size of the reward was shifted to 1 pellet for two of the groups and to zero (i.e. extinction) for the other two. For one each of these pairs of groups the changed reward condition was accompanied by a new stimulus (a light) in the goalbox. Subsequently all animals were tested to see whether they would acquire a hurdle-jumping response to take them out of the goalbox into a different compartment. On these hurdle-jumping trials the rats were placed in the goalbox with the same reward (1 pellet or none) and stimulus (light or no light) conditions which they had experienced in the reward-shift phase of the experiment. Four control groups were also used, two with 1 pellet at all stages of runway training and two with no pellets at any stage of training; each of these pairs of control groups were divided into one with a light in the goalbox and one without, like the equivalent experimental groups.

The results of Daly's experiment showed that, compared to the control groups (which did not differ from each other), all four experimental groups were faster to escape from the goalbox by jumping across the hurdle. Furthermore, there was a higher rate of hurdle-jumping if the downward shift in reward magnitude was to zero than if it was to 1 pellet; and a higher rate of hurdle-jumping when the light stimulus was added to the goalbox as part of the environment from which hurdle-jumping permitted escape. Finally, it should be noted that 1 pellet compared to a previously experienced reward of 15 pellets was able to maintain a higher rate of escape behaviour than 1 pellet if this was all the animal had ever experienced. Thus Daly's experiment permits the following conclusions: (1) a goalbox associated with extinction becomes aversive (confirming Adelman & Maatsch's findings); (2) the addition of an explicit CS (the light) paired with the omission of reward increases the level of aversiveness of the total situation (in agreement with Wagner's observations); and (3) a reduction in reward from an experienced level of 15 pellets to a non-zero but lower level (1 pellet) is also able to confer aversive properties on stimuli associated with this reduction, and thus appears to be functionally equivalent to (though less intense than) total removal of reward.

In the light of Daly's results a very well established phenomenon, known since the 1940s, can be seen to be evidence of the punishing effects of frustrative non-reward. This is Crespi's 'depression effect'. In this, animals are first trained to run down a runway for a reward of relatively large size and are then shifted to a low reward. A control group is run throughout with just the low reward. In such an experiment the animals running initially to the high reward will run faster than those running to the low reward. When the high reward is shifted to a low one, the shifted group reduces its running speed till it is eventually the same as the group that has always had the low reward. However, for a small number of trials after the shift, the shifted group runs significantly *slower* than the unshifted group (even though both groups now get the same reward), suggesting that a low reward experienced after a high one is capable of acting like a punishment, i.e. it suppresses performance of a response upon which it is contingent (in the example, running into the goalbox).

The Crespi depression effect shows how an animal's reactions to a non-zero reward depend upon its expectations. An experiment by Amsel & Surridge demonstrates that reactions to zero reward similarly depend upon the size of the expected reward that is not delivered.

The design of this experiment was as follows. Four groups of rats were trained in a runway with either 50% random partial reinforcement (food reward on one trial in two, randomly selected) or continuous reinforcement (reward on every trial), and with either large reward (a 500-mg pellet) or small reward (a 94-mg pellet). On half the trials, for all groups, a light came on over the goalbox, too late to affect running speed. This was always on the non-rewarded trials for the two partial reinforcement groups. After the running response was fully established, the conditions were changed so that the light came on just before the rat entered the last section of the alley. Running speed was affected by the light only in the partial reinforcement groups, which slowed down after the light came on. This observation by itself might merely indicate that the rat, informed by the light that there was no food in the goalbox, did not bother to keep running. However, Amsel & Surridge's critical observation was that there was a marked difference between the effects of the light in the partial reinforcement groups given the high and low rewards, respectively: the slowing-down effect was much greater for animals getting the 500-mg pellet than for those getting the smaller, 94-mg reward. The most reasonable interpretation of this result, given the other evidence that non-reward is aversive, is that zero reward for rats used to a large reward is more punishing than zero reward for rats used to small reward.

It seems then that stimuli associated with frustrative non-reward (and presumably therefore non-reward itself) are aversive in much the same way as punishment and stimuli associated with punishment. Recently rather simpler experimental designs have been developed to demonstrate this principle. Thus, just as Karpicke & Dout were able to show conditioned withdrawal from a stimulus paired with shock, so Bottjer demonstrated conditioned withdrawal from a stimulus paired with the omission of food that was otherwise automatically delivered. There is also physiological evidence that non-reward acts like a stressor: just as blood levels of corticosterone are increased by pain (Chapter 5), so are they increased when a previously rewarded response goes unrewarded.

Summation of drives

A second important feature of punishment is that it is able to increase the vigour of ongoing or immediately subsequent behaviour. We have so far said little about this aspect of punishment, and it is time to remedy the omission.

The invigorating properties of punishment have usually been studied as a way of testing the 'drive summation' hypothesis advanced by the important American learning theorist Clark L. Hull. Learning theories of the Hullian variety (and most of those which have been concerned with fear and frustration have been strongly influenced by Hull) suppose that behaviour at any moment occurs because of the muliplication of a habit (*sHr* in Hullian terminology) by the sum of existing drives (*D*). All drives (hunger, thirst, sex drive, etc.) summate to produce *D*, and the application of a punishment or of a conditioned fear stimulus is conceived as setting up one such drive, namely, the fear drive. The habit which is energised by *D* on a particular occasion is the one which has the highest momentary rank in the 'habit family hierarchy', and this rank depends above all on prior leaning and the particular stimuli (the '*s*' in the symbol '*sHr*') to which the organism is exposed in its current environment.

In considering approach–avoidance conflict in the previous chapter we were, in Hullian terms, considering the conditions determining which of two habits becomes dominant when the organism is exposed to conflicting stimuli. Thus, when we apply a punishment or a conditioned fear stimulus to an organism engaged on some form of appetitive behaviour (e.g. food-seeking, sexual behaviour), two things can happen: (1) we may, by changing the stimuli to which the organism is exposed, change the dominant habit, thus observing a *decrease in the proportion of time spent by the subject on the original appetitive behaviour*; (2) we may, alternatively, increase *D* and so observe an *increase in the vigour of the original appetitive behaviour*. And, of course, we may observe both these effects, decreased probability of emission of the appetitive behaviour being accompanied by increased vigour when it occurs.

We see, then, that predictions about drive summation are, in the light of Hullian theory, rather complex. If we can make the animal fearful while changing the stimulus situation to a sufficiently small degree for it to continue carrying out the appetitive behaviour, we should observe an increased vigour of this behaviour; but, if we change the stimulus situation to a degree which elicits strongly competing avoidance behaviour, we should see the kind of inhibition of appetitive behaviour discussed in Chapter 9. The task of making precise predictions in this area would obviously be easier if experimenters had clearly distinguished between measures of the probability of occurrence of the appetitive behaviour (which should show a fall) and measures of the vigour of the appetitive behaviour (which should rise); but for the most part this has not been done. A mathematical model which P. T. Smith and I have developed for the analysis of this kind of situation suggests that, if only measures of the overall amount of appetitive behaviour (probability multiplied by vigour) are available, then we are likely to observe an overall increase in the appetitive behaviour when the added aversive drive is low, but an overall decrease when the added aversive drive is high (Fig. 10.1).

The experimental findings on the effects of an added aversive drive on appetitive behaviour by and large fit this analysis well. They may be summarised by saying that with a relatively *low* intensity punishment or conditioned fear stimuli, and in an environment which is strongly associated with the appetitive drive, the addition of fear increases the overall appetitive behaviour observed, both instrumental and consummatory. If the shock used is of *high* intensity, or if the environment is not so strongly associated with appetitive behaviour, there is usually an overall reduction in the amount of appetitive behaviour. Finally, it is possible, under some conditions, to observe the simultaneous decrease in the probability of the appetitive behaviour and increase in its vigour which our analysis suggests must always occur.

A number of the experiments which have provided the evidence on which these conclusions are based have measured eating or drinking in the home cage (which is, of course, an environment which is strongly associated with these consummatory acts) after shock has been delivered in some other environment. The usual result, provided a shock of relatively low intensity is used, is for eating and drinking to be increased by such prior shock. If high shock is used, or if some novel element is introduced into the consummatory situation, a decrease in eating or drinking may be observed instead. Strongman, for example, exposed animals to three different intensities of shock outside

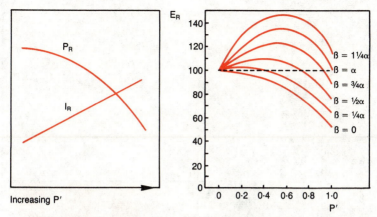

Increasing P′

Fig. 10.1. Outline of Gray & Smith's mathematical model for conflict and drive summation. The abscissa, P′, represents increasing degrees of the aversive drive, with appetitive drive held constant. That is, it represents increasing levels of input to the 'stop' or 'passive avoidance' mechanism in Fig. 9.2. When P′ = 1, the inputs to the approach and avoidance mechanisms are equal, and the animal is at the equilibrium point in Miller's equilibrium model (Fig. 9.3.). The figure on the left shows how, in the Gray & Smith model, probability (P_R) of the approach behaviour is expected to decrease, and intensity or vigour (I_R) of the approach behaviour, when it occurs, is expected to increase, respectively, as the level of the aversive drive approaches that of the appetitive drive. The figure on the right shows the resulting functions for $E_R = P_R \times I_R$, i.e. the overall amount of appetitive behaviour when separate measures for probability and vigour are not available. The parameter, β, of the curves in the figure on the right reflects the degree to which increases in I_R over- or under-compensate (for high and low values of β, respectively) for decreases in P_R as P′ increases. This parameter is expressed as a proportion of the weighting, α, which is attached to inputs to the 'approach' mechanism in Fig. 9.2. For further details, see Gray & Smith (1969).

the home cage and then, in the home cage, offered them either normal food pellets or pellets flavoured (for the first time in the animal's experience) with bitter-tasting quinine. His results (Fig. 10.2) show that only the lowest of the three shock intensities increased eating of the normal pellets, and all three reduced eating of quinine-flavoured pellets.

The most impressive evidence in favour of our analysis of the drive-summation experiments, however, comes from studies in which probability of appetitive behaviour and vigour of appetitive behaviour have been measured separately and simultaneously. There are two studies which meet these requirements.

Tugendhat looked at the way in which the feeding behaviour of the three-spined stickleback is affected by electric shock. The experimental situation was divided into a 'home area' and a 'food area', separated

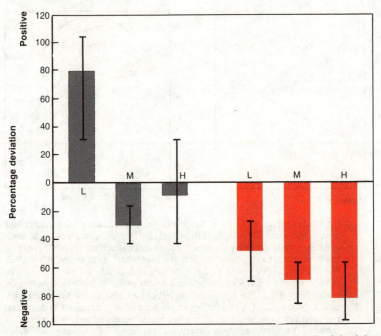

Fig. 10.2. Percentage change in food intake after three levels of shock-intensity – low (L), medium (M), or high (H) – when the rat is offered normal food pellets (grey columns) or pellets flavoured with quinine (red columns). The lines in the columns represent the ranges of scores for each group.

by a partition through which the fish were sometimes allowed to swim. Shock was only ever delivered in the food area. Tugendhat had previously taken careful measurements of the fish's feeding behaviour under normal conditions at different levels of food deprivation. She found that, as a result of shock, the fish spent less time feeding but, while they were feeding, their behaviour shifted in a manner appropriate to a more severe level of food deprivation (i.e. decreased probability of feeding accompanied by greater vigour).

Comparable results were obtained by Beach & Fowler, using very different techniques to examine the effects of fear on sexual behaviour in the male rat. The rats were first shocked in a cage for 5 successive days. They were then allowed to copulate with receptive females either in the same cage or in a neutral one. The rats which copulated in the environment previously associated with shock needed fewer intromissions to reach the point of ejaculation, that is, they behaved in a way appropriate to a higher level of sexual arousal. In an earlier study Beach had demonstrated this same phenomenon when rats were shocked actually during copulation. In this study the increased sexual arousal was accompanied (as we would expect) by a decrease in the total amount of time spent copulating. In line with the results of these experiments with rats, Barlow *et al.* have demonstrated a stronger penile erection response to an erotic film when young men watched it under the threat of electric shock.

Apart from the general theoretical significance of these experiments on the summation of drives, they are also of obvious importance for the understanding of such disorders as compulsive over-eating and *ejaculatio praecox* in Man: the clinical impression that these can be due to anxiety is given good experimental support.

In the light of the drive-summation experiments, then, it is reasonably certain that the fear aroused either by punishment or by stimuli associated with punishment may increase the vigour of appetitive behaviour occurring at the same time. Thus, on the fear = frustration hypothesis, we must suppose that the omission of an anticipated reward or stimuli associated with such frustrative non-reward can have the same effect.

The classic experiment in this field was reported by Amsel & Roussel in 1952. They used the situation shown in Fig. 10.3. The rat is trained to run down two alleys with two goalboxes placed successively. In the first goalbox it is on a partial reinforcement schedule, being rewarded in a typical experiment on a randomly chosen 50% of the trials; in the second goalbox it is rewarded on every trial ('continuous reinforcement'). If frustrative non-reward is invigorating we should expect that the rat will run faster in the second alley on trials on which it has *not* been rewarded in the first goalbox than on trials on which it *has* been rewarded there. This is exactly what Amsel & Roussel found. Furthermore, subsequent experiments have shown that this is genuinely due to an increase in speed after non-reward rather than a decrease in speed after reward; for animals which are either *always* rewarded in

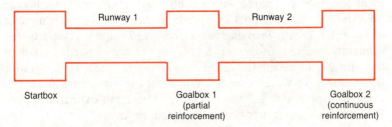

Fig. 10.3. Amsel & Roussel's double-runway situation. For a full explanation, see text.

the first goalbox or *never* rewarded there run at the same speed in the second alley as rats on a partial reinforcement schedule run after their *rewarded* trials. Thus the increased speed of running after frustrative non-reward appears to be a true 'frustration effect' and this is the term which is often used to describe the Amsel & Roussel phenomenon. Furthermore, using different but analogous procedures, other investigators have been able to demonstrate very similar frustration effects in pigeons, monkeys, and children.

Just as invigorating effects have been described for both punishment and stimuli associated with punishment, so they have been described, not only for frustrative non-reward itself, but also for stimuli associated with frustrative non-reward. Brown, Kalish & Farber showed in 1951 that a stimulus which had regularly been followed by shock could increase the magnitude of the startle reflex if it was presented to rats just before the presentation of the stimulus (a gunshot or loud tone) for the startle itself (Fig. 10.4). Using this technique Wagner, in the experiment described earlier in this chapter, showed that the stimulus associated with the non-rewarded trials of a partial reinforcement schedule was able to potentiate the startle reflex in exactly the same way.

Like punishment, then, frustrative non-reward – and stimuli associated with frustrated non-reward – are both aversive and invigorating. With these preliminary points established we can begin to seek more convincing evidence in favour of the fear = frustration hypothesis.

Testing the fear = frustration hypothesis

Some of the most interesting experiments providing such evidence have been concerned with the phenomena associated with partial reinforcement schedules, One of the best established findings in the whole study of animal and human learning is that a partial reinforcement schedule, compared with a continuous reinforcement schedule involving the same number of trials (and therefore a greater number of rewards), *greatly increases resistance to extinction* if reward is subsequently omitted altogether. Since, on the non-rewarded trials of the partial reinforcement schedule, the animal is exposed to frustrative non-reward, we would expect a satisfactory theory of frustration to offer an explanation for the partial reinforcement extinction effect. Such an explanation has been elaborated by Amsel.

Roughly speaking, Amsel's theory supposes that, as a result of its non-rewarded trials, an animal on a partial reinforcement schedule

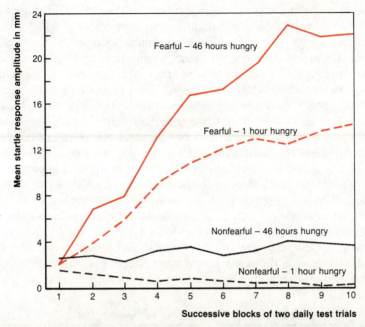

Fig. 10.4. Results of an experiment by Meryman, showing the amplitude of the startle response as a function of fear, no fear, intense hunger, weak hunger, and their combination. Startle responses to a pistol-shot were tested in four groups, two of the groups being rendered fearful by being given an electric shock in the test situation once a day. Both fear and hunger potentiate the startle response, and their effects summate.

comes to experience conditioned frustration during its performance of the instrumental response. This state of frustration in turn sets up distinctive internal stimuli (presumably those which allow us to say of ourselves that we 'feel frustrated' or 'disappointed') which, since they are followed by a reward on the rewarded trials, come to act as cues to continue performing the instrumental response. In effect, the stimulus feedback from the internal state of frustration becomes a signal to the animal to 'keep on trying'.

Now, if the effects of frustrative non-reward are very similar to the effects of punishment, it should be possible to substitute a punishment for frustrative non-reward (and *vice versa*) and get essentially the same results. We saw in the last chapter that just as a partial reinforcement schedule is a way of training an animal to continue performing a response in spite of frustration, so, by introducing a punishment initially at a very low intensity and then gradually increasing it and always following it by reward ('counter-conditioning'), it is possible to keep an animal responding in spite of punishment. Making use of this fact, Brown & Wagner conducted an experiment in which they trained three groups of rats: one on an ordinary continuous reinforcement (CRF) schedule, one on a partial reinforcement (PR) schedule with rewards on a random 50% of trials, and the third having half the trials rewarded and the other half receiving reward and punishment on the special counter-conditioning schedule just described. Each group was then divided into two, one half being put on a normal extinction schedule and the other being given CRF together with the highest intensity punishment reached by the counter-conditioning group. As would be expected from the fear = frustration hypothesis, the PR group showed greater resistance to *punishment*, and the counter-conditioning group greater resistance to *extinction*, than did the CRF controls (Fig. 10.5). Thus tolerance for non-reward carries with it tolerance for punishment and *vice versa*.

A somewhat surprising feature of the results shown in Fig. 10.5 is that training on a PR schedule was more effective in producing resistance to punishment than was training with punishment in producing

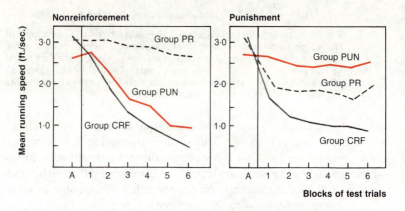

Fig. 10.5. Brown & Wagner's experiment on cross-tolerance between fear and frustration. The figure shows mean running speeds in a runway on the last day of acquisition (the point marked 'A' on the abscissa) and on subsequent daily blocks of six test trials. Group CRF was trained on a continuous reward schedule; group PR was trained on a partial reward schedule; group PUN was trained with continuous reward and gradually increasing intensity of punishment in the goalbox (a 'counter-conditioning' schedule). The left-hand graph shows the performance of the groups during ordinary extinction; the right-hand graph shows the performance of the groups exposed to continuous reward and high-intensity punishment. Group PR develops not only resistance to extinction but also resistance to punishment; and, similarly, group PUN develops not only resistance to punishment but also resistance to extinction.

resistance to extinction. Given that on any reasonable analysis foot-shock is more aversive than non-reward, one would have expected the reverse result. This feature of Brown & Wagner's findings still awaits explanation. However, it has recently been replicated by Halevy, Feldon & Weiner; indeed, these workers found cross-tolerance between non-reward and punishment only when animals were trained on the PR schedule and tested with punishment, not the other way round.

Brown & Wagner's result is presumably an instance of associative counter-conditioning (see Chapter 9): that is to say, the animal forms an association between the stimuli characteristic of an internal state (fear or frustration) and food reward for running, and then generalises from these stimuli to the (very similar) stimuli characteristic of the other state (frustration or fear). But it is also possible to demonstrate cross-tolerance between punishment and non-reward using a non-associative procedure. We have already encountered just such a demonstration, in the form of Chen & Amsel's experiment, described in the previous chapter, in which rats were given repeated foot-shock and were consequently more resistant to extinction of a food-rewarded running response tested many weeks later. Since the shock was deli-vered before the animals received any training in alley-running and in a different apparatus, it is difficult to see how stimuli arising from shock or from the emotional reaction to shock could have formed an association with any specific cues or responses related to alley-running. Presumably, therefore, there was some non-associative form of adap-tation to repeated shock which was able to affect later responding to a broad range of aversive stimuli, including non-reward. Again, there-fore, we must conclude that there exists some fundamental similarity between the mechanisms that mediate responses to shock and to non-reward, respectively.

Another experiment of Wagner's exploits Miller's finding that in an approach-avoidance conflict, the strength of the avoidance tendency falls off more rapidly with stimulus change than does that of the approach tendency. Wagner argued that on the fear = frustration hypothesis, the strength of extinction should similarly be reduced more than that of approach when the environment is changed. To test this prediction, rats were trained to run down a runway on CRF. Half of the subjects were trained and then extinguished in a 3-inch-wide black runway and half in a 6-inch white runway. Controls were run down the same runways but never given food in them. On each of the last 2 days of extinction each rat was given a single unrewarded trial in its usual alley and an equivalent trial in the other, novel alley.

If, like fear, frustration fails to generalise as widely as approach tendencies, there should be a revival of the running response in the novel alley. This is exactly what Wagner found, since the experimental animals ran faster in the novel alley, while the controls ran slower.

There are other experiments which exploit the theories of learning in which the terms 'fear' and 'frustration' figure to derive and successfully test predictions from the fear = frustration hypothesis. For example, Banks has reported a series of experiments showing that punishing animals on a proportion of rewarded trials (a 'partial punishment' schedule) affects subsequent resistance to continuous punishment in much the same way that a partial reinforcement schedule affects resistance to extinction.

A particularly interesting group of experiments has employed a 'transfer-of-control' design. Such an experiment is divided into a phase of Pavlovian classical conditioning and a phase of instrumental conditioning. The two phases are typically conducted in two different apparatuses and at two different times. They have in common a stimulus or stimuli to which some kind of motivational/emotional significance is given by pairing with an appropriate UCS in the Pavlovian phase and which are subsequently presented to the animal while it performs the instrumental response. The instrumental phase is used to assess the kind and degree of motivational and emotional significance that has accrued to the stimulus by the changes that it produces in instrumental performance. For example, Rescorla & LoLordo first trained dogs in a Sidman avoidance schedule (Fig. 9.4) in a shuttlebox and then exposed them to a number of Pavlovian conditioning procedures in a different apparatus but still using shock as the UCS (as in the Sidman avoidance schedule). After this the dogs were returned to the shuttlebox, where they resumed shuttling at a steady rate. The various stimuli which had been used in the Pavlovian conditioning sessions were now presented to the dogs while they were shuttling. This elegant experimental design provided a very sensitive measure of the significance which the Pavlovian stimuli had acquired for the animal. In particular, as would be expected, a CS which had been followed by shock in the Pavlovian phase acquired the capacity to increase fear and so to potentiate avoidance behaviour.

Grossen, Kostansek & Bolles took this experimental design a stage further, but using rats rather than dogs. The instrumental phase was essentially the same as in the Rescorla & LoLordo experiment. However, in the Pavlovian phase a tone was paired with the non-delivery of food that was otherwise automatically presented to the animal. Just like the CS paired with shock in the Rescorla & LoLordo experiment, this conditioned frustrative stimulus increased the rate of Sidman avoidance – that is, the rat reacted as if the tone increased its fear of the shock (see Fig. 12.3).

Yet other experiments have made use of a phenomenon originally described by Kamin and known as the 'blocking effect'. To demonstrate this phenomenon one pairs a first stimulus (CS_1) (e.g. a tone) with a UCS (e.g. shock), using standard Pavlovian techniques. One next presents this stimulus in compound with a second stimulus (CS_2) (e.g. a light), continuing to follow the compound with the shock UCS. CS_2 is a stimulus which, if presented on its own or in compound with a stimulus not previously paired with the UCS, would acquire considerable power to elicit CRs appropriate to a shock UCS (conditioned suppression, freezing, defecation, etc.). However, presentation of CS_2 in compound with a stimulus that has already been paired with the UCS blocks the normal process of conditioning: it is as though the animal knows already from the occurrence of CS_1 that shock will occur and so does not treat CS_2 as a useful additional predictor of this UCS. Indeed, Kamin's blocking effect has been widely interpreted as indicating that a UCS can support conditioning to a newly presented CS only to the extent that the UCS is surprising, i.e. not already predicted by other concurrently present stimuli; and there is much detailed experimental support for this interpretation.

Now, the degree to which CS_1 can block conditioning to CS_2 depends on the degree of similarity between the UCS used to condition CS_1 and the UCS paired with the CS_2–CS_1 compound. If the UCS is identical in the two phases of the experiment, blocking can be total; as the two UCSs become dissimilar, the amount of blocking is reduced; if the two UCSs bear no relationship to each other, blocking is absent; and if the two UCSs are opposite in sign (e.g. no shock paired with CS_1, shock paired with the compound), a phenomenon known as 'super-conditioning' may be observed – CS_2 develops greater conditioned strength than it does if presented on its own. It will be clear from this pattern of results that one may use Kamin's blocking design to assess the degree of similarity between two stimuli: the greater the power of one to block (as UCS) the other (also as UCS), the more similar they are.

Just this argument was used by Dickinson & Dearing in a study of the relationships between non-reward and shock. In this experiment rats initially received trials in which a clicker signalled the'delivery of food interspersed with other trials in which a light was added to the clicker and no food was delivered. This procedure established the light as a predictor of 'no food' (or, more technically, as a conditioned inhibitor with respect to the food UCS; see Fig. 11.3). It also gave the light the power to block the development of conditioned suppression to a tone when the light and tone were presented as a compound stimulus followed by shock. This result implies that, in agreement with the fear = frustration hypothesis, the rat treated a signal of no food as motivationally similar to a signal of shock.

Drugs which reduce fear and frustration: alcohol and amytal

In the previous section we considered a number of experiments that have made use of learning theory to derive and test predictions based upon the fear = frustration hypothesis. In general, these experiments have amassed an impressive body of evidence in support of this hypothesis. But some of the most interesting studies in this area have joined to learning theory the use of certain drugs in an effort to extend the investigation of the fear = frustration hypothesis beyond similarities of *function* to similarities of *physiology*. To understand these studies we must first discuss the important series of experiments conducted in Neal Miller's laboratory on the effects of two very common drugs, sodium amytal (one of the barbiturates) and alcohol, on behaviour in approach–avoidance conflicts. ('Sodium amytal' is a trade name; the generic name for this drug is 'sodium amylobarbitone' in England and 'sodium amobarbital' in the United States.)

In their influential book *Personality and Psychotherapy*, Dollard & Miller suggested in 1950 that these drugs produce their therapeutic effects by reducing the avoidance tendency in an approach–avoidance conflict more than they reduce the approach tendency. If this were so, it would account for the early findings made by one of Miller's students, Conger, that in a simple runway conflict, when shock intensity is set at a level which just inhibits the rat from going to the food in the goalbox, an injection of alcohol causes the animal to resume this behaviour. However, there are many other ways of accounting for this finding. The drug could make the animals hungrier; it could produce sensory changes (known to all of us who have had one whiskey too many) which, given the known greater steepness of the generalisation function for avoidance than approach, would result in a greater reduction in avoidance behaviour; it could cause forgetting of the habit which the animal learned most recently, which, in Conger's experiment as in most experimental investigations of conflict, was the avoidance habit; or it could affect discrimination, so that the animal could no longer recognise the danger signals provided by its environment. It is only as a result of a careful series of experiments that these alternative possibilities (and others) have been excluded to the point where the hypothesis that alcohol and amytal act by directly reducing fear, i.e. the avoidance component of an approach–avoidance conflict, is now rather well established.

We do not have space to do more than indicate the evidence on which this conclusion is based. Conger has shown with Brown's strength-of-pull technique (Fig. 9.3) that there is indeed a greater decrease in the strength of the avoidance than of the approach tendency after injections of alcohol. That there are sensory changes in the rat as a result of such injections has been shown by training animals to discriminate between being drunk and being sober. Conger gave rats food on some trials and food plus shock on others. The animals were run drunk on some trials and sober on others. For one set of rats, the shocks always came on drunk trials, for a second set only when they were sober. Both sets learned to run more slowly on the shocked trials, showing that they could tell when they were drunk and when they were sober; but the animals shocked when drunk had much more difficulty in learning to keep away from the food, showing a direct inhibition of fear by the drug. The same fear-reducing effects have been found in similar experiments with amytal; and both drugs have now been used in a wide enough variety of experimental situations for it to be clear that the results obtained do not depend on such factors as the nature of the approach drive used, the apparatus, the kind of warning signals used, the nature of the response, or whether it is the approach response or the avoidance response which is learned first. Dutch courage appears to be as universal a phenomenon in the rat as it is in Man.

The finding that alcohol is capable of reducing the avoidance component of an approach–avoidance conflict is able to account for the effects of this drug on social behaviour in human beings. As Miller remarks, 'alcohol has a perplexing variety of effects, making some aggressive, others amorous, some tearful, and others talkative.' If we suppose that the type of behaviour which is displayed after drinking alcohol was previously restrained by fear (which is in many cases obvious enough), all these effects can be attributed to a single mechanism of action – that of reducing fear.

The strong motivation which many human beings build up for alcohol (even before there is any physiological addiction to the drug) can be explained in a similar fashion. In general, animals will learn to do things which lead to a reduction in fear (Chapter 11). It follows that they should learn to take drugs which have this effect. In support of this deduction, it has been shown that cats offered a choice between alcohol and water in a situation in which they have been exposed to shock develop a preference for alcohol, even though their normal preference is for water. Another of Miller's experiments has shown that rats which are occasionally shocked will learn to press a bar which causes an automatic injection into them of a small quantity of sodium amytal, though unshocked rats do not develop this behaviour.

Given this wealth of evidence for the fear-reducing effects of alcohol and amytal, we are now in a position to use these drugs to test the fear = frustration hypothesis. If there is a true physiological overlap between the systems activated by punishment and frustrative non-reward, these drugs should also reduce frustration. The bulk of the evidence indicates that this is indeed so.

The most obvious derivation of this argument is that these drugs should increase resistance to extinction (by reducing the aversive effects of non-reward) if they are administered to the animal during extinction. This hypothesis has been confirmed in a number of experiments. An experiment of my own made use of the Adelman & Maatsch extinction technique described earlier in this chapter. One group of rats was extinguished in the usual way, being allowed to stop in the stem of the runway which led to the box in which they had previously been rewarded with food. Injections of amytal to this group retarded extinction, that is, animals under the drug continued to run to the old goalbox *faster* than control animals (Fig. 10.6). Another group of rats was extinguished by being trained to jump out of the now empty goalbox. Injections of the drug in this group *slowed down* the speed of jumping out compared to undrugged control animals. Thus rats given the drug were both less reluctant to get to the goalbox and less eager to leave it. A further group of control animals were trained to jump out of the goalbox on to the ledge for a food reward, and this group was actually speeded up by injections of the drug (Fig. 10.6). Thus the overall pattern of results obtained in this experiment was consistent with the notion that amytal reduces the aversive consequences of frustrative non-reward and would be difficult to explain in any other way.

Another simple prediction is that, in experiments on discrimination learning, the fear-reducing drugs should disinhibit responding in the presence of the negative stimulus (i.e. the one signalling non-reward for responding). This prediction has been confirmed by Wagner for alcohol and by Ison & Rosen for amytal.

A particularly interesting prediction concerns the partial reinforcement extinction effect. We have seen that according to Amsel's theory this occurs because the animals on the partial reinforcement schedule

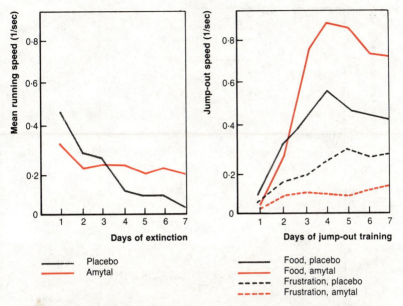

Fig. 10.6. Effects of 20 mg/kg sodium amytal on extinction in a runway (left-hand graph) and in Adelman & Maatsch's situation, in which the animal jumps out of the empty goalbox onto a ledge (right-hand graph). The drug reduces resistance to extinction in the runway and also reduces the speed with which animals jump to get out of the frustrating goalbox. Control rats rewarded with food for jumping out of the goalbox (right-hand graph) jump out faster if they have been drugged, showing that the drug does not impair the motor abilities involved in jumping.

learn to use the internal stimuli of frustration as cues for continued approach behaviour. Suppose, then, we reduce frustration by administering one of the fear-reducing drugs during training, but then remove the protective cover of the drug when the animals are put onto extinction. Clearly, this ought to attenuate or even abolish the difference between partially and continuously rewarded groups. Fig. 10.7 shows the results I obtained when I did this experiment: the partial reinforcement extinction effect, evident in groups trained and extinguished on placebo (i.e. with injections of physiological saline), is virtually abolished in groups trained on amytal and extinguished on placebo. Similar results have been independently obtained by Ison & Pennes.

However, the results depicted in Fig. 10.7 are open to another interpretation. The group of rats in which amytal abolished the partial reinforcement extinction effect received the drug in training but not in extinction. Thus it is possible that they had learned tolerance for frustration (i.e. to persist in the performance of an instrumental

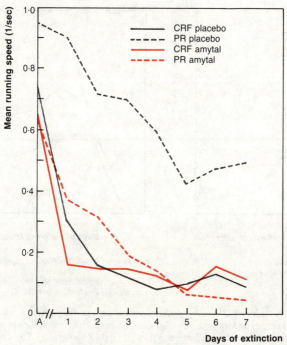

Fig. 10.7. Abolition of the partial reinforcement extinction effect by injections of 20 mg/kg sodium amytal during acquisition. The four groups represented by the curves in the figure received either continuous reinforcement (CRF) or partial reinforcement (PR) combined with injections of either amytal or a saline placebo during training. The point marked 'A' on the abscissa is the last day of acquisition. During extinction all groups were given placebo injections and no further reward. The PR placebo group is much more resistant to extinction than the CRF placebo group but the two amytal groups both extinguish rapidly.

response even when feeling frustrated) but that the expression of this learning had become limited to the state in which they were drugged. This general type of phenomenon (known as 'state dependence of learning') has been demonstrated many times. Just as an animal may learn to perform a particular response in one environment or apparatus but then fail to perform it in a very different one ('generalisation decrement'; see Fig. 2.5), so it may learn to respond in one drug state but not in another or in the absence of the drug that was present during learning. The control for state dependence is simple: you merely have to test the animal also when drugged. If you are dealing with state dependence the animal will express its learning when drugged but not when treated with an appropriate placebo.

When both Ison & Pennes and I did this experiment we found the partial reinforcement extinction effect to be intact in animals both trained and extinguished under amytal.

Now the particular conditions used in these experiments involved a relatively short inter-trial interval (about 5 minutes), i.e. the interval separating successive trials in the alley. Different results are obtained if the inter-trial interval is extended to 24 hours, i.e. the animal runs only one trial per day. To understand the significance of this change in the inter-trial interval requires us to consider a second theory of the partial reinforcement extinction effect, due to E. J. Capaldi. This theory was once seen as a rival to Amsel's; however, it is now clear that the two theories complement one another.

It will be recalled that Amsel's account of the partial reinforcement extinction effect supposes that the partially reinforced animal learns to make use of cues from the internal state of conditioned frustration (itself established by way of classical conditioning of the unconditioned frustration response to alley stimuli that regularly precede non-reward). These cues act as signals that performance of the running response will (eventually) be rewarded. Capaldi's account is similar in that it supposes that the partially reinforced animal comes to use certain cues as signals that running will be rewarded, but the cues are different: they consist in the after-effects or memory of the immediately preceding non-rewarded trial or trials. Now this theory is in no way incompatible with Amsel's. On the contrary, one can easily take Capaldi's 'after-effects of non-reward' to be Amsel's unconditioned frustration response; this 'after-effect' then becomes a 'memory of non-reward' as the directly perceived emotional turmoil caused by non-reward fades away. In this way, then, we have two potential mechanisms by which an animal might learn to persist in spite of irregular non-reward. Amsel's rat experiences a conditioned frustration reaction to apparatus cues and learns to continue running while feeling this way; Capaldi's rat continues to feel or remember the consequences of a recently experienced non-reward and learns to continue running while in *this* state.

In spite of their overall similarity these two theories of the partial reinforcement extinction effect make a variety of distinctive predictions, e.g. with regard to the effects of the sequences of rewarded and non-rewarded trials encountered during acquisition of the running response. These predictions have been exhaustively tested over the last 20 years or more, and neither theory has been able to triumph over the other. Rather, it seems that the partial reinforcement extinction effect can be produced by both Amsel's and Capaldi's mechanisms, though different conditions favour a predominant role for one or the other. Among these conditions is the length of the inter-trial interval. The role played by this variable is easy to understand. For Capaldi's mechanism to work, the animal must either feel or remember the consequences of a particular non-rewarded trial or trials from its recent past. Such feelings and memories fade with time, and they are subject to interference from other events that occur during the inter-trial interval. Amsel's mechanism is not affected in this way by time: Pavlovian conditioned reflexes (in the present instance, the CR of conditioned frustration) are largely independent of the intervals that separate successive exposures to the CS. Thus, once an Amselian rat has developed the reaction of conditioned frustration to apparatus cues, it will experience that reaction whenever it is placed in the apparatus, irrespective of the time since it was last put there. It follows from these arguments that, at short inter-trial intervals, the partial reinforcement extinction effect should be formed in two parallel ways, one Capaldi's, the other Amsel's; but, as the inter-trial interval is lengthened, the role played by Capaldi's mechanism should become progressively weaker. Experimental observations are in good agreement with this deduction.

It can now be seen that the experiment whose results are shown in Fig. 10.7 was ill-designed to test Amsel's theory of the partial reinforcement extinction effect. At an inter-trial interval of 5 minutes a very strong, perhaps a predominant, role would be played by Capaldi's mechanism; thus, even if amytal succeeded completely in blocking Amselian conditioned frustration, we would not necessarily predict abolition of the partial reinforcement extinction effect, since this could still be mediated by Capaldi's mechanism acting in parallel to Amsel's. For an adequate test of the susceptibility of conditioned frustration to amytal it would be necessary to run the experiment at a much longer inter-trial interval. For example, other evidence suggests that the role played by Capaldi's mechanism with a 24-hour inter-trial interval is at best minimal. Joram Feldon and I therefore ran an experiment at this inter-trial interval, with the results shown in Fig. 10.8. Under these conditions the partial reinforcement extinction effect was abolished by amytal given in training, whether the animal was tested in extinction under the drug or under placebo.

Taking this line of experimentation further, Feldon and I have shown identical results using a different fear-reducing drug, chlordiazepoxide (trade name, 'Librium'). This is a member of the class of benzodiazepines, now widely used in the treatment of human anxiety; we shall have more to say about these drugs later (see Chapter 12).

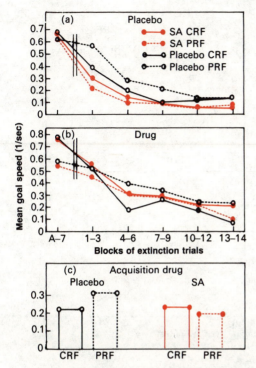

Fig. 10.8. Mean goal speeds during extinction (no trials rewarded) as a function of continuous reinforcement (CRF, every trial rewarded) or partial reinforcement (PRF, a random 50% of trials rewarded) during acquisition, and injection of sodium amylobarbitone (SA), 20 mg/kg, or placebo during acquisition. (a) Extinction under placebo; (b) extinction under drug; (c) mean goal speeds over all extinction trials averaged across extinction. A-7: Final acquisition trial. The experiment was run at 1 trial/day. From Feldon *et al.* (1979).

Like amytal, chlordiazepoxide (when given during both training and extinction) abolishes the partial reinforcement extinction effect at an inter-trial interval of 24 hours but leaves it unchanged at an interval of 5 minutes. Davis and I went on to use this drug in an examination of the partial punishment effect, i.e. the increased resistance to continuous punishment (punishment on every trial) that is produced by partial punishment during training (punishment on a randomly chosen proportion of trials). The identical pattern of results emerged: the partial punishment effect was abolished at an inter-trial interval of 24 hours but unchanged at one of 5 minutes (Fig. 10.9).

These results are a striking confirmation of the similarities between fear and frustration. Furthermore, if we argue by analogy to Amsel's

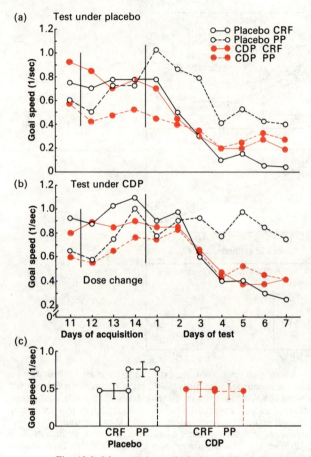

Fig. 10.9. Mean goal speeds during the final days of acquisition and during testing (every trial punished as well as rewarded) for groups of rats trained on a continuous reinforcement (CRF) schedule (all trials rewarded, none punished) or a partial punishment (PP) schedule (all trials rewarded, a random 50% also punished) and given either chlordiazepoxide (CDP, 5 mg/kg) or placebo (saline) injections. (a) The results when the test was carried out with saline injections; (b) the results when the test was carried out with injections of CDP; (c) the same data averaged across all trials of the testing period. Between Days 12 and 14 of acquisition the dose of CDP was gradually decreased or increased in those groups switched from drug to placebo or vice versa between acquisition and test. The experiment was run at 1 trial/day. From Davis *et al.* (1981).

and Capaldi's theories of the partial reinforcement extinction effect, they suggest that the partial punishment effect is also mediated by two parallel mechanisms. The effect of one mechanism (Amselian) is that the partially punished animal develops conditioned fear to alley stimuli that regularly precede shock and learns that running in the presence of the internal cues characteristic of fear is rewarded. The effect of the other (analogous to Capaldi's mechanism) is that the partially punished animal learns that reward follows upon running in the presence of the after-effect or memory of immediately preceding goalbox shock. And, it seems, chlordiazepoxide blocks both conditioned fear and conditioned frustration, but not the immediate consequences of either shock or non-reward: this is an important point which I shall develop below.

The experimental results illustrated in Figs. 10.7, 10.8, and 10.9 have important clinical implications. As we shall see in Chapter 14, there now exist powerful behavioural methods for the treatment of phobias and other neurotic conditions involving anxiety. The key ingredient of these methods of 'behaviour therapy' appears to consist in exposing the patient to the stimuli that make him anxious; as a result of such exposure the patient loses his fears. Now, this process bears many similarities to the manner in which a rat trained on a partial punishment or partial reinforcement schedule comes to tolerate anticipated pain or non-reward. But patients are also commonly treated for anxiety by administration of benzodiazepines. The data we have been considering suggest that this practice may have its dangers: benzodiazepines might reduce the impact of behaviour therapy, or even interfere with the tolerance for fear and frustration that might be expected to develop from exposure to relevant stimuli during the normal course of life. This is a theme to which we return in Chapter 14.

Drugs and arousal

We saw in the previous section that fear-reducing drugs appear to affect reactions to stimuli that signal the occurrence of shock or non-reward but not to the latter, unconditioned aversive events themselves. A similar principle seems to emerge when one examines the susceptibility to the action of these drugs on the invigorating or 'arousing' effects of frustrative non-reward.

Consider first Amsel & Roussel's double-runway frustration effect (Fig. 10.3): in this phenomenon rats run faster in the second of two sequential alleys if they have experienced non-reward (rather than reward) in the first goalbox. Now this effect appears to be a reaction to the unconditioned event of frustrative non-reward – in Capaldi's terms, an immediate after-effect of non-reward. Based upon the discussion in the previous section, therefore, we would not expect the double-runway frustration effect to be altered by treatment with one of the fear-reducing drugs. There are now several experiments, carried out with both barbiturates and benzodiazepines, that confirm this deduction.

The failure of fear-reducing drugs to alter the double-runway frustration effect stands in marked contrast to the results obtained in experiments utilising a different paradigm to measure the arousing effects of non-reward. This is the partial reinforcement *acquisition* effect.

The partial reinforcement acquisition effect consists in the fact that, when training in the runway has proceeded to the point at which the animals are running pretty well as fast as they ever will, PR animals run faster than CRF animals. This is accounted for by Amsel in terms of the invigorating effects of the conditioned frustration which the partially reinforced animals are believed to experience during their traversal of the runway. Abolition of this speeding-up effect of partial reinforcement has been reported for alcohol by Nelson & Wollen and for amytal by Wagner and by myself (Fig. 10.10).

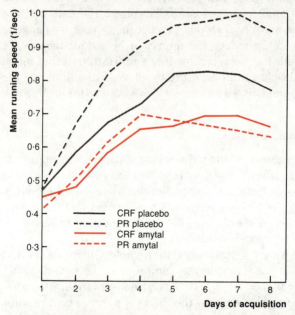

Fig. 10.10. Abolition of the partial reinforcement acquisition effect by injections of 20 mg/kg sodium amytal during acquisition. These are the same groups of animals as those in Fig. 10.7. The PR placebo group runs faster than the CRF placebo group but both drugged groups run at about the same speed as the CRF placebo group.

Now, unlike the double-runway frustration effect, the partial re-inforcement acquisition effect depends upon the arousing effects of *conditioned* frustrative stimuli, i.e. those that make up the startbox and stem of the alley in which running speed is measured. The topography of runway experiments allows one to make this distinction in an intui-tively simple spatial way. The double-runway frustration effect (Fig. 10.3) is measured by an increase in running speed that occurs *after* the point of non-reward (in the first goalbox). In contrast, the partial rein-forcement acquisition effect occurs *before* the point of non-reward, since it is measured while the animal is running towards the goalbox in which non-reward sometimes occurs. This apparently minor difference in experimental design makes all the difference: the double-runway frustration effect is immune to the action of fear-reducing drugs; the partial reinforcement acquisition effect is abolished by them. This pat-tern of results confirms the important principle that emerged in the previous section: reactions to conditioned but not to unconditioned aversive stimuli are susceptible to the fear-reducing drugs.

These arguments are based upon the premiss that the partial rein-forcement acquisition effect reflects the invigorating effects of expo-sure to stimuli associated with non-reward. If we add to this premiss the general fear = frustration hypothesis, we must predict that, in parallel to the data illustrated in Fig 10.10, fear-reducing drugs should reduce the invigorating effect of stimuli associated with punishment. Using Brown's potentiated startle reflex technique Chi has demonstrated just such an effect for amytal and M. Davis for two benzodiazepines; in neither case was the response to the uncon-ditioned stimulus used to elicit the startle reflex altered by these drugs. Thus the invigorating effects of conditioned fear stimuli and conditioned frustrative stimuli are similarly susceptible to the action of fear-reducing drugs, in agreement with the fear = frustration hypothesis.

Frustration and aggression

The similarity between fear and frustration comes out in a striking fashion when we consider the relation of both to aggressive behaviour. It will need very little experimental demonstration to convince the reader that pain is likely to result in aggression. If such a demonstration is needed, it can readily be supplied by the observation that if two rats are caged together, a shock to the feet will immediately and very reliably elicit fighting between them. The same is true of mice, hamsters, opossums, raccoons, marmosets, foxes, cats, snapping turtles, squirrel monkeys, ferrets, squirrels, roosters, alligators, crayfish and several species of snake – a striking catalogue of cross-species unanimity provided by Azrin. The hypothesis that frustration will also lead to aggression was advocated by Dollard & Miller in 1950, but only recently have we obtained good evidence for its validity.

Azrin describes an experiment which can leave little doubt of the relation between frustration and aggression. He trained pigeons to obtain food by pecking at a key. Another pigeon was in the experimental chamber, restrained and inactive. When the working pigeon was placed on extinction, so that its pecking was no longer rewarded, it turned and attacked the restrained pigeon viciously. In a similar experiment, Gallup trained pairs of rats to run simultaneously down two adjacent and parallel runways for food on a partial reinforcement schedule. After 30 seconds in the goalbox a door separating the two boxes was opened and the animals were allowed 1 minute together. The amount of aggressive behaviour shown by the paired animals was measured by means of a rating scale running from 1 to 7. After non-rewarded trials the mean aggression score was 6.83, compared to a score of 1.67 after rewarded trials, a highly significant difference.

Similar effects have been demonstrated in our own species. Nation & Cooney provided human subjects with two alternative responses with which they could escape from a loud noise: one (pressing a button) involved little effort and contained no aggressive elements; the other (striking a pad with a force of at least 25 lb) had the opposite characteristics. Concurrently, the subjects were rewarded with tokens for an independent manual response (moving a shuttle back and forth). While the latter response continued to be rewarded, the preferred mode of escaping from noise was to press the button; but when reward was discontinued (extinction), there was a shift to the pad-striking response.

The relationship between effort and aggression that was implicit in my description of Nation & Cooney's experiment has been demonstrated more formally in experiments with animals. The probable adaptive value of this way of reacting to non-reward will be obvious: at least some of the obstacles that prevent one from obtaining the necessities of life can be overcome by increased effort and/or aggressive behaviour. This intuition is supported by Thompson & Bloom's report that the increased rate of response observed in the rat during early extinction co-varied with an increased probability of attacking a second rat present in the experimental chamber (as in Azrin's experiment with pigeons).

The fear = frustration hypothesis reconsidered

There is still other evidence we could cite in support of the fear = frustration hypothesis. But enough has been said to show that despite its running counter to our intuitions, this hypothesis is worth taking seriously. It is time, therefore, to set out its content more formally than hitherto.

The data we have reviewed in this chapter have consistently supported both one equation and one inequality. The equation is between punishment and non-reward; the inequality is between unconditioned aversive stimuli (whether punishment or non-reward) and conditioned aversive stimuli. In consequence, we have not one but two fear = frustration hypotheses: (1) the unconditioned events of punishment and non-reward, respectively, set up closely similar and perhaps identical central states that give rise to closely similar responses; (2) conditioned stimuli associated with unconditioned punishment or non-reward, respectively, set up closely similar and perhaps identical central states that give rise to closely similar responses; furthermore, (3) the central states set up by unconditioned and conditioned aversive events, respectively, are different from each other and give rise to different responses.

We have already met clause (3) in this formulation, when we saw in Chapter 2 that painful stimuli and stimuli associated with pain elicit, not only different, but often diametrically opposed forms of response; and, indeed, that stimuli associated with pain (fear stimuli) may even inhibit the direct reaction to pain. Furthermore, when we come to deal with the brain in Chapter 13, we shall encounter evidence that unconditioned punishment and non-reward act upon the same system (which I have termed the 'fight/flight system'); that conditioned fear and conditioned frustrative stimuli act upon the same system (which I have termed the 'behavioural inhibition system'); and that the fight/flight and behavioural inhibition systems are quite distinct from each other. Now, if the same physiological system mediates responses to two different classes of stimuli (e.g. conditioned fear and conditioned frustrative stimuli), it is of comparatively minor importance that we are able in our daily lives consistently to apply different names (e.g. 'fear' or 'frustration') to the occasions of its operation, depending upon our knowledge of the conditions (threats of punishment or of failure) which have set it into operation.

We have also seen in this chapter evidence that three kinds of drug (alcohol, the barbiturates, and the benzodiazepines) all reduce reactions to stimuli associated with punishment or non-reward, but not reactions to the unconditioned events of punishment or non-reward themselves. These classes of drug are chemically quite distinct from one another yet they all affect behaviour, in both animals and Man, in essentially similar ways. As we shall see in Chapters 12 and 13, the common modes of action of these drugs (both behavioural and neuro-chemical) offer vital clues for our understanding of the central mechanisms that mediate reactions to conditioned fear and frustrative stimuli.

In addition, these drugs are able to offer us a somewhat less cumbersome vocabulary. In Man they are said to reduce anxiety; indeed, the benzodiazepines in particular constitute by far the most widely used treatment for conditions involving this emotion. Now the word 'anxiety' tends to be used as though it refers to a quite distinctive human emotion, only distantly related to the more basic fears that we perhaps share with animals. It will be clear by now that such a radical distinction between human and animal emotions runs quite counter to the spirit of this book. The evidence that human modes of emotional reaction are essentially the same as those of other animal species to which we are, phylogenetically speaking, closely related is overwhelming, and we shall continue to add to it in the chapters that follow. Thus I feel no trepidation in supposing that, if human beings experience a state of anxiety, so do animals such as the monkey or the rat. Indeed, if that were not the case, the fact that the anti-anxiety drugs (benzodiazepines, barbiturates, and alcohol) affect animal behaviour in the ways that they do would be a quite extraordinary coincidence. But, if animals experience a state of anxiety and this is reduced by the anti-anxiety drugs in them as in Man, then the arguments and data we have marshalled above provide us with a radically new perspective on this emotion: we may now define 'anxiety' as *that emotional state which is elicited by stimuli associated with either punishment or non-reward.*

This, indeed, is the way in which I shall henceforth use the word 'anxiety': to refer to the common state elicited by conditioned fear or conditioned frustrative stimuli. Correspondingly, we may now call these kinds of stimuli 'anxiogenic'; and the drugs that reduce their impact, 'anti-anxiety' or 'anxiolytic'. Note that since (in Chapters 12 and 13) we shall describe those parts of the brain which mediate responses to anxiogenic stimuli as a 'behavioural inhibition system', it follows that anxiety may also be defined as that central state which results from activity in this system; we shall have much more to say later about this way of analysing anxiety.

The implications of the fear = frustration hypothesis (understood now in the more elaborate manner set out above) for the understanding of human motivation and personality are quite far reaching.

We are at once able to extend the analysis of conflict behaviour, which we examined in the previous chapter, to the development, or lack of development, of those traits of persistence in the face of disappointment and frustration which are so important in every aspect of life.

We are also better placed to understand the widespread use of alcohol, benzodiazepines, and barbiturates by people faced with difficulties in their lives. Such difficulties rarely involve the kind of punishing stimuli which give rise to what is usually called 'fear'. Far more frequently, they consist of disappointed expectations or frustrated hopes. But, on the fear = frustration hypothesis, these are no less able to establish the avoidance component of an approach–avoidance conflict than is a painful blow; and, according to the experimental evidence, anti-anxiety drugs are able to alleviate the distress of conflict just as well when this arises from frustrative non-reward as when it arises from punishment. Under these conditions, animals will learn to increase their consumption of these drugs. Kraemer & McKinney, for example, showed that separation from their companions will increase alcohol consumption in rhesus monkeys (this is a highly social species, and we may plausibly construe such separation as a form of removal of reward). It seems, then, that the human need for alcohol and other anti-anxiety drugs should join the long line of other conditions which, as we have seen throughout this book, we share with our mammalian relatives.

Finally, the fear = frustration hypothesis casts a new light on the organisation of personality. If fear and frustration are the same, individuals highly susceptible to the one should be highly susceptible to the other. Guillamon and I tested this proposition, using a task in which the rat is alternately given water reward and non-reward on successive trials in the alley. On such a 'single alternation' schedule animals come to run slower on non-rewarded than on rewarded trials. This effect appears to involve anticipatory frustration, since injections of sodium amytal selectively increase running speed on non-rewarded trials. We can therefore predict that more fearful animals should show a greater reduction in running speed on non-rewarded (relative to rewarded) trials than less fearful animals. This is precisely what Guillamon and I found in comparisons between the Maudsley Reactive and Nonreactive strains (Chapter 4) and between male and female rats (Chapter 7). Were such a relationship between individual differences in susceptibility to fear and frustration, respectively, to hold for Man also, the task of describing the organisation of human personality would be considerably simplified. We shall do well therefore to look out for such a relationship when we consider fearfulness in Man, as we shall in our final chapter.

11 The learning of active avoidance

Throughout the short history of experimental psychology, the dominant approach to instrumental learning (i.e. learning which is instrumental in getting the animal *to* rewards or *away from* punishments) has been some form or other of 'reinforcement theory'. According to this view, there are for a given species a range of so-called 'reinforcing stimuli', that is, rewards (often called 'positive reinforcers') and punishments. The occurrence of a reward following a response increases the probability that this response will occur again under the same stimulus conditions; the occurrence of a punishment following a response decreases the probability that the response will recur under the same stimulus conditions. A punishing stimulus may normally also be used to increase the probability of a response, if the consequence of the response is to terminate, delay, or cause the omission of the stimulus. When used in this way, a punishing stimulus is correctly termed a 'negative reinforcer'. In the absence of an independent definition of rewards and punishments, these are, of course, not empirical statements at all: they merely define the use of the terms 'reward', 'punishment', 'positive', and 'negative reinforcer'. They manage to masquerade as a scientific theory of learning (and to take in some very well known psychologists in the process) only because we all have a very good idea of what constitutes a reward (food when you are hungry, water when you are thirsty, etc.) or a punishment (anything that causes pain).

However, given the reinforcement approach to the study of learning, there are certain empirical observations which we can go on to make, and certain empirical questions which we can go on to ask. For example, we can investigate the interactions between rewards and punishments when these are both contingent upon the same response, as we did when we considered approach–avoidance conflict in Chapter 9. Or we can ask questions about the way in which the brain carries out the operations which lead to the acquisition of new approach behaviour (to rewards) or new avoidance behaviour (away from punishment); this is the kind of question which will occupy us in Chapter 13.

But even when we make these limited demands upon the reinforcement approach to learning, reinforcement theorists have always found avoidance learning difficult to fit into their conceptual system. This difficulty is particularly acute in the case of *active* avoidance.

It will be recalled that at the beginning of Chapter 9, I drew a distinction between the two kinds of avoidance behaviour which psychologists call 'active' and 'passive'. In passive avoidance the individual abandons some activity or other because this is followed by punishment; any other behaviour which the individual chooses to engage in goes unpunished. In active avoidance, by contrast, the individual has to learn one particular action which will enable it to avoid punishment; any other behaviour is followed by punishment. Thus, in passive avoidance existing behaviour is suppressed, while in active avoidance new behaviour is acquired. The problem posed by active avoidance for a reinforcement theorist is to identify the (negative) reinforcing stimuli leading to acquisition of new behaviour.

One possible move is to deny that active avoidance behaviour depends upon instrumental learning at all. A view of this kind that has been put forward several times in different forms would treat avoidance as a product of classical conditioning. Currently the most influential version of this view is the theory proposed by Robert Bolles, according to which animals come equipped with a set of innate 'species-specific defense reactions', and only need to learn (by classical conditioning) which stimuli should be responded to in this way (i.e. stimuli associated with pain). Chief among the innate defense reactions for most species are immobility (freezing) and flight. I have no quarrel with this general position; indeed, it is identical to the one I adopted at the end of Chapter 2, in discussing conditioned freezing. However, in at least some formulations of his theory Bolles asserts that such classically conditioned responses account for *all* avoidance behaviour.

In support of this position Bolles points to the striking differences in the speed of acquisition of an avoidance response depending upon the response the experimenter chooses to reinforce by shock avoidance. If the designated avoidance response is closely similar to a species-specific defense reaction (e.g. running away from a box associated with shock into a safe box not so associated), a rat, say, learns very quickly; but if the response is quite dissimilar to any such reaction (e.g. pressing a bar), the rat learns very slowly and sometimes not at all. What is going on in these experiments, according to Bolles, is quite independent of the fact that running and bar-pressing are each followed by shock avoidance. Rather, running is learned easily because it is an innate response to stimuli associated with shock; bar-pressing, in contrast, is not an innate response to these stimuli so it is not learned. When bar-pressing is seen as an apparent avoidance response, Bolles claims, closer inspection will reveal that actually the rat is attacking the bar (a species-specific defense reaction) and only accidentally depressing it (and thus avoiding the shock).

Bolles' account of avoidance behaviour certainly captures something important about it – namely, that it is far easier to teach an animal an avoidance response that is compatible with its dominant response tendencies than to teach it an incompatible response. (Exactly the same is true about teaching animals any responses, avoidance or otherwise.) But, as a complete account of avoidance behaviour, it suffers from serious flaws (quite apart from the fact that aggressive behaviour is not normally elicited by stimuli associated with pain, so bar-pressing is most unlikely to occur, in the example cited above, as a by-product of conditioned attack). First, Bolles' theory strongly implies that if one compared animals trained on an avoidance schedule with animals trained according to a strict classical conditioning procedure, the former should not perform the avoidance response any more than the latter. But there have been several demonstrations of better learning by animals whose responding caused the omission of an aversive UCS than by classically conditioned controls (see, e.g., Fig. 11.1). Second, though it is harder to train animals to perform avoidance responses that are quite unlike any innate response to stimuli associated with shock, with care and experimental ingenuity it is certainly not impossible to do so. Thus, for example, Ferrari, Todorov & Graeff were able to train pigeons to peck a key to avoid shock, the birds eventually achieving a level of performance in which they received less than 10% of the programmed shocks; key-pecking is high in the pigeon's repertoire of responses to stimuli associated with food, but its innate response to a key paired with shock is to withdraw from it.

Examples such as these require us to seek some principle of learning that goes beyond classical conditioning if we are to offer a complete account of avoidance behaviour. So we turn to reinforcement theory. Now, the obvious thing to say is that active avoidance is reinforced by the *omission* of punishment. But the non-occurrence of an event can only affect an organism which *expects* the event; and workers in the Behaviourist tradition were very reluctant to make any move which admitted those all-too-mentalistic things, expectations, into the vocabulary of science. The oddity about this reluctance is that at exactly the same time, other psychologists equally in the Behaviourist tradition (and, indeed, they were often the same psychologists) were busy constructing the frustration theory of extinction of rewarded

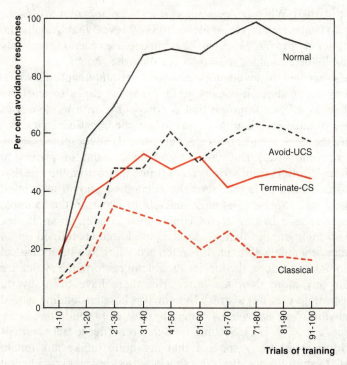

Fig. 11.1. Kamin's experiment on performance in the shuttlebox (Fig. 4.8) in groups of rats which could both terminate the CS and avoid the UCS (shock) by shuttling ('normal' group), terminate the CS only, avoid the UCS only, or do neither ('classical' group).

behaviour. Yet, as we have seen, frustration is due to the omission of an expected reward. Why the Behaviourists should have been able to stomach this, yet baulk at the omission of an expected *punishment* is a mystery; but baulk they did. And, being unwilling to recognise the omission of punishment, they therefore tried hard to find other reinforcements buried in the active avoidance situation. In doing so, they developed a theory of the role of fear in avoidance which has had important consequences, not only for our understanding of avoidance behaviour itself, but also for the general theory of learning.

The Miller–Mowrer theory of avoidance learning

As developed in particular by O. H. Mowrer and Neal Miller, this theory states, in effect, that what the animal learns in an active avoidance situation is not to avoid the punishment at all, but to escape from the stimuli which warn him of the imminent onset of punishment. (It should be noted that in the preceding sentence, 'to avoid' and 'to escape' are not synonyms, as in everyday English, but are used as part of the language of psychology. In this language, to 'escape' is to terminate one's exposure to the aversive stimuli after exposure has begun, while to 'avoid' is to take action which prevents one's exposure to the aversive stimuli from ever beginning.) In this way, the reinforcing event for active avoidance learning is not the omission of anticipated punishment, but the perfectly tangible termination of the warning signals. Now these warning signals are not usually themselves innately punishing stimuli, which the animal will escape from without special training. They are such things (in a typical experimental situation) as lights or tones or boxes from which the animal would not normally bother to escape. Thus this theory of avoidance learning requires some further account of how these innocuous stimuli acquire their 'secondarily aversive' properties. We have, of course, met these stimuli before: they are the classically conditioned fear stimuli of Chapter 2; for in order to turn a stimulus into a warning of impending punishment, we carry out the classical conditioning procedure of following the stimulus by the punishment.

It is in virtue of this introduction of a stage of classical conditioning into the processes which theoretically eventuate in instrumental learning that the Miller–Mowrer theory of avoidance learning, especially as it was developed by Mowrer, has its importance for the general theory of learning. For many other areas of learning are now treated according to the same 'two-process' model, in which a stage of classical conditioning first confers upon initially neutral stimuli a motivational and emotional significance, and then these stimuli guide instrumental behaviour to or from reinforcements (or 'goals'). We assumed as much in our treatment of conflict behaviour in Chapter 9, when stimuli along the runway which leads a rat to both reward and punishment were considered to become signals of both impending reward and impending punishment, and then to affect the strength of approach (via 'incentive' motivation) and passive avoidance (via 'fear'), respectively.

The essential assertions of the Miller–Mowrer theory are: (1) that initially neutral stimuli which are followed by punishment acquire secondarily aversive properties, in the sense that the individual exposed to them will take action to terminate this exposure; (2) that termination of these conditioned fear stimuli ('CS-termination') is the critical reinforcing event for active avoidance learning; and (3) that avoidance of the punishment ('UCS-avoidance') does *not* have any reinforcing effect. Using the language of the emotions, this theory is often stated by saying that: (1) the stimuli followed by punishment come to evoke conditioned fear; and (2) that it is reduction in this fear which reinforces avoidance behaviour.

First, what is the evidence that conditioned fear stimuli are aversive? In Chapter 2 we saw that it is possible to condition a particular symptom of fear, such as defecation, to an initially neutral stimulus, and also that it is possible for a conditioned fear stimulus to elicit passive avoidance. In our present context we must also show that an animal will take action to terminate its exposure to conditioned fear stimuli. Evidence that this is so was provided by some classic studies in the 1940s.

In one such study, conducted by Neal Miller, a rat was first shocked in a white compartment and allowed to escape from the shock by running into a black compartment. It was soon found that the rat would run from the white to the black compartment even in the absence of the shock. A stronger demonstration of the aversive properties acquired by the white compartment was provided by making the animal learn a totally new response to get out of it, even though it was never shocked again. This was done by interposing a barrier between the two compartments and arranging that the barrier could be removed if the animal turned a wheel. The rat duly learned to turn the wheel. The wheel was then made inoperative and a bar substituted which, when pressed, now opened the barrier. The rat abandoned wheel-turning and took up bar-pressing instead. Evidently the rat's initial experience with shock in the white compartment had left some fairly permanent residue which was sufficiently strong to motivate the learning of several new forms of behaviour.

Note that this result of Miller's, and the many others like it, compel us to adopt the view, adumbrated at the end of Chapter 2, that what is classically conditioned when a stimulus is followed by a punishment is not a particular fear *response* or set of responses, but rather a *change in the state of the organism which is then reflected in its subsequent behaviour*; for, in response to the shock, the rat had never turned a wheel or pressed a bar. A changed state of this kind can only be represented in the central nervous system.

The evidence is good, then, that stimuli followed by punishment do become secondarily aversive. What evidence is there that termination of these secondarily aversive stimuli acts as the reinforcement for active avoidance learning?

Kamin put the Miller–Mowrer theory to a stiff test by pitting CS-termination against avoidance of the punishment (electric shock) in the shuttlebox. He used four groups of rats. One was given the usual escape–avoidance conditioning procedure; that is, a response terminated the shock if it was already on, avoided the shock if it was not yet on, and terminated the CS (a buzzer) in either case. A second group was given the same buzzer–shock sequences, but could not affect either buzzer or shock by responding. As would be expected, the first group learned to shuttle back and forth between the two halves of the apparatus, while the second group did not (Fig. 11.1). The real interest, however, lies in the performance of the remaining two groups. One of these could terminate the CS by responding, but could not avoid the shock; the other could not terminate the CS, but could avoid the shock. Common sense has no difficulty in predicting that the second of these groups should learn to shuttle, while the first should not. The Miller–Mowrer theory (taken at its simplest) predicts with equal ease the reverse. The results satisfied neither prediction: both groups learned to some extent, but neither as well as the escape–avoidance group.

At first sight, then, Kamin's results offer some, but not complete, support for the Miller–Mowrer theory. There are two features of these results which offer difficulty for the theory.

First, there is the failure of the group which terminated the CS, but did not avoid the shock, to do even better. But this difficulty is more apparent than real. In this group, shuttling was actually *punished* by shock, and we would expect, therefore, that its probability of occurrence would be reduced. It is remarkable that in spite of this, CS-termination was able to maintain the rate of shuttling in this group as high as it did.

The second feature of Kamin's results which offers difficulty for the Miller–Mowrer theory is the ability of the group which avoided the shock, but did not (ostensibly) terminate the CS, to learn at all. However, it is not the case that this group did *not* experience CS-termination after making a shuttling response. Since CS-duration was fixed at 5 seconds, and since the rats tended to respond about 3 seconds after the onset of the CS, their response *was* followed by CS-termination, but with a delay of 2 seconds. That this is a point of substance is demonstrated by a subsequent experiment of Kamin's, also using the shuttlebox, in which delay of CS-termination after the response was varied, *all* responses avoiding shock. Learning occurred only if delay of CS-termination was less than 5 seconds; at delays greater than this, in spite of the fact that the response was sufficient to avoid shock, *no learning occurred at all*.

Thus Kamin's experiments offer even stronger support for the Miller–Mowrer theory than at first seems to be the case. At the very least, they demonstrate the remarkably powerful effects of CS-termination on the acquisition and maintenance of active avoidance behaviour. There are many other experiments which tend to the same conclusion; but there are also other data which the Miller–Mowrer theory would not predict and which make it clear that avoidance of punishment *per se* does have reinforcing effects.

The omission of anticipated punishment: safety signals

To begin with, Kamin's results cannot be generalised to tasks other than shuttlebox avoidance. Bolles, Stokes & Younger repeated his experimental design in both the shuttlebox and in a running wheel. In the latter apparatus they defined an avoidance response as a quarter-turn of the wheel in either direction. This offers the rat a much less complex problem than the response of shuttling, which requires the animal continually to return to the side of the box from which it has just fled, thus introducing an element of conflict. In addition, Bolles *et al*. investigated the effects of a contingency which Kamin had held constant: termination by the response of the shock if it was already on. For half the groups they eliminated this escape contingency by using a shock so short (0.3 seconds) that it was over before the rat had a chance to escape.

Bolles' results in the shuttlebox are shown in Table 11.1. It can be seen that Kamin's results were replicated in the groups which escaped shock, but that the escape contingency was itself partly responsible for maintaining shuttling both when combined with shock-avoidance and when combined with CS-termination. Shock-avoidance entirely on its own resulted in only 15 avoidances out of 100, compared to 14 when shuttling had no consequences at all. Thus this experiment confirmed Kamin's finding that in the shuttlebox, shock-avoidance *per se* is extraordinarily weak in maintaining avoidance behaviour. But it is also to be noted from Table 11.1 that CS-termination entirely on its own fared no better: 10 responses out of 100.

Bolles' results in the running wheel were quite different (Fig. 11.2). Avoidance alone (without assistance from either shock-escape or CS-termination) was now sufficient to maintain a respectable level of responding. This result, furthermore, was not due to delayed termination of the CS, as shown in a further experiment by making use of

Table 11.1. *Median conditioned avoidance responses (CRs) in 100 training
trials as a function of whether the CR
terminates the conditioned warning
stimulus (T), avoids the unconditioned
shock (A), or escapes the shock (E)*

Experimental condition			Median CRs
T	A	E	70
		–	37
	–	E	31
		–	10
–	A	E	40
		–	15
	–	E	9
		–	14

From Bolles, Stokes & Younger (1966).

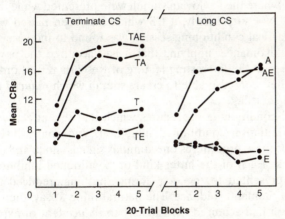

Fig. 11.2. Effects of CS-termination, shock-escape, and shock-avoidance on
performance in the running wheel. The acquisition of a running response in
a running wheel as a function of whether the warning signal (CS) could be
terminated (T) and of whether the shock could be avoided (A) or not and
escaped (E) or not. The group marked – affected neither CS nor shock by
running. From Bolles, Stokes & Younger (1966).

trace conditioning. In this procedure the CS was presented for a
period of time (half a second) too brief to allow completion of the
avoidance response, and the shock then followed 10 seconds later.
There could, therefore, be no question of apparent termination of

the CS by the avoidance response. Nevertheless, avoidance of the UCS alone was once more able to maintain responding.

These experiments show, then, that avoidance behaviour may be maintained by several distinct sources of reinforcement acting in concert: escape from the UCS, termination of a warning signal, *and* avoidance of the UCS. Evidently, the balance between these types of reinforcement can vary quite markedly depending upon the task the animal is set.

Other experiments also attest to the capacity of omission of anticipated punishment – like the omission of anticipated reward – to act as a reinforcing event in its own right. A particularly convincing demonstration has been reported by Rescorla & LoLordo, in experiments described in part in the previous chapter. These workers used a transfer-of-control design (Chapter 10) to test the motivational significance of stimuli treated according to the Pavlovian paradigms illustrated in Fig. 11.3. Dogs were first trained to avoid shock on a Sidman schedule in a shuttlebox. The Pavlovian phase of the experiment was then conducted in a different apparatus and the animals returned to the shuttlebox, where the Pavlovian stimuli were presented while they were shuttling. Not surprisingly, a CS which had been paired with shock in the classical conditioning session was found to increase the rate of Sidman avoidance, implying that it had acquired the power to increase fear. But the interesting feature of Rescorla & LoLordo's results is that they were also able to create stimuli which *reduced* the rate of Sidman shuttling.

The classical conditioning procedures which had this effect were those of differentiation conditioned inhibition, and contrast (Fig. 11.3). In each of these procedures one stimulus signals shock and one signals no-shock. Each of the latter kind of 'conditioned inhibitory' stimuli, when presented to the dog during shuttling, reduced the baseline rate of Sidman avoidance. The natural thing to say, then, is that these stimuli had acquired the capacity to *reduce* fear or expectation of shock. Similar results have been reported by Nieto, who showed in addition that a stimulus established as a signal for the omission of one of two kinds of aversive UCS (shock or a loud blast on a car horn) can alleviate the conditioned suppression elicited, not only by a CS paired with the same kind of aversive UCS, but also by a CS paired with the other. Thus a conditioned inhibitor of fear affects a rather general motivational state, not one that is narrowly linked to a particular aversive event.

Now there is nothing in the Miller–Mowrer theory which allows a stimulus to acquire the property of reducing fear. At first sight, this

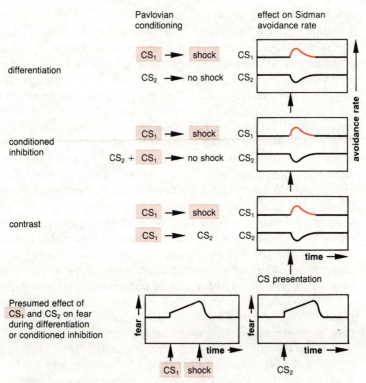

Fig. 11.3. Three Pavlovian conditioning procedures used by Rescorla & LoLordo to establish CSs which either increased or decreased the rate of Sidman avoidance responding (Fig. 9.4) when presented to dogs in a shuttlebox. The two panels at the bottom of the figure depict the changes in the level of fear which can be presumed to occur after presentation of the two kinds of CS during the Pavlovian conditioning phase of the experiment. For a fuller explanation, see text.

is merely a lack of inclusiveness in the theory. After all, it would be a natural extension to postulate that just as stimuli followed by a primary aversive stimulus come to be conditioned stimuli for fear, so stimuli followed by a reduction in aversive stimulation come, also by classical conditioning, to elicit a reduction in fear. But let us look a little more closely at the stimulus sequences which endowed originally neutral stimuli with this capacity to reduce fear.

Take the case of differentiation. In this CS_1 is always followed by shock, CS_2 is never followed by shock. There is, then, *no primary aversive stimulation to be reduced following presentation of CS_2*. But, one might argue, the conditioned rise in fear which is postulated to occur to CS_1 would surely generalise to CS_2 and, when shock failed to follow CS_2 there would be a reduction of fear which could become conditioned to CS_2. Certainly, on any view, it is necessary to suppose that there will be generalisation of fear to CS_2. But the second half of the argument cannot hold firm. Both CS_1 and CS_2 are deemed to elicit fear, the one by conditioning, the other by stimulus generalisation. In both cases (Fig. 11.3), we must presume that the intensity of fear rises to a maximum just before the time of shock onset. Furthermore, *in both cases*, there is a subsequent reduction in fear, the only difference being that in the case of CS_1 there is first interposed the primary aversive event, shock. There is no alternative, then, to the view that initially at least, the critical difference between the two stimuli, and the one which must be affecting the animal's subsequent behaviour, is that CS_2 is *not followed by shock*.

The omission of anticipated punishment, then, is an event that the animal can detect and which has effects in its own right. Let us call this event, in parallel to 'frustrative non-reward', 'relieving non-punishment'. What are the properties of a stimulus – a 'safety' signal – that has been paired with relieving non-punishment as in Rescorla & LoLordo's experiment? An answer to this question which appears to have some generality is that safety signals have properties that make them the mirror image of warning signals.

As we saw earlier in this chapter, animals will rapidly escape from stimuli that have been paired with shock. Analogously, Leclerc has shown that if a platform was regularly inserted into a chamber just after a rat was shocked there, the rat came to jump up onto the platform. So, just as warning signals elicit withdrawal, so safety signals elicit approach. Rescorla & LoLordo's experiment, as we have already seen, furnishes us with a second example of the opposing properties of warning and safety signals. A CS paired with shock, when presented to dogs while they performed a shuttlebox avoidance response, increased the rate of shuttling, presumably by increasing the fear of shock; stimuli paired with the omission of shock had the reverse effect, decreasing the rate of shuttling. A third way in which warning and safety signals are mirror images of each other is of particular significance for the understanding of active avoidance behaviour: just as warning signals are conditioned (or 'secondary') negative reinforcers (animals will work to terminate them), so safety signals are secondary positive reinforcers or rewards (animals will work to produce them). The Miller–Mowrer theory, incidentally, predicts this last result, once it is established (as, for example, in Rescorla and LoLordo's experiment) that safety signals reduce fear.

There are a number of studies which offer empirical support for the last claim enumerated above, namely, that safety signals act as rewards. For example, Kinsman & Bixenstine trained rats in a shuttlebox in which they could only escape (not avoid) shock. Just prior to the termination of the shock a distinctive stimulus (flashing light and buzzer) was presented to the animal, and this stimulus and the shock terminated together. Subsequently, the animals were tested in the same box to see whether they would learn to press a bar to turn this stimulus on; it was found that they pressed significantly more often when a response produced the stimulus then when it did not. Note that this is a fairly stringent test of the secondary reward hypothesis, since the subjects learned a totally new response (bar-pressing) to obtain the safety signal.

The secondary rewarding properties of safety signals offer a further possible source of reinforcement for active avoidance behaviour. For in nearly all instances of such behaviour the avoidance response is followed by stimuli associated with the omission of shock, even if these stimuli are composed only of the proprioceptive and kinesthetic feedback from the making of the avoidance response. Richard Morris has shown, with more explicit safety signals, that the contingencies that operate in a typical shuttlebox experiment are indeed able to turn safety signals into secondary rewards. A master group of rats was trained to avoid shock in the shuttlebox with white noise as a warning signal and a light as a safety signal (i.e. the light was turned on by successful avoidance responses). A second group of rats was yoked to the masters and received exactly the same sequences of noise, shock, omission of shock, and light, but having no opportunity yet to respond. Subsequently the yoked group was trained on the same avoidance task. As shown in Fig. 11.4 they learned very much faster than other yoked groups that had either been exposed initially to noise and shock, but not to the light, or which were not now offered the light as a response-contingent safety signal.

We should beware of considering the reinforcement offered by safety signals as a theoretical alternative to that offered by the termination of warning signals. On the contrary, there is every reason to suppose that these two sources of reinforcement complement each other. In support of this view, Owen *et al.* have shown that shuttlebox avoidance is learnt better if a response is followed by both termination of the warning and presentation of a safety signal than if it is followed by either one of these events on its own.

A further characteristic of safety signals emerges from an experiment by Lawler. He first trained rats to run into a distinctive box to escape shock. A buzzer was sounded at the same time as the animals were shocked. The animals were later tested to see whether the escape box had become a secondary reward. They were placed in a startbox,

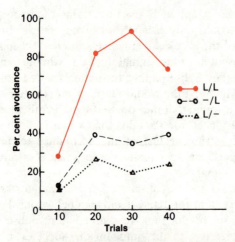

Fig. 11.4. Avoidance learning by three groups of rats in a shuttlebox. For Group L/L, the light had previously been established as a signal for the omission of shock and was now turned on by a successful avoidance response. For Group −/L, the light was similarly produced by each avoidance response, but had not previously been established as a signal for the omission of shock. Finally, for Group L/−, the light had been previously conditioned but was not presented during avoidance training. From Morris (1975).

as in the original training trials, and given the choice of two boxes to enter, one of which was the original escape box. For half the animals, the test trials used the same startbox as in the training trials and the buzzer was sounded; for the other half a different startbox was used and there was no buzzer. Controls were treated identically, except that they were not shocked during training; no animals were shocked during the test for secondary reward. All the animals shocked during training chose the original escape box more often than the controls, thus demonstrating that this box had indeed become a secondary reward. Furthermore, this tendency was no less (indeed it was stronger) in the group which was given a changed startbox and no buzzer during test trials and which, presumably, was made less fearful. Thus *the secondary rewarding effects of safety stimuli appear to be relatively independent of the degree of fear* aroused in the situation. Lysle & Fowler have similarly shown that the capacity of a safety signal to inhibit fear is little affected by quite wide variation in the strength of concurrent positive conditioned stimuli for fear.

This last point has some importance in view of a number of observations which show that there is a certain degree of independence between the amount of fear displayed by an animal and the efficiency of its avoidance behaviour. In the first place, once an animal has acquired successful avoidance behaviour, he shows no overt signs of being fearful: he performs the avoidance response quietly and efficiently and, between warning signals, carries on with the ordinary business of living. An experiment by Kamin, Brimer & Black provides more formal evidence that the signal for an avoidance response loses its fear-eliciting properties as avoidance behaviour becomes stable. They trained rats in a shuttlebox with a tone as the warning signal. The fear-eliciting capacity of the tone was independently evaluated by observing its disruptive effects on the animals' rate of bar-pressing for a food reward. This was done after 1, 3, 9, or 27 consecutive avoidances in the shuttlebox, while control rats were similarly tested after 1, 3, 9, or 27 pairings of the same tone and unavoidable shock. Suppression of bar-pressing by the tone was much greater in the control group; and, more importantly, suppression reached a maximum for the experimental group after 9 avoidances and then fell markedly again by the time 27 consecutive avoidances had been made.

This result has been replicated by Susan Mineka. In addition, she and Gino investigated the resistance to extinction (by shock omission) of the avoidance response with extinction commencing after either 9 or 27 consecutive avoidances. There was no difference in resistance to extinction, showing that even though fear of the CS had diminished after 27 avoidances the avoidance response had not weakened. The most obvious reason for the decline in fear of the CS as successful avoidances follow each other is that the CS is no longer followed by shock. However, Starr & Mineka have shown that this explanation is incorrect. They repeated the Kamin, Brimer & Black experiment, but also included yoked control groups which received the same patterns of CS and shock delivery as the master groups making the avoidance response. The animals yoked to the group making 27 consecutive avoidances showed no loss of conditioned suppression to the tone, though their masters did (see Fig. 11.5), in agreement with Kamin's findings.

Thus it is something about the actual making of the shuttlebox avoidance response that gives rise to the loss of fear of the CS. One possible factor is the controllability of the shock; this of course differed between the master and yoked groups in Starr & Mineka's experiment. As we saw in Chapter 9, with all parameters of shock delivery identical, inescapable shocks nonetheless have more deleterious effects on behaviour and physiology than do escapable shocks. It is not surprising, therefore, that (as demonstrated by Mineka, Cook & Miller) the fear conditioned to a situation in which the rat receives inescapable electric shocks is greater than the fear conditioned with escapable shocks of the same nominal intensity ('nominal' because, as described in Chapter 9, inescapable shocks have sometimes been observed to

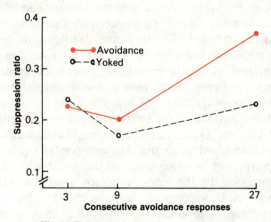

Fig. 11.5. Suppression produced by a tone that has either served as a warning signal for avoidance learning or been presented, with the same frequency of shocks, to yoked subjects. From Starr & Mineka (1977).

give rise to an analgesic response not seen after identical escapable shocks). Thus the very fact that the animal is successfully avoiding shocks may render him less fearful of the CS that signals the shocks.

Other experiments from Mineka's laboratory show that the provision of a safety signal can have the same effects as the making of a successful avoidance response. Recall that Starr & Mineka's master group that made 27 consecutive avoidances showed a loss of conditioned suppression to the CS but the yoked control group did not (Fig. 11.5). The addition, for both the master and the yoked group, of a safety signal indicating the completion of a successful avoidance response altered this pattern of results. Now both groups of rats showed a loss of conditioned suppression in the 27 avoidances condition. The same result was obtained if a safety signal was provided on non-avoidance trials after the occurrence of shock. In a parallel experiment, Mineka, Cook & Miller showed that the fear conditioned to a situation in which the rat received inescapable shock could be reduced if a safety signal was provided at the end of the shock. Thus the common ingredient of the three procedures that lead to loss of fear of the CS (making an avoidance response, presenting a safety signal on avoidance trials, and presenting a safety signal at the termination of shocks) may consist in the presence of a signal of shock-free time – that is to say, a safety signal.

It appears, then, that fear may first increase and then decrease while avoidance responses continue to be made with some efficiency;

and indeed that it is the making of the avoidance response and/or the clear safety signals that follow upon this response which give rise to the decrease in fear. But, if reinforcement for a successful avoidance response consists in fear-reduction on each trial (as claimed by the Miller–Mowrer theory), there should be some loss of reinforcement, and therefore some impairment of avoidance, as fear gets less from trial to trial. If, on the other hand, reinforcement consists in an approach to safety signals, and if (as we have seen in Lawler's experiment) these retain their secondarily rewarding property even when fear is low, this lack of close correlation between the degree of fear and the success of avoidance behaviour is less surprising.

The 'safety signal' view of active avoidance learning can also make sense out of another group of experiments which are an embarrassment for the Miller–Mowrer theory. As we have seen, this theory asserts that warning signals are secondarily aversive, and there is indeed evidence that this assertion is correct. Consider, therefore, an animal which is given the choice of being shocked in one environment without warning or shocked in another environment with a warning signal regularly preceding the shock. It is clear that the Miller–Mowrer theory must predict that the animal will prefer the environment containing unsignalled shock; for in the other environment he experiences not only the primary aversive stimulus but also the secondarily aversive warning signal. The evidence from many experiments, however, is unequivocal: animals will choose shock-plus-warning in preference to unsignalled shock. Katz has demonstrated the same phenomenon with human subjects.

There are several possible explanations of the preference for signalled over unsignalled shock. One is that the subject of the experiment can utilise the information about shock onset provided by the CS to prepare to receive the shock, perhaps by adopting some particular posture, and so reduce its impact. In its most literal sense this explanation is almost certainly wrong: direct measurement of current flow through the rat has shown that this is greater in the case of *signalled* rather than unsignalled shock, while grid contact time does not differ between these two conditions. However, a signal may lessen the impact of shock more indirectly. It has been known since Pavlov's time that the magnitude of a UCR may be diminished by a preceding CS. In agreement with this generalisation Katz has observed smaller skin resistance changes and less self-reported distress in response to predictable than to unpredictable shocks in his work with human subjects. One way is which a signal of shock may produce this effect is by triggering an endogenous analgesic response. In support of this hypothesis Fanselow has shown that naloxone (which blocks the opiate-mediated analgesic response to stimuli associated with shock) is able to eliminate the rat's preference for signalled shock.

While changed responses to the shock seem in part to underlie the preference for unsignalled shock, they are not the whole story. There is evidence implicating in addition safety signals. A role for such signals is readily intelligible. As the Rescorla & LoLordo experiment showed, a stimulus associated with no-shock in an environment in which shock is sometimes experienced acquires the capacity to reduce fear in its own right. Thus, in the situation where shock is regularly preceded by a warning signal, *absence* of the warning signal should become a safety signal of this kind. It follows that an animal choosing between an environment in which shock is unsignalled and one in which it is preceded by a warning is choosing between a steady high level of fear in the former case, and short periods of very high fear (during the warning) against a background of much lower fear (in the absence of the warning) in the latter case. It is no surprise, therefore, that shock-plus-warning is the preference normally displayed.

The role of safety signals is nicely demonstrated in an experiment by Badia, Abbott & Schoen. These workers offered rats a choice between two schedules of unavoidable shock, one with a 20-second warning, the other with a 5-second warning of shock. The intervals between shocks in the two schedules were the same. In addition, each schedule contained a safety signal filling the time when the warning signal was off. The rats preferred the schedule with the longer, 20-second warning. They were also tested for their choice of the two warning signals and the two safety signals. They preferred the 5-second warning to the 20-second warning (presumably because the 20-second warning increased the level of fear for a longer period), indicating that differential aversiveness of the two warning signals could not underlie their choice of shock schedules. However, they preferred the safety signal that occurred as part of the schedule involving the 20-second warning to the safety signal going with the 5-second warning. These results suggest two conclusions. First, safety acquires its value by contrast to fear: the 20-second warning was more aversive than the 5-second warning; correspondingly, the safety signal going with the 20-second warning was more rewarding than the one going with the 5-second warning. Second, at least under Badia *et al.*'s conditions, schedule choice was dominated by the rat's preference for one or other safety signal rather than by its preference for one or other warning signal.

The persistence of avoidance behaviour

There is one further feature of active avoidance behaviour which offers considerable difficulties for the Miller–Mowrer theory, but which is easy to fit in to the view that what the animal learns in an avoidance situation is to approach stimuli associated with relieving non-punishment. This concerns the extinction of avoidance.

Avoidance behaviour can be extremely persistent. Some of Solomon & Wynne's dogs tested in the shuttlebox continued to respond to the warning signal for 200 trials without a mistake after receiving only *one* shock and, when the experiment was terminated, there was no sign of any deterioration in their performance. Similar findings have been reported for other species and other forms of avoidance behaviour.

Now, according to Mowrer's theory, the reinforcement for avoidance behaviour consists in the reduction of conditioned fear. But, if the conditioning of fear follows the usual laws of classical conditioning, it should extinguish when the conditioned stimulus is presented without the unconditioned stimulus following it. Yet this is exactly what occurs when an animal is successfully avoiding the UCS on presentation of the CS. Ergo, fear should rapidly extinguish, there should be no fear to be reduced by a successful avoidance response, and avoidance should also extinguish.

Without major surgery, then, Mowrer's theory of avoidance learning cannot account for the persistence of avoidance behaviour. The alternative explanation of avoidance learning which we have been developing throughout this chapter does not experience this difficulty. According to this account, avoidance behaviour is reinforced by the safety signals (secondary rewarding stimuli) which accompany the successful avoidance of punishment. Such safety signals have two features which help to account for the persistence of avoidance behaviour. First, their effectiveness is relatively (but, as we shall see below, not absolutely) independent of the degree of fear aroused by warning signals (Lawler's experiment). Second, safety signals are themselves unusually resistant to extinction. Support for this last point comes from a series of studies, by Zimmer-Hart & Rescorla, of Pavlovian conditioned inhibition. Recall that a safety signal may be regarded as a Pavlovian conditioned inhibitor, as illustrated in Fig. 11.3; indeed, if the central claim of the Miller–Mowrer theory is correct, it is this property that allows safety signals to act as rewards. It is important, therefore, that Zimmer-Hart & Rescorla were unable to demonstrate any extinction of the conditioned inhibition of fear (as measured by alleviation of conditioned suppression of bar-pressing) when the conditioned inhibitor was repeatedly presented alone without further consequence.

This failure to extinguish conditioned inhibition of fear by simple repeated presentation of a conditioned inhibitor has been replicated several times. However, Lysle & Fowler have recently demonstrated one way in which such extinction can be brought about: by extinction of fear responses to the *positive* conditioned stimuli to which the animal is also exposed. They interpret this result to mean that 'inhibition is functionally dependent on excitation and thus is inoperative when the excitatory process is not viable'. In more familiar terms, safety signals lose their effectiveness when one is no longer afraid. Alas, while this inference is intuitively plausible, it brings us back full circle to the old problem: when fear extinguishes, so will safety signals and therefore the avoidance behaviour that they reinforce. Fortunately, an elegant exit (both theoretical and empirical) from this impasse has been described by Stefan Soltysik.

Consider an animal making successful avoidance responses in the familiar shuttlebox. A warning signal (say, a buzzer) sounds; the animal jumps to the other side of the shuttlebox, avoids shock, and terminates the buzzer. His jump may also turn on a safety signal (say, a light). But, even if no such explicit safety signal is built in to the experimental design, there are implicit stimuli to do the same job: feedback from jumping and landing on the other side of the box, the silence that follows termination of the buzzer, apparatus cues, etc. To make exposition easier, however, we shall suppose that a light provides an explicit safety signal. Once the animal has learned to make the avoidance response, then, he is exposed to long strings of the sequence buzzer–light (with no shock). *Ex hypothesi*, the light inhibits fear and serves as a reward for the shuttling response. Zimmer-Hart's findings suggest that repeated occurrence of the light will not extinguish these properties; some of Lysle & Fowler's results indicate that the intensity of the fear aroused by the buzzer can vary quite widely without compromising the capacity of the light to act as a conditioned inhibitor of fear; and Lawler's experiment similarly suggests that wide variation in the intensity of fear will not compromise the capacity of the light to act as a secondary reward. However, Lysle & Fowler's major result implies that, if fear extinguishes completely, the safety signal will also lose its fear-reducing and rewarding properties. Here is our problem. To get round it, Soltysik proposes that conditioned inhibition (elicited in our example by the light) preserves the inhibited response (i.e. conditioned fear) from extinction.

An effect of this kind was demonstrated a quarter of a century ago by Chorazyna, studying the standard Pavlovian conditioned salivary response. To this Soltysik has added the important further demonstration that the same phenomenon occurs with Pavlovian fear conditioning. In this experiment cats were conditioned to make several responses (paw-flexion and changes in respiration and heart rate) to two CS+s paired with shock. A third stimulus was established as a conditioned inhibitor (CI), being presented 2 seconds after the onset of the CS+ on trials when shock was omitted (so mimicking the typical sequence of events in an active avoidance task; see above). The CI came to inhibit the conditioned responses to both CS+s. Now came the critical part of the experiment. Both of the CS+s were presented for 180 consecutive trials without shock. One of them was always followed by the CI, to see whether this would protect the CS from extinction; the other was simply presented on its own. Soltysik's hypothesis predicts that the 'protected' CS should preserve, and the 'unprotected' CS lose, the capacity to elicit fear. When the two CSs were tested without shock or the CI at the end of the series of extinction trials, this is exactly the result that Soltysik obtained.

We seem, then, at last to have reached a satisfactory understanding of the resistance of active avoidance behaviour to extinction: safety signals (alias conditioned inhibitors of fear) reduce fear and provide secondary reward for the avoidance response; but at the same time they preserve fear from complete extinction and so ensure their own continued potency. This account has the added advantage that it fits comfortably into more general contemporary theories of learning, such as the one developed by Rescorla & Wagner. This theory holds that Pavlovian reinforcement consists in the discrepancy between the actual UCS and the expectation that the animal has for the nature (aversive or appetitive) and intensity of the UCS; a positive discrepancy (the intensity of the actual UCS being greater than expected) leads to excitatory conditioning, a negative discrepancy to inhibitory conditioning. Furthermore, the animal's expectancy is based on the algebraic sum of all the CS+s and the CIs to which it is exposed at the time the UCS occurs. Consider, then, what should happen on a successful avoidance trial in the shuttlebox, or on a 'protection-from-extinction' trial in Soltysik's fear-conditioning analogue of avoidance behaviour. Fear is elicited by the CS+ (or the buzzer in the shuttlebox), but then inhibited by the CI (or the making of the avoidance response). This should lead to a net expectation of aversive stimulation that is at or close to zero. No shock occurs, so there is no discrepancy with expectation. Thus, on the Rescorla–Wagner model, there should be no further conditioning, and both the CS+ and the CI should preserve their respective excitatory and inhibitory strengths.

The same point can be made in a slightly less formal way by considering the extinction of ordinary rewarded behaviour. Indeed, according to the safety signal theory, the extinction of avoidance can be regarded simply as a particular case of the extinction of rewarded behaviour. Now, as argued in the previous chapter, this type of extinction is due to the frustration caused by a mismatch between the expected reward and the reward actually obtained. But consider what the 'reward' is for active avoidance behaviour: it is non-punishment, i.e. the animal expects that *nothing* should occur after he has made the avoidance response. Thus, unless the experimenter starts to *punish* the avoidance response, there is no way in which a mismatch between the expected nothing and the actual nothing can arise, and so no reason for the animal ever to abandon the avoidance behaviour. And this, indeed, appears to be the case. Avoidance behaviour extinguishes spontaneously in rather complex situations, such as the shuttlebox, though even here, as we have seen, it is sometimes extremely persistent. But in simple one-way active avoidance situations just a few initial punishments are often sufficient to establish avoidance behaviour which the animal continues to display for as long as the experimenter has the patience to continue observing it.

A further reason for the persistence of avoidance behaviour has been pointed out by Levis & Boyd. When an animal has learned to perform an avoidance response to a warning signal it will usually respond soon after the warning signal commences. Every now and then, however, the animal will respond with a longer latency. Typically such a slow response is then followed by some more very fast ones, as though the animal had frightened itself by its own tardiness. This, for example, was the kind of behaviour observed by Wynne & Solomon in their study of the extinction of shuttlebox behaviour in dogs. Now one can express this intuition more formally by considering a warning signal as being made up of successive temporal segments. As the animal successfully avoids shock with short-latency responses he will repeatedly experience the initial segment of the CS followed by no shock. This would be expected to lead to a reduction in the amount of fear elicited by this segment of the CS, as demonstrated in experiments like Starr & Mineka's (Fig. 11.5), so that eventually the animal will make a long-latency response. But this response will expose the animal to later segments of the CS, whose fear-eliciting properties have not yet had much opportunity to extinguisl.. In consequence the animal will experience an elevated level of fear, as evidenced by its return to faster responding on subseqent trials. Levis & Boyd deduced from this type of analysis that, if the CS were not homogeneous (e.g. a single tone) but rather a sequence of discriminably different stimuli (e.g. as used in their experiment, a tone followed by white noise followed by a buzzer, for 6 seconds each), resistance to extinction would be increased, since there would be less generalisation of extinction of fear from initial to later segments of the CS. This indeed is the result they obtained: greater resistance to extinction of a one-way jump-up avoidance response with a heterogeneous than with a homogeneous CS. Given the complex nature of most real-life stimuli this result may have very wide application, especially in the kind of clinical cases discussed in Chapter 14.

There can be little doubt that active avoidance training is also effective in establishing extremely persistent forms of behaviour in Man. It is the basis of a large part of military discipline. It is also rather clearly involved in establishing the kinds of ritual behaviour involved in religion. Here, there is usually a certain set of actions which must be performed, failure to do so being punished early in life and leading to extreme anxiety in the adult. Similar kinds of ritual behaviour are sometimes developed, not in social groups as is the case for religions, but by a single individual. In this case, the behaviour is classified by psychiatrists as 'obsessional' or 'compulsive', and the syndrome is called 'obsessional–compulsive neurosis'. In this syndrome, the patient feels compelled to go through some meaningless set of actions whenever he is exposed to certain kinds of stimuli. If he omits to go through the ritual (repeated washing of the hands, repeated checking that all doors are locked, repeated thinking of certain thoughts, are common examples), he experiences considerable anxiety, though no objective harm can come from this omission. In both religious ritual and in obsessional neurosis, as in the case of animals which have been drilled in an active avoidance response, the way in which the actions are performed becomes increasingly stereotyped with repetition. This is in great contrast with behaviour which has been learned on the basis of reward, which usually remains quite variable in spite of repeated occurrence.

Neither warning nor safety signals are necessary for avoidance

So far two-process theory, especially once Mowrer's warning signals were supplemented by safety signals, has made quite a good job of handling the varied data reviewed in this chapter. Both warning and safety signals are used by two-process theory to provide immediate, response-contingent stimulus change that is correlated with eventual avoidance of the aversive UCS and thus able to guide the animal's behaviour into appropriate channels. But there are now a number of demonstrations of successful avoidance under conditions in which neither warning signal termination nor safety signal presentation are available to play the theoretical roles allotted to them.

Consider first an experiment reported by Herrnstein & Hineline. These workers, in effect, offered rats a choice between two frequencies of being shocked. If they pressed a lever they were shocked at unpredictable intervals with the mean frequency of shocks equalling, say, m times a minute. If they did anything else at all, they were shocked at equally unpredictable intervals with a frequency, n times a minute, which was greater than the frequency associated with lever-pressing, i.e. $n>m$. Under these conditions, nearly all the rats tested learned to reduce the frequency with which they were shocked by pressing the lever at a steady rate and, moreover, this rate was proportional to the amount of reduction in shock-frequency with which lever-pressing was reinforced. Yet, if the learning of avoidance behaviour is reinforced by the termination of conditioned aversive stimuli, as required by the Miller–Mowrer theory, these animals should have been unable to learn at all, for the experimental situation contained *no conditioned stimulus for the rat to terminate*. Nor could safety signals help, for there were none of them either.

The same is perhaps also true of the widely used Sidman avoidance schedule (see Fig. 9.4), but in this schedule there is at least a fixed time interval between the animal's last response and the delivery of punishment. Thus it is plausible to argue that certain internal stimuli related to the passage of time (and there is abundant evidence that animals possess such an internal 'clock') can become conditioned warning (or safety) signals for an animal trained according to a Sidman avoidance schedule. But in Herrnstein & Hineline's experiment, since shocks are delivered at randomly varying intervals of time, only the mean frequency of occurrence being affected by the animal's response, this argument cannot apply. As Herrnstein points out, if a proponent of the Miller–Mowrer theory wished to postulate any kind of internal stimuli for the Herrnstein–Hineline random-shock schedule, the only attribute it could have is its statistical relation to shock-frequency; and, to detect these 'stimuli', the animal would have to be reacting to shock-frequencies in the first place. The conclusion is inescapable, therefore, that animals can detect a reduction in the frequency of occurrence of a punishment.

Inescapable or not, this conclusion conflicts with a central principle that has guided virtually every theory of learning since Pavlov: namely, that associations are formed at particular moments in time when particular events come in some way into contact with each other. Since shocks occurred randomly in time whether Herrnstein & Hineline's rats had pressed the lever or not, the occurrence (or non-occurrence) of a shock at any particular point in time could not by itself convey any information to the animal. Only the mean frequencies of shock after responding and not responding, respectively, could convey the needed information; and these frequencies are properties of sequences of events that are extended in time. Thus Herrnstein & Hineline's results imply, in contradiction of the principle stated above, that the rat can:

(a) measure the average frequency of shocks it receives when not responding;

(b) store this frequency;

(c) measure the same quantity during periods of time after it responds;

(d) compare this with the stored frequency for shocks when not responding; and

(e) choose that behaviour which is associated with the lower shock frequency.

The random nature of the two shock schedules used by Herrnstein & Hineline ensured that shocks often occurred shortly after bar-presses, as well as after not bar-pressing. Nonetheless, at any given interval of time after bar-pressing the probability of shock was lower than at the same time after "not bar-pressing" (assuming that the animal could define the latter event for himself). This is a subtle difference, but it does provide the animal with some kind of immediate consequence of its behaviour which is correlated with the general, long-term reduction in shock frequency that bar-pressing produced. A later experiment, by Mellitz, Hineline *et al.*, goes one step further and eliminates all immediate consequences of avoidance responding without eliminating the animal's ability to learn the relevant response.

In this experiment rats were first trained on a Sidman avoidance schedule in which presses on either of two levers were equally effective in postponing programmed shocks by 20 seconds. When the animals were proficient at this task, having achieved at least 85% shock avoidance, a further contingency was introduced, affecting only the non-preferred lever (i.e. the one on which the animal was responding less). This contingency reduced the overall session length (initially set at 152 minutes) by 1 minute for every response on the non-preferred lever. The contingency was discontinued 2 minutes before the end of the session, ensuring that there could be no immediate reinforcement by removal from the experimental chamber. Thus, by shifting responding to the non-preferred lever, the rat produced absolutely no change in the events occupying the remainder of the session, but he did shorten the session. Remarkably, the animals showed clear evidence of learning, not only shifting their responses to the non-preferred lever, but then shifting them back to the other lever when the contingencies were reversed.

By any token this is a surprising result, and we would do well to wait until it is replicated before building too much upon it. Nonetheless, it is consistent with a number of other experiments which have shown that avoidance responding can be maintained in both rats and pigeons even when it has no immediate consequences but serves only to cancel later shocks. In such experiments it is difficult – I would say 'impossible', except that one should never underestimate the ingenuity of the dedicated theoretician – to detect anything capable of playing the role of either warning signal or safety signal.

Hineline's experiments eliminate warning and safety signals by the way in which the animal's task is constructed. A cruder way of arriving at the same end employs surgery to interrupt the sensory pathways along which the information constituting such signals must travel. Taub & Berman studied rhesus monkeys which had the responding limb de-afferented by section of the appropriate dorsal roots in the spinal cord. In this way the proprioceptive stimuli which normally indicate to the central nervous system that a limb has moved in a certain way were eliminated. The monkeys were trained, in some cases before this operation and in some cases after, to avoid shock upon receipt of a warning signal (a buzzer) by flexing the de-afferented limb. View of the limb was impeded, thus eliminating visual feedback. Whether the monkey was trained before or after de-afferentation, it performed the task perfectly well. In a further refinement of the experiment the warning signal was changed to a click of fixed duration, too short for there to be any question of its being apparently terminated by the monkey's response. Again the animals were able to learn and perform the required avoidance response. It is difficult to quarrel with Taub & Berman's own conclusion from these experiments: 'avoidance responding proceeded in a situation in which secondary negative reinforcement (sc. termination of warning signals) could not be presented over either proprioceptive, interoceptive or exteroceptive pathways'. To which one might add that secondary positive reinforcers (safety signals) were equally thoroughly eliminated.

It is extremely difficult to maintain, in the face of this evidence, that the two processes of two-process theory are *necessary* for avoidance learning to occur. Taub & Berman's experiments show that the rhesus monkey, deprived of all information other than the issue from the central nervous system of one or other command to the motor system and the occurrence or otherwise of shock, is able to correlate these two kinds of event and determine which motor commands are followed by shock and which are not. Thus the only items which can develop secondary aversive or secondary rewarding properties are the motor commands themselves. But to push the language of classical conditioning and secondary reinforcement this far is to take two-process theory beyond the point at which it remains useful. It seems safer to suppose that Taub & Berman have been able to demonstrate pure instrumental avoidance learning: a motor command which is followed by the non-occurrence of shock increases in probability.

Does this conclusion mean that our excursion into two-process theory (and the excursion, much more laborious, made by two-process theorists themselves) has been a waste of time? I think not. Animals can sometimes display pure instrumental learning, when experimenters make it impossible for them to solve an avoidance problem in any other way. But, more often, they will make use of all the diverse forms of support the environment offers them, be this in the way of escape responses that resemble the effective avoidance response, termination of signals of aversive UCSs, or production of signals correlated with safety (and we have seen examples of all of these earlier in the chapter). Similarly, their behaviour will sometimes be determined by distant consequences of their actions that override more immediate contingencies. But, more often and more strongly, it is the immediate consequence of behaviour that determines what is learnt and what is done.

In illustration of this last point, let us finish this section with a description of one more of Hineline's experiments. He trained rats on a schedule made up of recurring fixed 20-second cycles. Each cycle began with the insertion of a retractable lever into the chamber. If the rat pressed the lever within the first 8 seconds of the cycle, the lever was at once withdrawn and shock occurred 18 seconds into the cycle; if the rat did not press the lever, shock occurred 8 seconds into the cycle and the lever was withdrawn 2 seconds later. Thus the only effect of lever-pressing was to postpone shock by 10 seconds – there was no reduction in the overall frequency of shock. All the rats learned to press the lever. Similar experiments by Gardner & Lewis have shown that rats will learn in this way even if the consequences of lever-pressing include, besides postponement of shock, the delivery of several shocks later in the cycle, thus actually increasing the overall rate of shock. The factors that determine the relative effectiveness of immediate and remote consequences of actions in promoting the learning and maintenance of avoidance behaviour remain obscure. Mackintosh, in a review of the relevant literature, concludes that remote consequences provide 'at best a weak source of reinforcement for avoidance responding and can hardly be the major contingency affecting responding in most standard situations'.

We may conclude, therefore, that neither the warning nor the safety signals that figure so heavily in the baggage of the two-process theory of avoidance learning are necessary for such learning to take place. But all the particular effects predicted by two-process theory (reinforcement by termination of warning signals; reinforcement by presentation of safety signals; increased performance of avoidance responses in the presence of warning signals; decreased performance in the presence of safety signals) have been demonstrated in the laboratory. And it is reasonably certain that it is these signals, in human beings as well as in animals, which provide the emotions (both fear and relief) and the motivational drive that power most avoidance behaviour.

The elimination of avoidance behaviour

Obsessional neurosis and similar forms of neurotic behaviour offer a considerable challenge to our ability to eliminate avoidance behaviour. What help can the experimental psychologist offer to his colleagues in the psychiatric clinic who have to tackle this problem?

In considering the attempts which have been made in the laboratory to hasten the extinction of active avoidance behaviour, we must first draw a distinction between the degree of the individual's fear of the object (with its attendant warning signals) which he is avoiding, and the strength of the avoidance behaviour by which he avoids it. We have already seen that active avoidance behaviour is relatively independent of the degree of fear aroused by warning signals, this being one of the chief arguments against the Miller–Mowrer theory. The same point has emerged during studies of the extinction of avoidance behaviour. We have already dealt with the conditions which affect the degree of fear of the avoided object, in our treatment of conflict (Chapter 9). At present, we shall concentrate on the elimination of the avoidance behaviour.

Essentially, the methods of hastening extinction of avoidance which have been tried in the laboratory consist either in: (1) breaking the connection between making the avoidance response and presentation of the full safety environment, considered in this section; or (2) actually punishing the avoidance response, considered in the next. (In both cases, of course, the experimenter ceases to punish failure to make the avoidance response.)

One way of breaking the connection between the avoidance response and the safety environment is to prevent the avoidance response from occurring at all, by physically blocking it. Since failure to make the avoidance response is now not punished, the animal is given the opportunity to associate *not* making this response with non-punishment. In an experiment of this kind, Black trained dogs to avoid an electric shock by touching a plate with their heads and then exposed them to the sequence, warning signal – no shock, after they had been injected with a paralysing drug, curare. This treatment, which effectively prevented performance of the avoidance response, hastened extinction of the response as judged by the animals' behaviour in subsequent sessions in the undrugged state. The same kind of result can be obtained by interposing a barrier between the animal and the safety environment it has been trained to approach.

Now, of course, from a common-sense point of view, it is hardly surprising that these procedures should hasten extinction, since they give the animal the opportunity (which he cannot have so long as he continues to perform the avoidance response) to learn that conditions have changed and that the warning signal is no longer followed by shock even if he fails to make the avoidance response. However, we should remember the obsessional patients who are fully aware that failure to make their avoidance response will have no unpleasant consequences and yet continue to make it. If awareness of consequences can do so little for human beings, we ought perhaps to be careful about attributing much effectiveness to it in rats or dogs.

In fact, other experiments which have blocked the avoidance response during extinction tend to discredit the idea that this result is due to the animal's having learned that the warning signal is no longer to be feared. Indeed, the reverse appears sometimes to be the case. Page, for example, trained rats to avoid shock by running from one box to another. During extinction, the door between the two boxes was blocked off for the first five trials for one group, but not the other. The former group extinguished the avoidance response more quickly. But the two groups were then trained to approach the shock box to obtain a food reward. It was the group which had *not* had the avoidance response blocked which learned this approach behaviour faster. Thus, blocking the avoidance response had eliminated running out of the shock box more quickly, but appeared to have left a more lasting fear of this box. No doubt this was due to some further conditioning of fear which took place on the blocked trials; for it is usually the case that blocking the avoidance response leads to an immediate and steep rise in the fear displayed by the subject. This is to be expected, given the account of active avoidance developed in the preceding pages. As we have seen, one of the effects of a safety signal is to reduce fear, presumably as a result of Pavlovian inhibition (Fig. 11.3). If, then, we remove safety signals (by blocking the avoidance response) fear should at least temporarily increase.

Blocking the avoidance response leaves the animal exposed to the danger signals of the environment in which it has been punished. Another way of achieving the same end is to allow the avoidance response to be performed, but at the same time to prolong the warning signal after it has been made. An experiment carried out in Warsaw by Soltysik bears this out. Dogs had been trained to avoid shock by flexing the paw on presentation of a warning CS. During training, this was terminated as soon as the response was made. On extinction trials, the CS was prolonged when the response was made, and shock was no longer given whether the animal responded or not. As in Page's experiment, this led to more rapid extinction of the flexion-response than when CS-termination continued as before. Also as in Page's experiment, there were signs that this procedure, while eliminating the avoidance response, had left the fear of the CS intact; for, even after paw-flexion had been abandoned, the CS elicited a conditioned increase in heart rate.

The type of experiment in which the avoidance response is blocked or the termination of warning signals is discontinued has received a great deal of attention in recent years because of the analogy it offers to one widely and successfully used method of behavioural treatment for phobias and other neurotic conditions (Chapter 14), namely, 'flooding'. Essentially, this treatment consists of exposing the patient to intense and prolonged stimulation from fear cues while preventing him from engaging in his usual pattern of avoidance behaviour – the analogy will be obvious. It would be alarming, therefore, if the outcome obtained by Page – rapid extinction of avoidance behaviour but lasting fear of cues associated with shock – were invariably seen after flooding procedures in experiments with animals.

In fact, this is not the case: quite diverse results have been obtained in experiments in which both extinction of avoidance and residual fear have been measured after blocking of the avoidance response. One factor that Shipley, Mock & Levis have shown to be important is the total time of exposure to the warning signal. In many experiments this has differed widely between the group of animals whose avoidance response is blocked and the various control groups to which this group is compared. Shipley's results show that when care is taken to equate CS-exposure time between such groups, the differences between them become much smaller and even disappear. This finding suggests, as the Miller–Mowrer theory would predict, that extinction of avoidance is chiefly determined by the extinction of fear that should ensue from exposure to the warning signal without presentation of the UCS.

As we shall see in Chapter 14, the same principle has emerged in studies of the extinction of human neurotic behaviour: the dominant variable, across treatments that differ widely in detailed procedures, is the total time the patient is exposed to the stimuli he fears. But, while this is a comforting parallel for those who seek communalities between human and animal behaviour, we should be wary of supposing that total CS-exposure time is the only factor determining the extinction of fear and avoidance behaviour. As we have already seen, Starr & Mineka's experiment (Fig. 11.5) has shown that with total CS- and UCS-exposure controlled during the acquisition of avoidance behaviour, animals that could control the CS and UCS came to display less fear of the CS than yoked animals that could not. This finding is the natural complement to results like Page's, suggesting as they do that the removal of control (by blocking the avoidance response) *increases* fear of the CS.

Unfortunately, this principle too is not universal. In one of their experiments, Mineka & Gino obtained exactly the opposite result. They blocked a shuttlebox avoidance response for 30 trials and compared the group treated in this way to a control group given equal CS-exposure time while continuing to make the avoidance response. On a measure of subsequent extinction of avoidance, the two groups did not differ, supporting Shipley's findings. But on a measure of conditioned suppression to the warning signal, fear was reduced *more* in the group for which avoidance responding was blocked.

If we seek guidance from these studies in the design of therapies for human neurotic behaviour, the conclusions they permit are far from satisfactory. We know several ways of eliminating avoidance behaviour: by blocking the avoidance response; by prolonging the CS so that it continues after the avoidance response is made; or by exposing the animal to the warning signal without presentation of shock. But the best we can say about the relationship between the extinction of avoidance and the extinction of fear (which is likely to be at least as important a goal for therapy) is that it is not close. Some procedures appear to reduce avoidance more than fear, some do the reverse, and we are a long way from understanding why. Thus caution should be used in applying flooding techniques to the treatment of human behaviour: one may eliminate avoidance behaviour but leave fear intact or even inadvertently increase it.

Vicious circle behaviour

The perils attendant on the second way of hastening the extinction of an active avoidance response – punishing it – are even greater. There are by now rather a large number of experiments on record which show that punishing an avoidance response may actually *strengthen* it. Mowrer has termed this phenomenon, for obvious reasons, 'vicious circle behaviour' (VCB).

A striking example of vicious circle behaviour was reported by Solomon, Kamin & Wynne in their shuttlebox experiments with dogs. The dogs were showing extreme resistance to extinction already, having carried on shuttling for 200 trials with no shock. At this point, shock was put on the side of the box into which the animal was jumping: now if he stayed still nothing happened, but if he jumped he got shocked. Nevertheless, most of the dogs continued jumping *into shock* for another 100 trials (at which point the experiment was terminated), and some of them actually jumped faster into shock than they had been jumping before punishment was introduced. Remember that these were animals which had learnt the initial response at the cost of only one or two shocks. To make their behaviour still more mysterious, all the dogs developed anticipatory reactions before jumping, indicating that they 'knew' they were to be shocked.

This is not always the effect obtained when an avoidance response is punished. Indeed, in this same experiment, three out of the 13 dogs abandoned shuttling quite suddenly once punishment was introduced. But it is surprising how often VCB has been reported. For example, an early experiment by Gwinn obtained the phenomenon using a completely different apparatus, and rats instead of dogs. The animals were trained to run round a circular runway so as to escape from shock by entering the last section. They were then divided into two groups: the experimental group was now exposed to shock in the third quadrant of the runway only, while the control group had no shock in any section of the runway. The experimental group continued running round the whole runway for more trials, and ran *faster* over the section from the startbox to the quadrant in which only they were shocked, than did the controls. About a dozen other experiments have obtained VCB, and another half-a-dozen have looked for it and failed to find it. It is not easy to see anything which distinguishes the experiments obtaining the phenomenon from those which do not.

This mysterious phenomenon becomes a trifle less mysterious when we realise that, if the animal is approaching safety (i.e. rewarding) signals in making an avoidance response, he is in the same position as the rats in Fowler & Miller's experiment (discussed in Chapter 9) which ran to shock-plus-food more eagerly than others ran to food alone. As we saw in that experiment, the critical determinant of the outcome of this conflict was the compatibility of the response to punishment with the nature of the approach response: rats which lurched *forward* as a result of shock to the hindpaws showed the paradoxical behaviour described above; those which lurched *backward* from shock to the forepaws showed a more normal inhibition of the punished response. Now, in a VCB experiment, the initial avoidance training effectively suppresses the innate *passive* avoidance responses to warning signals and substitutes for them some form of approach behaviour. Thus, when the avoidance response is suddenly punished, the animal's inital reaction is to continue running forward or jumping to the other side of the shuttlebox. Since the shock used in the VCB experiments is always of fixed duration, this continuation of the old avoidance response is followed eventually by shock-offset, i.e. by a reward. In this way, the old safety signals, in spite of the fact that they are now also associated with punishment, can become even more powerfully reinforcing than before.

Further experiments are needed to decide whether this account of vicious circle behaviour has any merit. In any case, it seems certain that it is a phenomenon of some importance. It might well be yet another strand in the causal web which leads to masochistic, and especially to compulsive masochistic, behaviour in human neurotics. At a practical level, it is something the behaviour therapist (Chapter 14) should be wary of when he attempts to treat neurotic avoidance behaviour (e.g. phobias or obsessional rituals) by punishing it. By doing this he could make matters worse.

This completes the survey of the organisation of fearful behaviour which we began in Chapter 9. In this survey, we have looked at passive avoidance and conflict, at the similarities between the effects of punishment and those of frustrative non-reward (the 'fear = frustration' hypothesis), at helplessness, and at the way in which organisms learn an active avoidance response. As we did in the case of the *origins* of fearful behaviour, which occupied us throughout the first half of the book, we shall want to proceed to an examination of the physiological basis of the *organisation* of fearful behaviour. This basis can only be found in the brain, or, better, in the neuroendocrine system. But we shall be better placed to see how the brain is involved in the organisation of fearful behaviour if we first form a map of the '*conceptual* nervous system' which is implied by the principles we have seen at work in the last few chapters. This is what we shall try to do in the following chapter.

12   A conceptual nervous system for avoidance behaviour

In the long run, any account of behaviour which does not agree with the knowledge of the nervous and endocrine systems which has been gained through the direct study of physiology *must* be wrong. But this is not to say that in the short run the psychologist must wait for the physiologist to amass this knowledge before venturing on the task of constructing a science of behaviour. It does not even mean that in the short run the psychologist need be too unhappy if the theories he constructs to account for behavioural data do not entirely agree with the picture of the neuro-endocrine system which the physiologist is currently able to offer him. Indeed, it is possible that the structure of the neuro-endocrine system *inferred* by the psychologist from purely behavioural data may be at certain points in advance of physiological knowledge. Thus it is that at present there are two kinds of nervous system, which are sometimes related only by their common abbreviations, CNS: the Central Nervous System studied by the physiologist, and the Conceptual Nervous System (a happy phrase coined by D. O. Hebb) inferred by the psychologists.

It should not be thought that this kind of relation between psychology and physiology is in any way unusual. On the contrary, it is one that has frequently arisen in the history of science. A very good parallel is the treatment of the atom by chemists and physicists. The atomic theory of matter was proposed by Dalton on the basis of chemical knowledge well before physicists were in a position to make any direct observations of atomic structure. Let us hope that the psychologist's conceptual nervous systems – or a few of them, at any rate – will be similarly fruitful.

In the present chapter I wish to propose a conceptual nervous system (or 'model') for the organisation and control of avoidance behaviour, both active and passive. For the most part this is based on purely behavioural data, of the kind we have been examining for the last three chapters, though account is also taken of one or two experiments in which the relations between brain and behaviour have been studied. The principles upon which this model is based have all been introduced earlier in the book, and it is possible to regard this chapter as a summary of them. What is offered is a personal synthesis which is bound to be wrong in part and maybe wrong completely;

but, at the very least, it will provide a frame of reference in which to place the known facts about the real brain when we come to deal with them.

Of course, there is not one brain to deal with avoidance behaviour and another to deal with everything else. Thus it is impossible to construct a conceptual nervous system for avoidance behaviour without also considering how other kinds of behaviour are organised. Both in this chapter and the next, therefore, we shall apparently stray from our topic of fear in order to place this topic in broader perspective.

Reward and punishment from electrical stimulation of the brain

Let us take as our starting point the phenomenon known as 'electrical self-stimulation of the brain'. Knowledge of the basic neurophysiology of motivation has been greatly increased by experiments, pioneered by James Olds in the United States, in which animals (usually rats), with electrodes (which are small pieces of wire, insulated except at the tip) implanted deep in the brain, are allowed to press a bar which either causes a small electric current to flow in their own brains or, alternatively, turns off a current which the experimenter has caused to flow in their brains. It has been found that with certain placements of electrodes the rat will press a bar to stimulate its own brain electrically for hours on end; and, with certain other placements, the rat will be equally eager to terminate or prevent the occurrence of the electrical stimulation. Now it takes no great leap of the imagination to suppose that these results indicate the existence in the brain of two fundamental motivational systems, a 'reward' mechanism and a 'punishment' mechanism. That is to say, the common denominator of the heterogeneous class of events which an animal finds rewarding (e.g. food, water, copulation) is that they cause an increase in the activity of the reward mechanism of the brain, while the common denominator of such diverse punishments as electric shock, loud noise, sudden loss of support, and so on, is that they cause neurons to fire in the brain punishment mechanism.

If we now bring in the principles of classical conditioning, which we first encountered in Chapter 2, we have most of what we need to guide the organism towards its rewards or away from its punishments.

We suppose that stimuli which regularly precede the occurrence of a reward acquire the capacity to activate the reward mechanism; and that the closer in time to the innately rewarding stimulus they occur, the stronger is this capacity. The reward mechanism is so constructed that, via connections with the animal's 'motor' system (i.e. those parts of the brain which issue commands to the limbs), it strives to maximise such conditioned or 'secondary' rewarding stimulation. In this way, given a stable environment in which sequences of stimuli recur with a degree of regularity, it is able to guide the animal towards the innately rewarding stimulus. We could, in fact, liken the reward mechanism to a homing or 'approach' device, of the kind used by a guided missile to aim up a heat gradient at the hottest spot around.

In a similar manner, stimuli which regularly precede the occurrence of a punishment acquire, by classical conditioning, the capacity to activate the punishment mechanism; and again this capacity is greater the closer in time to the innately punishing stimulus they normally occur. The punishment mechanism is so constructed that via its connections with the motor system it strives to minimise its own inputs. It does this by putting a brake on behaviour which leads to an increase in its own activity. That is, it is a mechanism for passive avoidance, giving the command 'Stop!' to the motor system.

In short, in the language of feedback control systems, the reward mechanism is a 'positive feedback' device, while the punishment mechanism is a 'negative feedback' device. It is easiest to think of their operation in a spatial framework. Let us take as an example one of the conflict experiments discussed in Chapter 9. (Indeed, so far, we have done no more than re-state, in more mechanistic terms, Miller's equilibrium model for conflict – see Fig.9.2.) We may suppose that in this kind of experiment, as the animal runs towards a goalbox in which he has experienced both reward and punishment, there is an increment in the activity of both the reward mechanism, increasing the approach tendency, and the punishment mechanism, increasing the tendency to halt. Note, however, that it is not necessary to think of the operation of the reward and punishment mechanisms in spatial terms. The reward mechanism will home in on, and the punishment mechanism try to decrease, any kind of stimulation. Among the most important classes of stimuli for this purpose are those which result from the organism's own motor activities (feedback from the muscles, etc). As pointed out by Mowrer, an approach mechanism which homes in on the feedback from the organism's own responses, provided these are followed by reward, offers a completely general system for learning *any* response which is followed by reward, *Mutatis mutandis*, this is equally true for a passive avoidance mechanism which includes among its inputs feedback from the animal's own responses.

I have so far glossed over a major distinction that needs clarification before we proceed further. As we have seen before (Chapters 2 and 10), animals respond differently to conditioned and unconditioned aversive stimuli. To put the point most simply, in response to a painful electric shock, animals will typically show increased activity, run, jump, scream, hiss or attack a suitable target (e.g. another animal) in their vicinity; but, in response to a stimulus associated with shock, the animal will most likely freeze and remain silent. As we shall see in the next chapter, the brain mechanisms that mediate these two kinds of reaction are quite distinct, as are the drugs to which they are sensitive. In recognition of this distinction I have termed the mechanism that reacts to unconditioned aversive stimuli, the 'fight/ flight system'; and the mechanism that reacts to conditioned aversive stimuli, the 'behavioural inhibition system'. It will be clear from the description of the punishment mechanism given in the preceding paragraphs that this is equivalent to the behavioural inhibition system, i.e. it is concerned with the suppression of responses that lead to punishment, not with the execution of active flight or fight behaviour. We return below to the fight/flight system; for the moment, let us consider further the relations between the punishment and reward mechanisms.

The distinction between reactions to conditioned and unconditioned stimuli is relevant also to our understanding of the reward mechanism. Unconditioned rewards (food, water, a sex partner) elicit a variety of specialised consummatory responses, each appropriate to the particular UCS employed (chewing, licking, copulation, etc.). Such consummatory responses may also appear in a conditioned form; but in addition conditioned rewarding stimuli elicit a pattern of behaviour (involving increased activity, approach, and exploration) that is relatively independent of the particular UCS with which they have been paired (this is Konorski's 'preparatory reflex'). Now, just as I have supposed that the punishment mechanism mediates responses specifically to conditioned aversive stimuli, so I suppose that the reward mechanism mediates preparatory responses to conditioned appetitive stimuli. The relation between this kind of conditioned response and the type of learning ascribed to the reward mechanism will be readily apparent.

In the case of both reward and punishment, then, the principal contribution of unconditioned reinforcing stimuli is to act as Pavlovian UCSs for the conferral upon associated stimuli of the power to act as secondary reinforcers. A simplifying assumption would be to allot this Pavlovian function to the same system that mediates other effects of unconditioned stimuli; in the case of unconditioned punishing stimuli this would be the fight/flight system. Given that this Pavlovian conditioning has taken place, the stimuli that actually activate the reward and punishment mechanisms, and which thereby allow the animal to optimise its transactions with the environment, are *conditioned* reinforcing stimuli.

At this point, we may start making a diagram of our conceptual nervous system (Fig.12.1). As soon as we do this, it is clear that besides the reward and punishment mechanisms and motor system with which we have already provided the animal, it is necessary to give him a 'decision mechanism' to decide between approach and passive avoidance under conditions of conflict. Without such a decision mechanism, conflicting commands will be issued direct to the motor system, with probably disastrous consequences. This is not an appropriate place to go into the workings of the decision mechanism. The interested reader is referred to the article in which P. T. Smith and I originally proposed this general model. In it, we show that, by giving the decision mechanism a very simple mathematical rule to operate on, it is possible to predict the general lines of many of the behavioural phenomena disclosed by investigations of conflict, partial reinforcement, and discrimination learning. For our present purposes, all we need note is that the decision mechanism closes *either* the switch on the route from the reward mechanism to the behaviour command box

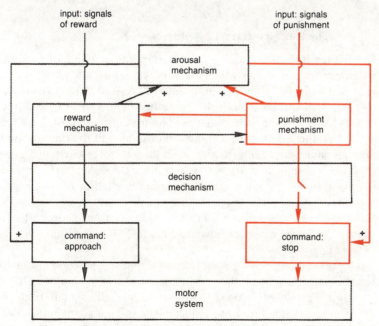

Fig. 12.1. Part of the Gray & Smith model for conflict and discrimination learning. For a full explanation, see text.

'Approach!' *or* the switch on the route from the punishment mechanism to the behaviour command box 'Stop!' but not both. (The general conditions under which approach or passive avoidance behaviour will in fact predominate were discussed in Chapter 9.)

The remaining item in Fig. 12.1 is called an 'arousal mechanism'. It receives inputs from both the reward and the punishment mechanisms and in turn sends outputs to both command boxes, the one for approach and the one for passive avoidance. In this way, whichever of these boxes is switched in by the decision mechanism has its activity increased in intensity by an amount which depends on the magnitude of the inputs to the arousal mechanism from both the reward and the punishment mechanisms. This will lead to the phenomena of drive-summation which we considered in Chapter 10: when approach behaviour occurs, it will occur more vigorously if the animal is simultaneously exposed to threats of impending punishment. It is the joint operation of the arousal and the decision mechanisms which leads to the predictions that, as threats of punishment for approach behaviour are increased, the probability of approach behaviour goes down, while the vigour of approach behaviour, if it nevertheless occurs, increases (see Fig. 10.1).

The fear=frustration hypotheses

In Chapter 10 I discussed the evidence that the omission of an anticipated reward has effects which are very similar to, and perhaps even identical with, the effects of punishment. This similarity gives rise to the fear=frustration hypothesis, or rather (as noted in Chapter 10) to two fear=frustration hypotheses. The first of these postulates that the states elicited by unconditioned punishing and frustrative events are the same; according to the second, the states elicited by conditioned punishing and frustrative stimuli are the same; but (as already discussed in the present chapter) the states elicited by unconditioned and conditioned aversive (punishing and frustrative) events are *not* the same. The fear=frustration hypotheses can be incorporated in our model in the manner shown in Fig. 12.2. This is a fuller version of Fig. 12.1.

The reward mechanism now sends a signal to the 'comparator for reward' informing it of the kind and amount of reward which can shortly be expected to occur (on the basis of the signals from the environment which the reward mechanism is itself receiving). That is to say, it sends to the comparator for reward a copy of the stored memory that, whenever these conditioned stimuli (the secondary

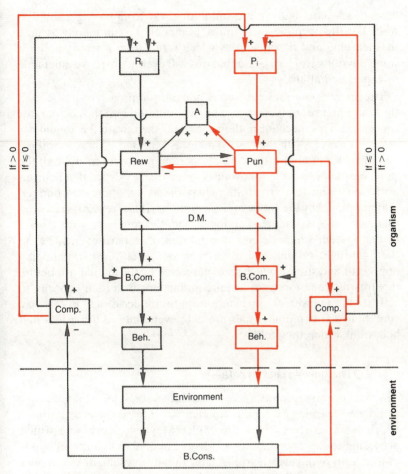

Fig. 12.2. Full version of the Gray & Smith model. R_i and P_i inputs to the reward and punishment mechanisms; Rew and Pun. D.M., the decision mechanism; A, the arousal mechanism; B Com., behaviour command to 'approach' (on the reward side) or to 'stop' (on the punishment side); Beh., the observed motor behaviour; B Cons., the consequences (rewarding or punishing) of the behaviour that occurs; Comp., comparator mechanisms which compare the actual consequences of behaviour with the expected consequences and make appropriate reward or punishment inputs. R_i consists of conditioned stimuli associated with reward or non-punishment; P_i consists of conditioned stimuli associated with punishment or non-reward. The reward and punishment mechanisms are connected by reciprocally inhibitory links.

rewarding stimuli) have occurred in the past, they have been followed at such-and-such an interval of time by this unconditioned stimulus (the expected reward). How such memories are established, stored, and made use of is, of course, the central problem of the physiological basis of learning; and we are still a long way from having a solution to it. For the purposes of the present model, we can merely assume that learning and the memories which result from learning do have some physiological basis, and that this will eventually prove amenable to experimental analysis.

The animal now goes ahead and makes its approach response, and the consequences of its response are also entered into the comparator for reward. This mechanism then compares the reward the animal has actually received with the reward that was expected. (There is nothing very mysterious about this kind of operation; computers do it all the time. We shall see in the following chapter good evidence that neurons do it too.) The fear=frustration hypotheses are then incorporated into our model by the following assumptions. If less reward is received than was expected, there is an input from the comparator for reward to that system which detects unconditioned aversive events, i.e. to the fight/flight system. The usual processes of classical conditioning now occur and stimuli which were perceived by the animal just before this frustrative non-reward acquire an increment in their capacity to activate the punishment mechanism on future occasions. In this way, conditioned frustrating signals are able eventually to bring to a halt behaviour which they follow.

The hope=relief hypothesis

The final process we must incorporate into our model is that of active avoidance learning (Chapter 11). The conditioned aversive stimuli which were so stressed by the Miller–Mowrer theory, and which, according to the experimental evidence, undoubtedly play an important role in avoidance learning, are already present in our model, since we have postulated that stimuli which are regularly followed by punishment acquire the capacity to activate the punishment mechanism. However, the experimental evidence also demonstrates the effects on avoidance behaviour of what I have termed 'relieving non-punishment', and so our model must make provision for this too. This we can do in an exactly symmetrical manner to the way in which the model detects frustrative non-reward.

We suppose that the punishment mechanism is linked to a 'comparator for punishment', to which it signals the kind and amount of punishment which, in the past, have followed the stimuli which are currently being received from the environment and which, therefore, may be expected to occur again. The animal now makes some response or other and the consequences of its response are also fed into the comparator for punishment. If the animal has in fact made a successful response, the comparator detects the resulting mismatch between expected and received punishment and sends an input to a system that detects unconditioned reward. (It is difficult at present to be more specific than this. Unconditioned rewards other than relieving non-punishment come in many different kinds – food, water, sex, etc. – and each of these appears to depend upon a relatively independent mechanism, mostly focussed on the hypothalamus. There is no good reason to associate relieving non-punishment with any one of these more concrete forms of reward rather than any other, nor any strong evidence for a separate brain system mediating only active avoidance behaviour.) Again, the usual processes of classical conditioning are presumed to occur and stimuli which were perceived by the animal just before this relieving non-punishment acquire an increment in their capacity to activate the reward mechanism on future occasions. Thus, stimuli of this kind (conditioned 'relieving' or 'safety' signals) are approached by the animal when he is subsequently placed in the avoidance situation: which is to say that the animal learns the active avoidance response.

We see, then, that this hypothesis for the basis of active avoidance learning incorporates the successful part of the Miller–Mowrer theory – the aversive nature of stimuli which have been followed by punishment – as a necessary step in the final acquisition of an active avoidance response. For relieving non-punishment can be detected only if the animal has first learned to expect punishment after such stimuli. But the final establishment and maintenance of active avoidance behaviour depend on the animal's learning to approach safety stimuli, including those which result from his own behaviour. We also see that, just as we can think of the extinction of approach behaviour as being based on the activity of the *punishment* mechanism (by the fear=frustration hypothesis), so the view of active avoidance learning proposed here is that it is based on the activity of the *reward* mechanism. If we follow Mowrer and call the activity caused in the reward mechanism by secondary rewarding stimuli 'hope', and the activity caused by stimuli associated with non-punishment 'relief', we can describe this view as the 'hope=relief' hypothesis.

Adoption of both the fear=frustration and the hope=relief hypotheses allows a very simple classification of reinforcing events. Such a classification is set out in Table 12.1; it should be read as stating that presentation of a reward, or omission or termination of a punishment, increase the probability of recurrence of behaviour which has preceded these events, while presentation of a punishment, or omission or

Table 12.1. *Change in probability of response as result of four basic reinforcing events*

	Presentation	Omission or termination
Reward	increase (approach learning)	decrease (extinction)
Punishment	decrease (passive avoidance)	increase (active avoidance)

termination of a reward, decrease the probability of recurrence of behaviour which has preceded these events. Adoption of the fear=frustration and the hope=relief hypotheses has the effect of reducing the fourfold classification of reinforcing events set out in Table 12.1 to a two-way classification; for these hypotheses assert that each of these four events is functionally equivalent to the corresponding event in the diagonally opposite corner of the table.

As we saw in Chapter 10, there is now a substantial body of evidence in support of the fear=frustration hypothesis. In contrast, the hope=relief hypothesis is as yet virtually innocent of contact, favourable or unfavourable, with the harsh world of experimental fact. Indeed, apart from the demonstrations of the general point that safety signals and secondary rewarding stimuli can both function as positive reinforcers (Chapter 11), only a few relevant experiments have been performed.

One kind of experiment uses the transfer-of-control design described in Chapter 10. In this design, Pavlovian conditioning is used to confer motivational significance upon an initially neutral CS, and the nature of the motivational effect is then assessed by the capacity of the CS to alter instrumental behaviour. Figure 12.3 illustrates the results obtained in two such experiments. In both, the instrumental behaviour employed was Sidman avoidance (Fig. 9.4) and the stimuli tested had been established as predicting the occurrence (CS+) or omission of a UCS (CS−) or they had been presented (as a control) in a random temporal relationship to the UCS (CS_0). In Weisman & Litner's experiment (left-hand panel of Fig. 12.3), the UCS was shock; in the experiment by Grossen *et al.* (right-hand panel), it was food. A comparison between the results of the two experiments shows that Sidman avoidance was reduced by presentation of both a CS− for shock (i.e. a safety signal), presumably because it reduced fear, and a CS+ for food. Thus, as predicted by the hope=relief hypothesis, the effects of stimuli predicting shock omission and food presentation

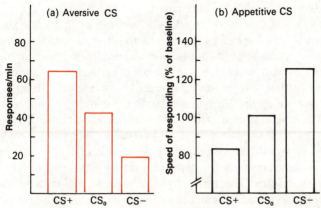

Fig. 12.3. The effects of superimposing various conditioned stimuli (CSs) on a baseline of unsignalled avoidance responding. The CS+ had previously been paired with an unconditioned stimulus (UCS); CS_0 had been uncorrelated with the presentation of the UCS (and thus provides a control condition against which to assess the effects of other CSs); and CS− had signalled the omission of the UCS. (a) UCS = shock; after Weisman & Litner (1969); (b) UCS = food; after Grossen *et al.* (1969). This figure is based on one in Mackintosh (1983).

were the same. (We have already remarked in Chapter 10 upon the other major feature of the results depicted in Fig. 12.3: the similarity of effect between the CS− for food and CS+ for shock, supporting the fear=frustration hypothesis.)

A second kind of experiment has made use of Kamin's blocking paradigm, described in Chapter 10. In this, CS_1 is first paired with one UCS, and then a compound of CS_1 and a second CS is paired with a second UCS. The degree of conditioning to CS_2 (measured subsequently to the compound-conditioning trials) provides a measure of the motivational similarity between the two UCSs. The more similar the two UCSs, the more CS_1 blocks conditioning to CS_2 (the blocking effect); conversely, if the animal treats UCS_1 as affectively opposite to UCS_2, one observes superconditioning – a greater degree of conditioning to CS_2 than under control conditions. (This pattern of findings is generally treated as indicating that the reinforcing value of the UCS depends on the degree to which it is surprising or unpredicted. Blocking occurs when the animal already expects UCS_2 or something very like it in the compound-conditioning phase; superconditioning occurs when the animal expects something quite unlike UCS_2.)

In a study using this paradigm, Goodman & Fowler first paired a flashing light with either food or the omission of food. Subsequently the light was combined in a serial compound with a clicker (20 seconds of clicker with the light accompanying the last 5 seconds), and the compound was followed by shock. The amount of fear conditioned to the clicker was assessed by its capacity to suppress licking for water. Acquisition of conditioned suppression was unaffected by compounding the clicker with the light; but resistance to extinction of suppression (when no further shocks were delivered) was changed. Relative to controls for which the light was presented for the first time during the compound-conditioning trials (so having no association with food), *greater* resistance to extinction was observed in the group for which the light had signalled food (superconditioning) and *less* resistance to extinction in the group for which it had signalled food omission (blocking). The latter effect is consistent with a large body of data reviewed in Chapter 10: animals treat the omission of food (frustration) as affectively similar to the delivery of shock. Conversely, the superconditioning effect suggests that the rats in Goodman & Fowler's experiment treated the delivery of food as affectively *opposite* to the delivery of shock, as the hope=relief hypothesis predicts. In a similar experiment Dickinson has also observed superconditioning, now affecting the acquisition of conditioned suppression: conditioning to a tone was faster if the tone was compounded with a light that had previously signalled food.

Goodman & Fowler's and Dickinson's experiments demonstrate superconditioning with an appetitive UCS_1 (food) and an aversive UCS_2 (shock). Superconditioning under the opposite conditions (aversive UCS_1 and appetitive UCS_2) has also been shown. Krank found that conditioning of the rabbit's response to water (jaw movements) proceeded faster when the test CS was compounded with a CS previously paired with paraorbital shock than under control conditions. This result suggests that a reward (water) is experienced as particularly rewarding if it occurs at a time when punishment (shock) is expected. An effect of this kind – perhaps we should call it the 'thrill effect' – might underlie the well-known propensity of people to find pleasure particularly acute when it is accompanied by a mild degree of fear (consider the popularity of downhill skiing).

In general terms Krank's observation of superconditioning is consistent with the existence of two opposing motivational systems, one subserving fear and frustration, the other subserving hope and relief. However, a second feature of his findings is less encouraging for this model. Given that a CS+ for shock increased conditioning to a compounded CS for water, the hope=relief hypothesis must predict that a CS− for shock (i.e. a stimulus predicting shock omission) will block such conditioning. But when Krank conducted this symmetrical experiment, he found no effect of a CS− for shock compared to a stimulus previously presented in random association with shock.

With the exception of this last, negative result then, the data described in this section (and in Chapters 10 and 11) support the postulate that there exist two basic motivational systems with generally opposing properties, one that mediates approach behaviour to secondary rewarding or safety signals, the other mediating behavioural inhibition in response to secondary punishing or frustrative stimuli.

Fight or flight

So far, we have discussed the conceptual nervous system for avoidance behaviour almost entirely from the point of view of learning. But we must also devote a little thought to the organisation of the unconditioned behaviour which an animal displays when it is actually exposed to a punishment. Broadly speaking, the animal will engage in either fight or flight. But, before discussing the conditions under which fight or flight occurs, we must first distinguish between two kinds of fight: the 'defensive attack' which is delivered by a frightened animal against a member of its own species or against a predator, and the 'predatory attack' delivered by a predator against its prey.

The distinction seems intuitively plausible, and some species adopt characteristically different modes of attack in the two cases. It is strongly supported by studies in which aggressive behaviour has been elicited by direct electrical stimulation of the brain. In the cat, stimulation of the lateral part of the hypothalamus produces a typical predatory attack on a rat: a silent, stalking approach, with hair sleeked back, and no sign of fear. Such stimulation is found to be rewarding by Olds' technique, and the aggressive behaviour produced is closely connected with the act of eating, which can also be elicited by stimulation of this part of the brain. Thus predatory attack seems to be essentially *approach* behaviour of the same kind as food-seeking or water-seeking. Defensive attack, in contrast, can be elicited by electrical stimulation of the medial part of the hypothalamus in both cat and rat. In the cat, the aggressive behaviour observed is accompanied by hissing and erection of the hair. This stimulation may also produce unconditioned escape behaviour and is found by Olds' technique to be punishing.

The difference between the two kinds of aggressive behaviour is well brought out in an experiment by Adams & Flynn. Cats were trained to escape from shock to the tail by jumping on to a stool. This response occurred immediately when electrical stimulation of the medial hypothalamus (which had produced defensive attack in earlier tests) was substituted for shock to the tail, but did not occur at all when stimulation of the lateral hypothalamus (which had produced predatory attack in the earlier tests) was used instead of tail-shock. Thus defensive attack is connected with fear and avoidance behaviour, while predatory attack is not.

The unconditioned response to a punishment, then, may be either defensive attack or attempts at escape. Now, as I mentioned at the end of Chapter 2, the classical conditioning of fear displays a peculiar feature in that the behavioural effects of conditioned and unconditioned stimuli are often diametrically opposed to each other. In particular, whereas the unconditioned stimulus produces a great increase in activity, either directed towards escape or attack, the conditioned stimulus usually elicits a pattern of tense, silent immobility. I suggested there that animals may come equipped with an innate response to stimuli which have been followed by unconditioned aversive stimuli and only need to learn, by conditioning, *which* originally neutral stimuli should be responded to in this way. In effect, we have now expanded this suggestion into the hypothesis that this innate response consists in switching in the punishment mechanism for passive avoidance which has been described earlier in this chapter. But it should be noted that this must mean that there is a separate punishment mechanism (the fight/flight system) for organising the unconditioned response to a punishment.

Now, if it were the case that some punishments tended to elicit fight and others flight, we would obviously not wish to speak of a single fight/flight system. However, experimental work on this problem makes it clear that whether fight or flight occurs does not depend on the actual punishment delivered but on other stimuli present at the time of punishment.

In particular, a large number of experiments conducted by Azrin and his collaborators show that (under conditions where effective escape is impossible) an electric shock to the feet produces unconditioned escape behaviour if there is nothing suitable in the animal's environment to attack; but if a suitable object is present, the most likely response to shock is an attack on it – even though it is in no way causally connected with the delivery of the shock. The most suitable object for attack is, of course, another animal; but attacks are also delivered against a model of an animal, a toy doll, or even a tennis ball. Thus, given the choice between ineffective flight and equally ineffective fight, animals appear to prefer fight: an experimental demonstration of the well-known fact that a cornered animal is a very dangerous beast indeed. The preference for fight under these conditions is so great that animals will even work to provide themselves with suitable stimuli to attack. For example, in one of Azrin's experiments, monkeys were able to pull a chain to present themselves with a ball: when and only when they were being shocked they pulled the chain at a high rate and bit the ball. However, if the experimenter provides the animal with the opportunity to perform an escape response which *is* effective in terminating shock, the animal rapidly comes to prefer this response to aggressive behaviour. If the escape response is made ineffective again, there is a return to aggressive behaviour.

There are many other factors which affect the probability of aggressive behaviour in response to punishment; but they all tend to the general conclusion that whether fight or flight occurs depends largely on the particular stimuli which are present in the animal's environment when punishment is delivered. Thus, rather than thinking in terms of two systems for reaction to different classes of punishment, it makes better sense to imagine a single fight/flight mechanism which receives information about all punishments and then issues commands *either* for fight *or* for flight depending on the total stimulus context in which punishment is received.

The conceptual nervous system and human emotion

We now have three separate mechanisms in our conceptual nervous system: one (the reward system) responds to signals of reward or non-punishment by activating approach behaviour; a second (the fight/flight system) responds to unconditioned punishment or non-reward by activating escape or aggressive behaviour; and the third (the behavioural inhibition system) responds to signals of punishment or non-reward by suppressing behaviour. These three mechanisms show an intriguing parallel to the modes of response that have emerged from studies of the autonomic nervous system in different emotional states. As we saw in Chapter 5, in spite of many attempts to find greater specificity, the best that appears possible is to distinguish, on the one hand, between positive and negative emotional states and, on the other, between negative emotional states that involve relatively more active or passive behaviour. It is interesting to speculate that positive emotional states, as reflected in the autonomic nervous system, may be due to the activity of the reward system, active negative emotional states to that of the fight/flight system, and passive negative emotional states to that of the behavioural inhibition system.

A number of more ambitious attempts to relate human emotions to these different CNS mechanisms have also been made. We shall limit our attention here to attempts to analyse fear or anxiety. These attempts may usefully be distinguished in terms of their reliance upon the behavioural inhibition or the fight/flight system: Jaak Panksepp identifies fear and anxiety largely with the activity of the fight/flight system; I identify them with the activity of the behavioural inhibition system; and the Brazilian scientist Federico Graeff has proposed what is in effect a synthesis of these two views.

Many of the arguments both in favour and against these different approaches to the modelling of human anxiety have turned upon the behavioural effects of the anti-anxiety drugs. It will be helpful, there-fore, if we first summarise briefly the effects of these drugs. (We have already considered some of their effects in earlier chapters, especially Chapter 10.) We shall limit ourselves to animal experiments, taking it for granted that these drugs reduce anxiety in Man (see Chapter 14). It is, of course, the latter aspect of their action that makes these drugs such important tools for the testing of animal models of human anxiety. If we suppose (a) that animals experience a state of anxiety homologous to the human state, and (b) that the anti-anxiety drugs reduce this state in animals as in Man, then we may argue from the behavioural effects of these drugs in animals to the kind of behaviour that constitutes anxiety in Man. These are strong assumptions, but we shall see below that there are reasons for accepting them.

The behavioural profile of the anti-anxiety drugs

The substances covered by the term 'anti-anxiety drugs' include three major classes: the benzodiazepines, the barbiturates, and alcohol. I have reviewed some hundreds of experiments in which these drugs have been given to animals under a wide variety of behavioural con-ditions, and there are few if any systematic differences between the effects of the three different classes of drug in the tasks I shall discuss here. The experiments have made use of many different species, ranging from goldfish to chimpanzees; again, there appear to be no important systematic differences in the results depending upon the species employed.

The behavioural profile of the anti-anxiety drugs, as this emerges from experiments with animals, is important (as we have already found) for several different arguments pursued in this book; it will become particularly important in the next chapter, where we shall see that certain kinds of damage to the brain produce a closely similar profile. For that reason, I have summarised the effects of the anti-anxiety drugs in a form (Table 12.2) that is more detailed than can be followed in the text.

Central to this profile is the impairment produced by the anti-anxiety drugs in passive avoidance behaviour, which we encountered in Chapter 10. In contrast, escape behaviour or simple, one-way active avoidance (e.g. to a safe box at the end of a straight alley) is unaffected by these compounds unless the dose is raised to a level at which non-specific sedation or motor incoordination occurs. In certain kinds of active avoidance tasks the anti-anxiety drugs even produce a facilitation of responding. This is seen for example when the avoidance response consists in pressing a bar in a Skinner box. Here, it appears that the drug counteracts the general inhibition of movement that is elicited by apparatus cues that have come to be associated with shock.

The clearest case of avoidance facilitated by the anti-anxiety drugs is in the shuttlebox (Fig. 4.8). This is almost certainly because the shuttlebox puts the animal into conflict between an active and a passive avoidance tendency. To avoid shock he must run to the now-safe side (active avoidance); but, because of its previous association with shock, this same side also elicits a tendency to inhibit running (passive avoidance). Anti-anxiety drugs therefore increase the rate of shuttling by weakening the opposing passive avoidance tendency. This analysis is supported by several lines of evidence, including some from the study of individual differences in fearfulness. Thus male rats are poorer at shuttlebox avoidance, but better at passive avoidance, than females (Chapter 7). Similarly, the Maudsley Reactive strain of rats, bred on a criterion of Open Field defecation to be especially fearful (Chapter 4), is worse at shuttlebox and better at passive avoidance than the low-fearful Nonreactive strain. This association in the Maudsley strains between defecation and shuttlebox avoidance could be accidental, due to particular combinations of genes present in the original population from which these strains were bred. That this is not so is shown by the results of two other breeding experiments. Bignami and Brush each bred strains of rats to be either good or bad at shuttlebox avoidance. In both cases, the poor avoidance strain turned out also to obtain higher defecation scores in the Open Field test. In addition, Bignami's 'Roman' low avoidance strain shows greater passive avoidance and shock-induced suppression of drinking than the Roman high avoidance strain.

Further light is cast upon the opposition between passive avoidance and shuttlebox avoidance by the results of an analysis of the genetics of shuttlebox avoidance behaviour conducted by Wilcock & Fulker.

Table 12.2. *Comparison between the behavioural effects of septal (S) and hippocampal (H) lesions and anti-anxiety drugs*

	1	2	3	4	5	6
	Section		Syndrome common to S and H lesions	Features in S, not H, Syndrome	Features in H, not S, syndrome	Effects of anti-anxiety drugs
Ch. 2	Ch. 6	Task				
1	1	Rewarded running, CRF	0			0
1	1	Rewarded bar-pressing, CRF		+		0
7	1	Rewarded bar-pressing, intermittent reinforcement	+			+
2	2	Passive avoidance	−			−
3	3	Classical conditioning with aversive UCS	0			0
3	3	On-the-baseline conditioned suppression	−			−
3	3	Off-the-baseline conditioned suppression		?	?0	?0
3	3	Taste aversion		?	−	?
4	4	Agonistic escape	+			?
4	4	Skilled escape	0			0
5	5	One-way active avoidance		−		0
5	6	Non-spatial active avoidance	+			+
5	7	Two-way avoidance	+			+
6	8	Threshold of detection of shock	0			0
	8	Movement elicited by shock	+			?
	9, 10, 13	Movement elicited by novel stimuli	+			?
	10	Distraction	−			?
	11	General activity		?	+	?
10	12	Exploration of novel stimuli	−			−
10	12	Spontaneous alternation	−			−
10	12	Open-field ambulation		−/+	+	+/−
10	12	Rearing	−			−
10	13	Habituation rate	?−			?
10	14	Open-field defecation		−		?
10	14	Emergence time		−	?	−
6	15	Shock-induced aggression		+	−	?
	15	Hyperreactivity syndrome		+		0
6	15	Social aggression	−			?+
10	16	Social interaction		+	?	?+
7	17	Resistance to extinction	+			+
7	17	Partial reinforcement extinction effect	−			−
7		Partial reinforcement acquisition effect		?0	?	−
7	18	Performance on DRL schedule	−			−

Table 12.2 (*cont.*).

1	2		3 Syndrome common to S and H lesions	4 Features in S, not H, Syndrome	5 Features in H, not S, syndrome	6 Effects of anti-anxiety drugs
Section Ch. 2	Ch. 6	Task				
7	18	Fixed-interval scallop		−	?+	−
8	19	Simultaneous discrim−ination	0			0
8	20	Successive discrimination	−			−
8	20	Single alternation		?	?−	−
	21	Spatial discrimination	−			?
8	22	Reversal learning	−			−
9	23	Double runway frustration effect	0			0
7	23	Crespi depression effect		?	?−	−

Note: +, Facilitation; −, impairment; 0, no consistent change; ?, insufficient data, If an entry in column 4 or 5 is accompanied by a question mark in the other column, the difference between the septal and hippocampal syndromes cannot be established because of insufficient data.

Source: This comparison is based on the information in the various sections of Chapters 2 and 6 in Gray (1982), as listed to the left of the table, together with the more detailed reviews given by Gray (1977) and Gray & McNaughton (1983).

These workers found that two separate genetic factors affect performance in the shuttlebox, both with directional dominance, indicating that they have probably been important for Darwinian survival. One of these factors promotes good shuttlebox performance late in acquisition, the other favours *poor* performance early in acquisition. Given the analysis of avoidance behaviour presented in Chapter 11 and developed further in this chapter, it is plausible to suppose that the latter factor reflects the activity of the behavioural inhibition system (mediating responses to warning signals early in acquisition), while the former factor reflects the activity of the reward system (mediating approach to safety signals late in acquisition).

In sum, insofar as escape and avoidance behaviour are concerned, the anti-anxiety drugs selectively impair passive avoidance, and their other effects are secondary to this change. The impairment in passive avoidance behaviour is not due to a loss of effectiveness of the punishing UCS itself. Measurement of the threshold electric current at which flinching or jumping is provoked has failed to reveal any sign of analgesia. At one time it was claimed that the benzodiazepines reduced aggressive behaviour. However, more recent data suggest that this effect is not reliably seen at doses without influence on general motor behaviour. At low doses, especially when given chronically, the benzodiazepines have even been reported to facilitate aggression, a change also seen after low doses of both alcohol and the barbiturates. Lynch *et al.* have suggested that this effect of chronic administration of low doses of benzodiazepines may be responsible for some cases of baby battering. It is possible that increases in aggressive behaviour, like the improvement in shuttlebox avoidance, are secondary to blockade of passive avoidance (i.e. to reduced fear of the opponent or of the consequences of aggression).

If the loss of passive avoidance is not due to changes in the effectiveness of the UCS, it must reflect some change in the behavioural control exercised by conditioned punishing stimuli. The process of classical conditioning is not generally affected by the anti-anxiety drugs, especially when discrete autonomically controlled responses (e.g. heart rate, defecation) are measured. However, more global measures of fear (such as conditioned suppression, the potentiated startle response, or reluctance to enter a box in which shocks have been received) have yielded a more confused pattern of results. Sanger & Joly, for example, recently found that chlordiazepoxide given to mice before they were shocked in a dark box reduced the time to enter the box on a test trial 24 hours later. Thus the data do not exclude an effect of the anti-anxiety drugs on the conditioning of fear, but this is unlikely to be the whole story. In addition, these substances reduce the expression of conditioned fear (that is, responses to secondary punishing stimuli) even when conditioning is complete. Rawlins and co-workers, for example, have reported a clear alleviation of conditioned suppression when chlordiazepoxide is administered to rats well after asymptotic training levels have been reached.

A similar picture emerges when we consider the effects of the anti-anxiety drugs in tasks involving non-reward. Neither the acquisition nor the performance of rewarded behaviour is systematically affected by these agents. The immediate response to unconditioned non-reward (as we saw in Chapter 10) is also unaffected. The best replicated example of this immunity is the double-runway frustration effect (Fig. 10.3). Thus, as in the case of paradigms using painful stimuli, the many effects of anti-anxiety drugs seen in experiments involving non-reward (see Chapter 10) are due to an alteration in the control of behaviour by secondary aversive (frustrative) stimuli. There are no reported experiments which allow one to judge whether the anti-anxiety drugs alter the conditioning process whereby secondary frustrative stimuli are formed. But there is clear evidence that these drugs can impair the control over behaviour exercised by such stimuli after they have been formed. Guillamon and I, for example, trained rats in the alley on a single alternating schedule of reward and non-reward, with the consequence that they came to run fast on rewarded trials and slow on non-rewarded trials ('patterned running'). After they had been trained for over 700 trials and had displayed consistent patterned running for over 300 trials, the rats were injected with sodium amytal. The drug increased running speed on non-rewarded trials without altering speeds on rewarded trials.

So far, then, we may conclude that anti-anxiety drugs selectively reduce responses to secondary aversive (punishing and frustrative) stimuli. To this simple summary we need to add only that these drugs also reduce responses to novelty. In elaborating upon this point we need to distinguish between two kinds of effect of novelty.

In the first place novel stimuli may act simply as general stressors. Thus a bright light in Broadhurst's Open Field test (Fig. 4.1) is not necessarily any more novel than a dim one, but it suppresses ambulation to a greater extent. Inhibitory effects of this kind can be seen in many situations. Thus an unfamiliar situation suppresses eating, drinking, and social interaction; and Sandra File has shown that, like Open Field ambulation, social interaction is suppressed more, the brighter the illumination. All these effects are reversed by the anti-anxiety drugs. File has in addition shown that novelty gives rise to an increase in plasma corticosterone levels (Chapter 5). This, too, is greater, the brighter the illumination and blocked by treatment with benzodiazepines.

The second way in which novel stimuli act is by drawing the animal's attention directly to themselves. Consider, for example, a task used by Jim Ison. The rat is first placed in the stem of a T-maze, the arms being blocked by transparent sheets of Perspex. One arm of the T is black, the other white, and the animal is allowed to inspect them for 3 minutes. The rat is then removed, the arms are changed so that both are now white or both black, the sheets of Perspex are taken away, and the rat is returned to the maze. The normal animal will enter the changed arm (thereby demonstrating an investigatory response to novelty) about 75% of the time. Animals given an anti-anxiety drug, however, choose between the two arms at random. Other experiments along similar lines (though they are not abundant) reinforce the conclusion drawn from this one: anti-anxiety drugs reduce the specific exploratory response to novelty.

Ison's experiment on the rat's response to stimulus change suggests that anti-anxiety drugs affect attentional processes. This has been confirmed in an elegant experiment by McGonigle and his co-workers.

They trained rats on a random 50% partial reinforcement schedule (see Chapter 10) to choose the positive cue (black or white) in a choice-box, choice of the negative cue never being rewarded. The rats were then shifted to a combined-cue discrimination (black vs. white *and* horizontal vs. vertical stripes, both cues being presented together) in which the old positive cue remained positive. During this stage of the experiment half the animals received sodium amytal and half placebo. Note that, since the problem could still be solved by attending simply to black vs. white, the animals did not need to learn anything about orientation of stripes. Whether they had done so or not was tested in a final transfer stage, conducted with only horizontal vs. vertical stripes and no drug. Other research has shown that a partial reinforcement schedule increases the spread of attention, that is, the animal learns about more separate features of its environment than does an animal given only rewarded trials. Thus McGonigle expected that the undrugged animals would show in the transfer test that they had learned to use the orientation of stripes to guide their choice response; and this is what he found. But the animals that had been drugged in the combined-cue phase of the experiment showed no learning about the added cue. From this and other related results one may infer that exposure to non-reward increases attention to novel cues and that this effect is blocked by the anti-anxiety drugs.

The behavioural inhibition system

If we add this inference to the other two behavioural effects of the anti-anxiety drugs which we noted in Chapter 10 – disinhibition of suppressed responding and increased level of arousal – we arrive at the model presented in Fig. 12.4. This figure may be read in three ways.

First, it is a convenient summary of the behavioural effects of the anti-anxiety drugs in animals. It is remarkable that it is possible to summarise in so simple a manner the great variety of results that have been obtained in very diverse experimental situations and with animals of many species.

Second, Fig. 12.4 presents a psychological model of the action of the anti-anxiety drugs. This model states: (1) that there is a behavioural inhibition system, which (2) responds to signals of punishment, signals of non-reward, or novel stimuli by (3) inhibiting ongoing behaviour, increasing readiness for action (arousal level), and increasing attention to environmental stimuli, and that (4) the anti-anxiety drugs reduce anxiety by antagonising the activity of the behavioural inhibition system. In common sense terms this is equivalent to defining anxiety

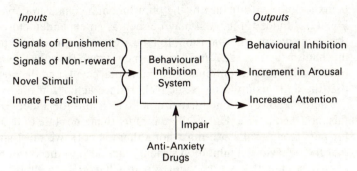

Fig. 12.4. The behavioural inhibition system.

(i.e. the state caused by activity in the behavioural inhibition system) as an emotion that is elicited by threats of punishment or failure and by novelty, and which causes the animal to stop, look and listen, and prepare for vigorous action. This, I think, is a definition which would meet with ready acceptance from the man in the street, especially if he has been lucky enough to escape the Freudian infection. Note that Fig. 12.4 is based entirely on experiments with animals: the fact that it contains a plausible model of *human* anxiety is strong evidence for the assumptions which, as noted above, must underlie any attempt to base a theory of anxiety upon animal experiments involving the anti-anxiety drugs: namely, that animals and human beings experience a similar state of anxiety, and that in both cases this state is reduced by these drugs.

Third, the model shown in Fig. 12.4 is a guide to the kind of neural system which would be expected to mediate anxiety – a guide that we shall find useful in the next chapter. For, if there *is* a behavioural inhibition system, it can only exist in the brain. Note also that since our model of anxiety appears to be equally applicable to species ranging from higher fish to higher primates, anxiety and its neural substrate must be phylogenetically old and stable. This inference is consistent with the evidence from neurochemical studies of the mode of action of the benzodiazepines. Studies of this kind have shown the existence of a specialised receptor, located on neuronal membranes, with high affinity and specificity for these drugs. This 'benzodiazepine receptor' (considered in more detail in the next chapter) is present in essentially the same form in higher fish and all mammals, including Man.

This, then, is the behavioural inhibition system as inferred from experiments on the behavioural effects of the anti-anxiety drugs. It bears a close resemblance to the punishment mechanism depicted in Fig. 12.2. Indeed, Fig. 12.4 may be seen as an expanded close-up of the punishment mechanism. To the features attributed to this mechanism in Fig. 12.2 it adds one new input (novelty), one new output (increased attention to environmental cues), and the postulate that the anti-anxiety drugs selectively reduce activity in this system.

There is, however, one further input which must be added to make this model complete. Unless there is more than one state of fear (and, for parsimony's sake, we must hope there is not), we must suppose that the behavioural inhibition system reacts also to the various types of fear stimuli that we encountered in Chapter 2, notably, stimuli associated with special evolutionary dangers (Seligman's 'prepared stimuli') and stimuli arising during the course of social interaction (see also Chapter 3). Since both these kinds of stimuli largely (though not entirely) derive their fear-eliciting properties from innate factors, I have labelled the appropriate input in Fig. 12.4 'innate fear stimuli'. Note, however, that this input cannot be inferred from the behavioural effects of the anti-anxiety drugs, because the relevant experiments appear not to have been done. If at some time in the future they are, and responses to innate fear stimuli turn out not to be affected by the anti-anxiety drugs, parsimony and the behavioural inhibition system would face a problem.

Modelling human anxiety

Having taken a look at the behavioural effects of the anti-anxiety drugs, let us return to the question of how best to model human anxiety. The reasons that motivate my own answer to this question will, I think, be clear from the preceding section: if anti-anxiety drugs reduce anxiety by impairing the activity of the behavioural inhibition system, then it is the behavioural inhibition system that must give rise to the state of anxiety. What of the arguments used by Panksepp to motivate his answer, namely, that anxiety is due to activity in the fight/flight system?

The basis of Panksepp's approach is to identify fear and anxiety as the emotion corresponding to flight and escape behaviour. More particularly, Panksepp starts from the fact that it is possible to elicit such behaviour by electrical stimulation of a number of regions of the brain, including the hypothalamus (we shall examine these regions in greater detail in the next chapter). He then proposes that activity in the brain circuits that traverse these regions, and whose electrical stimulation produces the observed behaviour, constitutes anxiety.

Now, in the absence of other information, the choice between response suppression (mediated by the behavioural inhibition system) and escape behaviour (mediated by Panksepp's hypothalamic circuits) as the basis for an analysis of anxiety is arbitrary. But we do have other information: we know that the drugs that reduce self-reported anxiety in Man impair response suppression but do not impair escape behaviour (when these are elicited by conventional environmental stimuli rather than brain stimulation). This information must rather strongly bias our choice in favour of the behavioural inhibition system as the basis of human anxiety.

However, a further possibility has emerged from Graeff's important experiments. He too has studied escape and defensive aggression elicited by brain stimulation, concentrating on the central grey in the midbrain as well as the hypothalamus (see Figs. 13.22 and 13.23). In contrast to the effects of anti-anxiety drugs on this type of behaviour when it is elicited by conventional environmental stimuli (Table 12.2), Graeff (confirming several earlier reports) finds that these compounds reliably reduce both aggressive and learned (though not unconditioned) escape behaviour elicited by brain stimulation. Furthermore, he has been able to identify the probable reason for the discrepancy between studies using environmental and brain stimulation, respectively. It appears that the brain systems subserving response suppression include a serotonergic projection (i.e. a pathway using serotonin as its neurotransmitter) that inhibits the output from the central grey fight/flight neurons. Other evidence indicates that the anti-anxiety drugs reduce activity in this serotonergic pathway. Thus, when conventional stimuli are used to elicit fight/flight behaviour, these drugs produce two effects that tend to cancel each other out: by reducing serotonergic inhibition of the central grey they increase fight/flight behaviour, but by a direct action on the central grey they decrease it. When brain stimulation is used, in contrast, the first of these two effects is largely by-passed, so that the anti-anxiety drugs act only to decrease fight/flight behaviour.

With these results in mind, let us consider again the way in which the behavioural inhibition system must operate. The primary function of this system is to suppress behaviour that threatens to produce an unwelcome outcome. It follows that the system can only usefully be put to work if some other system is producing behaviour that needs to be suppressed. We have identified two systems that can do this: the reward and fight/flight systems. To these should be added a number of systems that mediate various forms of unconditioned appetitive behaviour (eating, drinking, copulation, etc.). These considerations imply that at any one time when the behavioural inhibition system is active, the total emotional experience will be an amalgam of the emotional effects of activity in this system and those of activity in the system whose output is under inhibition. It is reasonable to suppose that such an amalgam will be maximally negative in affective tone when the system inhibited is the fight/flight system.

The Maudsley strains

It is instructive to apply this analysis to the differences that selective breeding for Open Field defecation has wrought in the behaviour of the Maudsley Reactive and Nonreactive strains of rats (Chapter 4). For the most part these map very well onto the differences that one sees between undrugged rats and rats that have been injected with one of the anti-anxiety drugs. Thus Reactive rats show greater passive avoidance and conditioned suppression but less ambulation and poorer shuttlebox avoidance than Nonreactive rats (of Table 12.2). This pattern of results is consistent with the hypothesis (adumbrated above in the discussion of shuttlebox avoidance) that the Reactive rat has a more effective behavioural inhibition system than his Nonreactive counterpart. At the same time, the fact that the contrasts which distinguish the two Maudsley strains distinguished in addition drugged from undrugged animals must increase our confidence that the behavioural profile listed in Table 12.2 is not just an artefact due to the existence of multiple but functionally unrelated drug effects. Confidence in this point is further increased by the fact that the same pattern of contrasts applies also to the two sexes (Chapter 7).

However, the differences between the Maudsley strains are in important respects wider than those that distinguish drugged from undrugged animals. Most critical is the observation that Reactive rats escape from shock faster than Nonreactive animals. This is not matched by the effects of the anti-anxiety drugs, which reduce escape speed only at sedative doses unless escape is motivated by brain stimulation (see the discussion of Graeff's work above). This type of result is also obtained in experiments using frustrative non-reward. Savage & Eysenck measured the size of the double-runway frustration effect (Fig. 10.3) in the Maudsley strains: Reactive rats escaped from non-reward in the first goalbox faster than Nonreactives. (Note that yet again fear=frustration: Reactives are faster at escaping from both shock and non-reward.) But anti-anxiety drugs do not alter the double-runway frustration effect, in spite of many attempts to show that they do. These findings suggest that whereas (except when aversive brain stimulation is used) the anti-anxiety drugs affect only the behavioural inhibition system, selective breeding on the criterion of defection in the Open Field has generated differences in both the behavioural inhibition system and the fight/flight system.

This inference is supported by two further considerations. First, as we saw in Chapter 7, there are striking parallels between the behavioural and physiological contrasts that distinguish, on the one hand, the two Maudsley strains and, on the other, male and female rats. Female rats appear to have a weaker behavioural inhibition system than males (less passive avoidance, less freezing, better shuttlebox avoidance, etc.). But, in addition, they are less aggressive than males (whereas anti-anxiety drugs do not reliably reduce aggressive behaviour unless this is elicited by aversive brain stimulation). This suggests that the female rat also differs from the male in having a weaker fight/flight system. Recall, however, that the pattern of sex differences observed in Man (and probably other primates) differs from that seen in the rat (Chapter 7). The sex difference in aggressive behaviour continues to take the same direction (females less aggressive than males), but the direction of the difference in fearfulness is reversed (female primates appear to be more fearful than males). This variation in the pattern of sex differences with species emphasises the fundamental separation between the behavioural inhibition and fight/flight systems. The fact that the breeding of the Maudsley strains and the sex difference in the rat have affected both systems should not be allowed to obscure this separation.

Second, it is unlikely that the behavioural inhibition system exercises control over the defecation response which was used to establish the Maudsley strains and which also differentiates the two sexes. It is difficult to investigate this issue with the anti-anxiety drugs, since these have direct constipatory effects that hamper interpretation of defecation scores. However, as we shall see in the next chapter, there are close parallels between the behavioural effects of anti-anxiety drugs and those of lesions to two brain regions, themselves closely related to each other, the septal area and the hippocampal formation. This communality of effect has given rise to the hypothesis that the septo–hippocampal system forms a major part of the neural substrate of the behavioural inhibition system. Now, in most relevant tasks the behavioural effects of septal lesions closely resemble those of hippocampal lesions. However, Open Field defecation constitutes a clear exception to this rule: septal lesions reduce this, but hippocampal lesions leave it unchanged. It is unlikely, therefore that Open Field defecation scores directly reflect the activity of the behavioural inhibition system.

From these various arguments, then, it would seem reasonable to conclude that: (1) the behavioural inhibition and fight/flight systems are separate from each other; (2) the anti-anxiety drugs primarily affect the former (except when aversive brain stimulation is used to elicit fight/flight behaviour); and (3) the Maudsley strains and the two sexes (in the rat) differ in both these systems.

Interactions between systems

We now have a plethora of systems: a number of mechanisms for dealing with the unconditioned effects of rewards (food, water, etc.) that we have not specified further, since they are not germane to the concerns of this book; a fight/flight system for dealing with the unconditioned effects of punishment and non-reward; and the behavioural inhibition and reward systems depicted in Fig. 12.2 for dealing with the effects of conditioned aversive and appetitive stimuli, respectively. We conclude this chapter by considering various ways in which these systems are likely to interact with one another.

First, it will be seen that Fig. 12.2 includes reciprocally inhibitory links between the reward and behavioural inhibition systems: that is, activity in each of these systems tends to suppress activity in the other. These links are necessary to account for the pattern of change observed over successive trials in an approach–avoidance conflict of the kind we examined in Chapters 9 and 10. In particular, in Gray & Smith's mathematical treatment of the ideas contained in Fig. 12.2 (see also Fig. 10.1), they are used to explain the partial reinforcement acquisition effect (Fig.10.10). More informally, when an animal is in an approach–avoidance conflict it is typically observed that with continued exposure to the situation approach behaviour comes to dominate over passive avoidance or vice versa. The reciprocally inhibitory links between the reward and behavioural inhibition systems can generate this type of effect by gradually increasing the degree to which the initially more powerful input (reward to punishment or vice versa) comes to dominate the less powerful.

Other data that are consistent with such reciprocal inhibition are the results of the superconditioning experiments described earlier in the chapter. The occurrence of superconditioning implies that a CS for shock depresses the activity of the reward mechanism below its baseline level (Krank's experiment), and that a CS for food has a similar effect upon the activity of the punishment mechanism (Goodman & Fowler's experiment).

A third type of experiment that provides relevant data employs the so-called 'retardation' paradigm. In this, one first trains the animal to associate either an aversive or an appetitive UCS with a given CS and then attempts to condition the same CS to a UCS of the opposing kind. The initial expectation for this type of experiment was that prior conditioning with, say, shock would retard subsequent conditioning of the same CS to a food UCS and vice versa (hence 'retardation experiment'). Early results supporting this prediction were reported by Konorski & Szwejkowska, who showed (but in a single dog) that development of the salivary response to a CS paired with food was retarded if the CS had previously been paired with shock to the paw; and conversely that development of the paw-flexion response was retarded if the CS had previously been paired with food. Subsequently, however, it has become clear both that prediction for this type of experiment is more complicated and that the results do not fully agree with those reported by Konorski.

In predicting the outcome of a retardation experiment one can follow three separate lines of argument. First, on the general view that there exist two opposing motivational systems, one can make the original retardation prediction, as did Konorski. Second, one can start from the hypothesis that UCSs reinforce more strongly, the less they are already expected. On this view the presentation of food when shock is expected (as a result of prior conditioning of the CS), or of shock when food is expected, should be particularly surprising, and so conditioning with the new UCS should be particularly effective. (This is the same argument that predicts Goodman and Fowler's super-conditioning result in the compound-conditioning experiment described in a previous section.) Third, one can start from the view that prior conditioning of any kind will increase attention to the CS and/or its ease of associability: the animal knows already that the CS is a significant event. The latter two arguments predict, in opposition to Konorski, that conditioning to a shock UCS will proceed faster if the CS has previously been paired with food (or vice versa).

These equally plausible arguments are not necessarily incompatible with each other. As it happens, recent results fit a definite pattern which fails to support one to the exclusion of the others. When initial conditioning is to a shock UCS and later conditioning to a food UCS, results like Konorski's are obtained: later conditioning is retarded. But when initial conditioning is to a food UCS and later conditioning to shock, the opposite is found: later conditioning is improved relative to controls. It is as though fear conditioning takes precedence: if a CS is first associated with pain, the animal takes a long time to overcome the initial conditioning; while if the second association is with pain, the animal is quick to detect the change in the significance of the CS. We saw a similar pattern of results in Chapter 9, in our discussion of Lovibond & Dickinson's experiments on counter-conditioning. The general adaptive value of this mode of behaviour will be apparent, but the mechanism by which it is achieved is unknown. It may perhaps depend upon a shift in the balance between the different processes emphasised in the three hypotheses outlined above.

Results of this kind imply that though we may wish to retain the notion that the reward and behavioural inhibition systems exert reciprocal inhibition upon each other, the inhibitory link from the behavioural inhibition to the reward system is stronger than the reverse link.

Reciprocally inhibitory links appear also to connect the systems that deal with unconditioned aversive and appetitive stimuli. As we saw in Chapter 9, counter-conditioning experiments have shown that when food is followed closely by shock, eating is strongly suppressed; while if the order of these events is reversed, the shock may lose some of its capacity to act as a punishment and come instead to elicit food-related behaviour. Indeed, these effects seem to be stronger than those that are obtained when shock is presented earlier in the chain of food-seeking behaviour, or when food is used to counter-condition stimuli associated with shock rather than shock itself (see the sections in Chapter 9 dealing with the effects of punishment and resistance to punishment). Thus, depending upon the temporal sequence in which they are activated, the systems mediating consummatory behaviour appear to be able to inhibit the fight/flight system or vice versa.

A final set of links that must be added connects the behavioural inhibition system to both the unconditioned mechanisms mediating consummatory behaviour and the fight/flight system. These links are both inhibitory and (via the arousal mechanism; Fig.12.2) excitatory. Consummatory behaviour (eating, drinking, etc.) may be either reduced or increased in the presence of fear stimuli (see the section on drive summation in Chapter 10). Inhibition of responses to painful stimuli by conditioned fear stimuli was described in Chapter 2; as we saw, this may be mediated by the conditioned release of endogenous opiates. Facilitation of the response to unconditioned punishment by conditioned fear stimuli is demonstrated in the potentiated startle reflex: a rat presented with a tone previously associated with shock

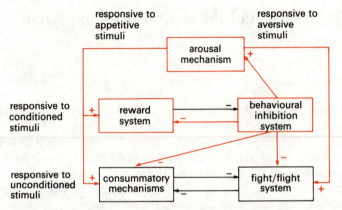

Fig. 12.5. Proposed inhibitory (−) and excitatory (+) interactions between the various components of the conceptual nervous system identified in this chapter. The reward and behavioural inhibition systems respond to conditioned reinforcing stimuli; the fight/flight system and consummatory mechanisms respond to unconditioned reinforcing stimuli. The reward system and consummatory mechanisms respond to appetitive stimuli; the behavioural inhibition and fight/flight systems respond to aversive stimuli. The behavioural inhibition system and the associated arousal mechanism are shown in red. For further explanation, see text.

will show a greater startle reflex to shock than under control conditions.

These various relationships between the different systems outlined in this chapter are illustrated in Fig. 12.5. This figure completes our catalogue of aids for the excursion which we shall take next into the real brain. It has proved possible to compress quite a large number of phenomena, as well as a variety of principles of learning, into two small diagrams (Figs. 12.2 and 12.5), though, to be sure, these diagrams presuppose the laws of classical conditioning, first established by Pavlov during the first few decades of this century, and still the most important part of the theory of learning.

13 Fear and the central nervous system

In one sense the psychologist's job is finished when he has constructed a conceptual nervous system of the kind which we drew up in Chapter 12. It is then a matter of comparatively minor interest to him (though not, of course, to the physiologist or the neurosurgeon) *which* part of the real brain performs the various functions which must be performed. But, of course, we can never be sure that our conceptual nervous system has been deduced correctly until we have demonstrated that the real nervous system is constructed along the same lines. And, even at an early stage in the construction of the conceptual nervous system, the tentative identification of one of the components of our model with a particular structure in the real brain may help us to devise powerful experimental tests of the model. In this chapter we shall consider what is known about the involvement of various structures in the brain in fear, avoidance learning, and related forms of behaviour, and see if we can get any clues about the functions discharged by these structures. (By the word 'structure', I do not necessarily mean one particular anatomical locus in the brain; a structure which discharges a function of interest to us might well consist of a number of diverse anatomical loci with extensive neural interconnections with one another.)

What, then, does the brain need to have if it is to match our conceptual nervous system?

First, it must obviously have mechanisms to receive and analyse the stimuli which are emitted by the world around it. In general, stimulus reception involves an 'end organ', specialised for response to a particular form of physical energy (e.g. the eye and the retina in the case of visible light) and pathways for conduction of the resulting nervous impulses to a particular part of the cerebral cortex (which is where the most complex forms of sensory analysis are carried out). The major route by which these impulses reach the cortex is via a relay station in a part of the brain known as the thalamus.

The arousal mechanism

There is, however, a second major route by which sensory stimuli affect the activity of the brain. This route goes by way of the ascending reticular activating system (ARAS), a network of short fibres which interconnect extensively with one another and run through the centre of the lower part of the brain from the bulb at the top of the spinal cord to the thalamus (Fig. 13.1). All levels of this system receive inputs branching off from all the main sensory tracts. As a result, sensory stimulation is conveyed upwards to the cortex by this route as well. However, the function discharged by this stimulation is not to convey precise information about the quality of the sensory input (this is done by the direct thalamo–cortical route), but apparently to vary the general level of alertness and excitement.

Stimulation of the part of the ARAS in the midbrain (the 'reticular formation') via implanted electrodes wakes up a sleeping animal and produces behavioural signs of alertness in a relaxed one. At the same time, there are striking changes in the electroencephalogram (EEG), which records the electrical waves which are broadcast by the brain. The frequency and amplitude of these waves change according to what the animal is doing. The pattern which is seen in an alert animal consists of a mixture of frequencies (therefore it is 'desynchronised'), mostly rather fast, and of comparatively low amplitude. In a state of relaxation, by contrast, the EEG shows a rather regular ('synchronised') pattern of slower and higher-amplitude waves; the 9–11 Hz 'alpha' waves which can be recorded from the occipital cortex of a relaxed human being are of this kind. As the animal gets drowsy, this pattern in turn is replaced by still slower and larger-amplitude

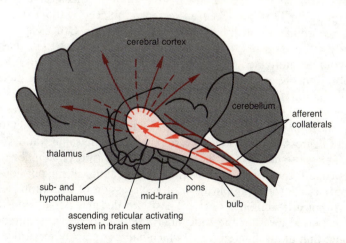

Fig. 13.1. The ascending reticular activating system. Outline of a cat's brain showing the distribution of afferent collaterals to the reticular formation.

waves, this time in an irregular pattern. Stimulation of the midbrain reticular formation via implanted electrodes replaces an EEG pattern of drowsiness or relaxation with the desynchronised fast waves of alert attention. Lesions in this area, on the other hand, can produce a permanently comatose animal.

The general theory that behaviour varies along a dimension of arousal or intensity (independent of the *direction* of behaviour) between, say, sleep and extreme excitement had been developed on purely behavioural grounds by such writers as Freeman and Duffy during the 1930s and 1940s. The Hullian notion of 'general drive' is a version of such a theory. As we saw in Chapter 10, there is good evidence that both punishment and frustrative non-reward can increase the vigour of ongoing appetitive behaviour, as the general drive notion leads us to expect. In Chapter 12 this notion was translated in the block diagram of our conceptual nervous system (Fig. 12.2) into an 'arousal' mechanism which receives inputs from the 'punishment' mechanism and is then capable of facilitating either approach behaviour or passive avoidance behaviour. We now take one further step and suppose that this arousal mechanism lies in the ARAS and especially the midbrain part of the ARAS.

Many of the experiments which we regarded in Chapter 10 as showing that punishment does in fact add to the vigour of ongoing appetitive behaviour were based on measurements of running speed. It is encouraging for our identification of the arousal mechanism with the reticular formation, therefore, that experiments by Sterman & Fairchild have shown that electrical stimulation of this structure in cats running for food produced an increase in speed without disrupting the coordination or direction of their behaviour. Other experiments have also supported the view that the ARAS can act to increase the vigour of motivated behaviour. Fuster found that monkeys reacted more quickly in a visual discrimination task as a result of electrical stimulation of the reticular formation; Isaac observed a similar decrease in reaction time in the unconditioned foreleg flexion response to shock in the cat; and Sheard & Flynn were able to facilitate the cat's aggressive behaviour towards the rat in the same way.

The dense network of short interconnected fibres that make up the reticular formation is recalcitrant to both anatomical and physiological analysis. However, application of a succession of ever more refined anatomical techniques over the last two decades has demonstrated that buried away in the reticular formation there are a number of long, thin unmyelinated fibres which had escaped earlier notice. It is

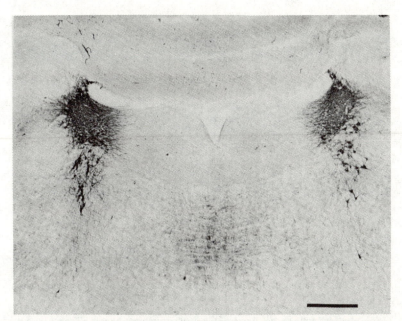

Fig. 13.2. The locus coeruleus. The photograph shows a transverse section through the rat brainstem at the level of maximal cross-sectional extent of the locus coeruleus, stained (dark cells) for the enzyme, dopamine-β-hydroxylase (which catalyses the synthesis of noradrenaline from dopamine). Bar = 500 μm. From Grzanna & Molliver (1980).

possible that at least some of the arousal functions that I have attributed to the ARAS are in fact mediated by these longer-axoned cells.

Among the principal candidates for this role are noradrenergic neurons (i.e. neurons that use noradrenaline as their transmitter) which originate in the locus coeruleus, a brainstem nucleus illustrated in Fig. 13.2. There are only about 1500 cells in this nucleus in the rat, but these innervate widespread regions of the brain, including much of the neocortex and most of the limbic system (Fig. 13.11). To achieve this feat, each cell body gives rise to several bifurcating axons destined for separate regions of the brain. The locus coeruleus fibres which target the forebrain travel in the dorsal ascending noradrenergic bundle (Fig. 13.3). In addition, other locus coeruleus fibres travel to the cerebellum and down into the spinal cord (Fig. 13.3).

A number of investigators have shown that electrical stimulation of the locus coeruleus, or direct application of noradrenaline to a

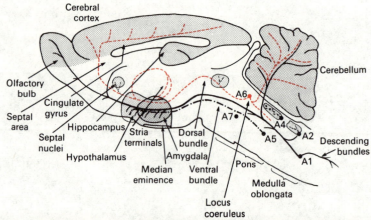

Fig. 13.3. Sagittal representation of the rat brain, showing the principal ascending and descending noradrenergic pathways. Cell bodies in the locus coeruleus (A6) give rise to pathways (red dashed lines) innervating all cortical areas of the brain. The dorsal bundle arising from A6 also innervates areas of the amygdala and anterior hypothalamus, while a short descending pathway innervates lower brainstem nuclei. Cell group A1 (solid lines) gives rise to a descending bulbo-spinal pathway and a major ascending ventral pathway, which follows the course of the medial forebrain bundle all the way to the olfactory bulb. This ventral bundle, which comprises ascending fibres from cell groups A5 and A7 (dots and dashes) as well as from A1 and A2, gives off branches to the lateral mammillary nuclei, to the lateral and ventral hypothalamus (the latter providing terminals to the median eminence and infundibulum), and to large parts of the limbic forebrain (including the anterior medial amygdaloid complex, the ventral medial septum, and cingulum). Shaded areas indicate regions of noradrenergic terminals. From Livett (1973).

target brain region served by the dorsal noradrenergic bundle, has two effects: the spontaneous firing rate of neurons in the projection area of the locus coeruleus is reduced, but their response to stimulation of other afferent pathways to the same region is increased. In consequence, the signal-to-noise ratio of the target organ with respect to the non-noradrenergic afferent is increased. This kind of effect is exactly what one would expect of an arousal system, that is, the locus coeruleus appears to boost the activity of other, widely distributed brain systems rather than itself subserving some specialised function. We shall see later in the chapter that other evidence implicates the locus coeruleus particularly in those aspects of arousal that are important in states of anxiety.

Pain

As we saw in Chapter 2, the notion (proposed by Watson) that pain is one of the innate stimuli for fear is not tenable. Thus the sensory mechanisms that give rise to the perception of pain do not require detailed attention in this book (for an excellent introductory account, see Melzack & Wall's book *The Challenge of Pain*). Pain does concern us, however, at two points. First, it is a primary stimulus for fight/flight behaviour; thus we need to enquire how the sensory mechanisms for pain perception make contact with the mechanisms that subserve this type of behaviour. Second, as noted earlier in the book, there is evidence that fear may inhibit pain, or at least those reactions by which we assess pain; thus we need also to enquire how the behavioural inhibition system makes contct with the mechanisms for pain perception and/or the fight/flight system. Both these issues will occupy our attention later in the chapter. Here, we must consider briefly the manner in which pain signals reach the brain.

Consider the case in which a painful stimulus is applied to the surface of the body. The skin contains six anatomically distinct specialised receptors, as well as much larger numbers of free nerve endings (Fig. 13.4). These different kinds of end organs transmit nerve impulses to the spinal cord by way of three types of nerve-fibres: the A-beta fibres are large in diameter, myelinated, and fast-conducting; the A-delta (small and myelinated) and C (small and unmyelinated) fibres are much slower in conduction speed. It used to be thought that some free nerve endings act as specialised pain receptors, responding only to intense stimuli and activating in particular the slower-conducting afferents to the spinal cord. From there, pain sensation was supposedly sent on unchanged to pain centres in the brain. To service this latter part of the communication system, the spinal cord possesses no less than six ascending channels. Three of these (the spinoreticular, paleospinothalamic, and propriospinal systems) are phylogenetically old, are slow-conducting, and course medially through the brainstem (Fig. 13.5); the other three (the neospinothalamic, spinocervical, and dorsal-column postsynaptic systems) are phylogenetically newer, are fast-conducting, and course laterally in the brainstem (Fig. 13.6). The three slower-conducting pathways project in the main to the reticular formation in the brainstem, and particularly to an area around the cerebral aqueduct known as the central or periaqueductal grey (Fig. 13.5), though some fibres also penetrate as far as the thalamic portion of the reticular formation. The three faster-conducting pathways mainly target the ventrobasal complex of the thalamus (Fig. 13.6),

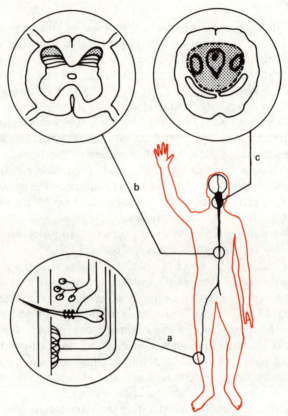

Fig. 13.4. Schematic representation of the receptors and projection pathways of the somatic sensory system. (A) The diagram of the skin shows widely branching free nerve endings (which produce overlapping receptive fields) as well as some specialised end organs. The fibres project to the spinal cord. (B) The cross-section of the spinal cord shows the laminae (layers) of cells in the dorsal horns which receive sensory fibres and project their axons toward the brain. The cross-hatched area represents the substantia gelatinosa (laminae 1 and 2). (C) The brainstem (lower part of the brain) receives a large somatosensory input, and it projects to higher as well as lower areas of the central nervous system. The cross-hatched area represents the reticular formation. Below it on each side is the medial lemniscus. The spinothalamic projections – which are shown within the reticular formation – lie above the lemniscal tracts. From Melzack & Wall (1983).

but some fibres also project to the midbrain and thalamic reticular formation.

We now know that this relatively simple picture is wrong on several counts. Much of the new knowledge in this field has been organised in the 'gate-control' model of pain perception proposed in 1965 by

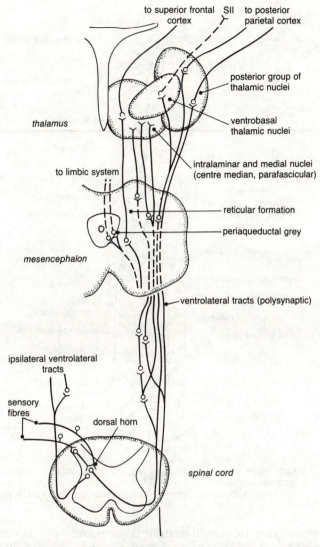

Fig. 13.5. The slowly conducting somatosensory projection pathways. The breaks in the projection lines represent multi-synaptic connections. The proprioscinal fibres are not shown, but consist of short fibres which are distributed throughout the cord. From Melzack & Wall (1983).

Ronald Melzack and Patrick Wall. These workers have emphasised two aspects of pain perception. First, pain depends upon the total pattern of sensory stimulation, not principally upon the firing of specialised pain receptors. Second, the sensation of pain is in addition

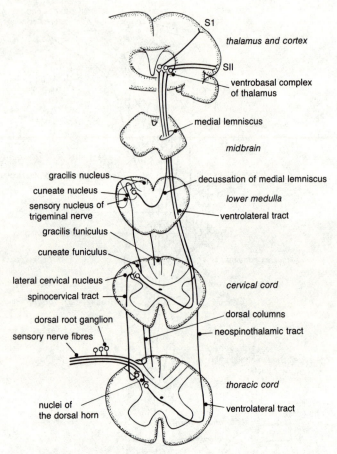

S1

thalamus and cortex

SII

ventrobasal complex
of thalamus

medial lemniscus

midbrain

gracilis nucleus — — decussation of medial lemniscus
cuneate nucleus — *lower medulla*
sensory nucleus of
trigeminal nerve — ventrolateral tract
gracilis funiculus
cuneate funiculus
lateral cervical nucleus — *cervical cord*
spinocervical tract
— dorsal columns
dorsal root ganglion — neospinothalamic tract
sensory nerve fibres
thoracic cord
nuclei of — ventrolateral tract
the dorsal horn

Fig. 13.6. The rapidly conducting somatosensory projection pathways. The three main projection pathways are the dorsal column–medial lemniscal pathway, the dorsolateral tract (of Morin), and the neospinothalamic tract. The lower sections are shown on a larger scale than the upper sections. From Melzack & Wall (1983).

subject to powerful central controls that descend from the brain into the spinal cord. The most recent version of their theory is illustrated in Fig. 13.7.

According to this model, the sensation of pain depends upon the net output from spinal 'pain transmission' cells (labelled 'T' in Fig. 13.7). The output of these cells is in the first instance determined by the total pattern of impulses reaching the spinal cord. This pattern is determined both by the specific characteristics (tactile, thermal, pressure, etc.) of the applied stimulus and by its noxious features (signalled

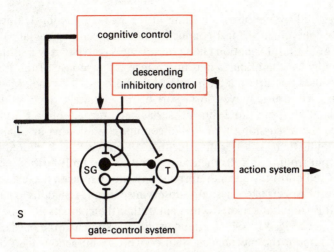

Fig. 13.7. Melzack & Wall's gate-control theory, Mark II. The model includes excitatory (open circle) and inhibitory (filled circle) links from the substantia gelatinosa (SG) to the transmission (T) cells as well as descending inhibitory control from brainstem systems. The round knob at the end of the inhibitory link implies that its action may be presynaptic, postsynaptic, or both. All connections are excitatory, except the inhibitory link from SG to T cells. From Melzack & Wall (1983).

in part by receptors that respond to stimulation intensity). Both types of stimulation, specific and noxious, excite spinal cells that receive inputs from the centre of the stimulated region; but in addition, cells that receive inputs from a wide area surrounding the centre are inhibited. This 'inhibitory surround' probably provides the basis for techniques, such as acupuncture, in which relatively gentle cutaneous stimulation, applied simultaneously with a second stimulus that normally causes pain, is used to lessen this pain. This complex pattern of messages, both excitatory and inhibitory, is conveyed to the cord by small-diameter (S) and large-diameter (L) fibres (see Fig. 13.7). The S fibres, known to be strongly activated by injury, excite the T cells directly; but in addition they tend to open a gate (so enhancing pain transmission more generally) located in the substantia gelatinosa of the cord ('SG' in Fig. 13.7; and see Fig. 13.4). The L fibres, which convey principally specific information about the stimulus, also excite the T cells; but they tend to close the gate, by exciting cells in the substantia gelatinosa which themselves inhibit transmission in the T cells.

In addition to these inputs, direct and via the substantia gelatinosa, from S and L fibres, the output of the T cells is subject to descending controls from the brain. These controls are of two kinds. The first is triggered by impulses travelling in the fast-conducting L fibres and subsequently in the fast-conducting lateral ascending spinal pathways (Fig. 13.6). Information taking these routes is able to travel sufficiently rapidly for the brain to bias in advance the response of the T cells to impulses that arrive later by way of the S fibres. This type of 'feed-forward' control may well form part of the circuitry which reduces an animal's response to a shock that has been turned into a CS for food (see the discussion of Pavlovian counter-conditioning in Chapter 9). Second, there is a negative feedback loop which is triggered by the output from the T cells themselves; this acts to limit further increases in T cell output. We shall return later in the chapter to these and other central aspects of the systems subserving pain and reactions to pain.

It is stimuli associated with pain, rather than pain itself, which elicit fear (see Fig. 12.4). Unfortunately, little is known at the physiological level about the mechanisms that convert an innocuous light or tone or other environmental stimulus into a warning of pain. Much progress has been made by Eric Kandel in the analysis of the neural basis of classical conditioning in relatively simple invertebrate species, such as *Aplysia*, but it is too soon to tell whether any of the principles uncovered will also be applicable to the more complex nervous system of the mammal. Other important advances have emerged from experimental analyses (conducted by Richard Thompson's group in California and by Yeo & Glickstein in London) of the circuitry that gives rise to the rabbit's conditioned nictitating membrane reflex in response to a CS associated with an airpuff or paraorbital shock. This research has localised the critical circuits to the cerebellum (Fig. 13.3). However, it is likely that the conditioned nictitating membrane response is an instance of what Konorski has called 'consummatory conditioning', not the 'preparatory conditioning' that has been central to our analysis of conditioned fear (see Chapter 10 for a discussion of this distinction). Indeed, Thompson's group has shown that the same cerebellar lesions that abolish the conditioned nictitating membrane response leave intact conditioned heart-rate deceleration (plausibly, an index of conditioned fear) to the CS. Thus it would be inappropriate to go further into the details of these experiments. We shall offer some speculations about the neural structures involved in preparatory or emotional conditioning later in the chapter.

Novelty

A second type of stimulus that is important for fear (see Fig. 12.4) is novelty. Responses to novelty were studied in great detail in the 1950s and 1960s by a group of workers in Moscow under the leadership of E. N. Sokolov. They described a series of 'orienting' responses which are elicited by novel stimuli in any sensory modality and which, with repetition of the stimulus, die out or 'habituate'. One of the most important of these orienting reactions is the EEG arousal response which, as we have seen, depends on mechanisms in the midbrain reticular formation. Many of the other orienting responses, among them the skin conductance response, also have important organising centres in this part of the brain.

Habituation of the orienting responses to the repeated stimulus is extremely specific. It is sufficient to change any one of a dozen parameters of the stimulus (its intensity, modality, duration, quality, repetition rate, combination with other stimuli, etc.) for the orienting reflex to recur. Omitting the stimulus at a time when (as a result of past regularities) it would be expected to occur is also sufficient to have this effect. This specifity of habituation convinced Sokolov that it is necessary to suppose that the brain builds up a 'neuronal model' of the repeated stimulus and compares this model, in all its aspects, with the actual stimulation received on any particular occasion. It is then able to signal 'match' (familiarity) or 'mismatch' (novelty) (Fig. 13.8). Sokolov further suggested that such a process of model-building and comparison could be carried out by the joint actions of three kinds of neurons: (1) afferent neurons, which always respond to an appropriate stimulus; (2) extrapolatory neurons, which respond only when a particular stimulus has been repeated on a number of occasions; (3) comparator or novelty neurons, which compare the firing patterns of the first two kinds and fire if the patterns do not match, thus signalling 'novelty'.

Sokolov put forward this scheme in 1963. In 1966 he and his co-worker, Olga Vinogradova, were able to report that they had discovered both novelty neurons and comparator neurons by recording directly from individual nerve-cells during the repeated presentation of originally novel stimuli. (The afferent neurons, of course, offered no difficulty: they can be found readily in the sensory cortex and in the sensory nuclei of the thalamus.) Novelty neurons were found in the visual cortex, the reticular formation and the caudate nucleus, but the most important concentration of them was in the hippocampus (Fig. 13.9). Extrapolatory neurons have to date been found only in the hippocampus.

The hippocampus is part of a series of interlinked nuclei and tracts, known as the 'limbic system' (Fig. 13.10), which surround the thalamus and lie beneath the cerebral cortex. During periods of arousal, when the cortex and ARAS display the desynchronised electrical rhythms described earlier, the hippocampus by contrast shows regular, high-vol-

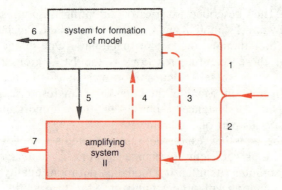

Fig. 13.8. Sokolov's model for the orienting response. (I) Modelling system. (II) Amplifying system. (1) Specific pathway from sense organs to cortical level of modelling system. (2) Collateral to reticular formation (represented here as amplifying device). (3) Negative feedback from modelling system to synaptic connections between collaterals from specific pathway and reticular formation (RF) to block input in the case of habituated stimuli. (4) Ascending activating influences from the RF to the modelling system (cortex). (5) Pathway from modelling system to amplifying system (this is the pathway through which the impulses signifying non-concordance between input and existing neuronal models are transmitted from cortex to RF). (6) To specific responses caused by coincidence between the external stimulus and the neuronal model (habitual responses). (7) To the vegetative and somatic components arising from the stimulation of the RF.

tage, slow 'theta waves', with frequencies about 6–12 Hz in the rat and 4–8 Hz in cats and dogs (see Fig. 7.2). Similar rhythms (at about 6 Hz) have been seen in Man in recordings from the temporal lobe of the neocortex, which overlies the hippocampal region; these tend to occur in states of emotion, especially disappointment and frustration.

Sokolov's model of the orienting response originally placed the 'system for the formation of the model' (fig. 13.8) in the neocortex. In the light of Vinogradova's subsequent observations of extrapolatory neurons in the hippocampus, it would seem more plausible to locate the neuronal model in this structure. As we shall see, the idea of the hippocampus as a comparator (i.e. a system that compares actual with expected stimulation) has proved to be extremely influential in shaping theories of hippocampal function.

The mode of action of anti-anxiety drugs

The analysis of the brain mechanisms mediating the perception of novelty in the preceding section has directed our attention towards

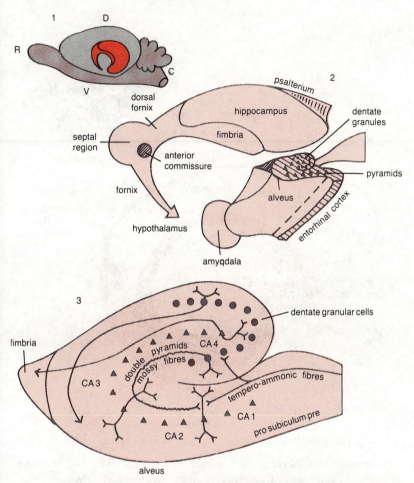

Fig. 13.9. The hippocampus, (1) as it lies in the rat brain, side view. D, dorsal (towards the back); V, ventral (towards the stomach); R, rostral (towards the nose); C, caudal (towards the tail). (2) A more detailed side view of the hippocampus, with a section through the middle portion. (3) A close-up of the section in (2). The description of the different 'cyto-architectonic fields' (CA1, 2, 3, and 4) of the hippocampus depends on the particular types of cells seen under the microscope. (CA stands for '*cornu Ammonis*', or 'Ammon's horn', a synonym for the hippocampus.)

the hippocampus. This structure turns up again when we ask the question, 'by what action upon the brain do the anti-anxiety drugs reduce anxiety?'.

One can set about answering this question in two different ways. The first is relatively direct: one investigates the direct action of the drugs concerned, asking how they affect neuronal firing, receptors

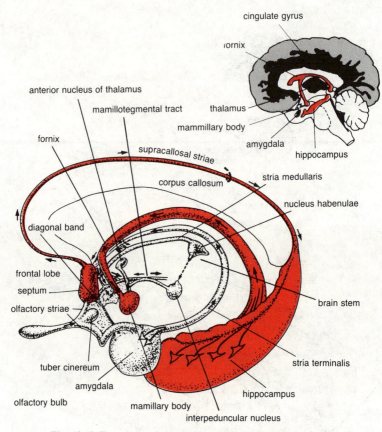

Fig. 13.10. Two views of the limbic system in the primate brain.

for neurotransmitters, neuronal membranes, etc. This is the route usually taken by physiologists and biochemists. In contrast, psychologists have tended to take a more indirect route, seeking for interventions in the brain which systematically reproduce the behavioural effects of the anti-anxiety drugs (as summarised, say, in Fig. 12.4) or, alternatively, interventions that produce effects that are systematically opposed to those of the anti-anxiety drugs. On the face of it, the direct approach is the superior one. However, all drugs have multiple effects, and the anti-anxiety drugs are no exception. These agents, for example, are muscle-relaxants, anti-convulsants, and sedatives and give rise to motor incoordination. There is no simple way of telling whether some observed change in, say, neuronal firing or receptor function is related to one of these effects or to the anti-anxiety effect which is our own central concern. Thus the direct approach needs supplementing with the information derived from the indirect approach. It is the latter kind of information which has pointed to the hippocampus as a structure with a central role in anxiety. But we shall first review the information that has been gleaned using the more direct approach favoured by neurophysiologists and neurochemists.

This approach has recently been very productive, especially in the case of the benzodiazepines. It has been shown that these drugs bind strongly to a highly selective receptor located on neuronal membranes in the brain. This discovery has given rise to the belief that, as in the analogous case of the opiate receptor, there must be an endogenous ligand for the benzodiazepine receptor, i.e. a transmitter-like substance that is secreted by neurons and acts upon this receptor. So far, however, in spite of many efforts to isolate it, the endogenous ligand has escaped capture. If it exists, it is as likely to be an anxiogenic as an anxiolytic substance.

The discovery of the benzodiazepine receptor undoubtedly marks an important milestone. Its relevance to the issues treated in this book is underlined by several reports that strains of rats (including the Maudsley strains) and mice differing in fearfulness also differ in the density of brain benzodiazepine receptors. Sex differences (cf. Chapter 7) in the density of benzodiazepine receptors have also been observed. Nonetheless the significance of research in this field is sharply limited by the fact that the benzodiazepine receptor does not bind either the barbiturates or alcohol; yet the behavioural effects of these drugs are closely similar to those of the benzodiazepines. Thus, occupancy of the benzodiazepine receptor is not in itself a *sine qua non* for anxiolytic drug action.

However, the next step in the chain of neurochemical events brings us onto stronger ground. The benzodiazepine receptor turns out to be closely coupled to a second receptor, one for the well-known inhibitory transmitter γ-aminobutyrate (GABA) (Fig. 13.11). Activation of the benzodiazepine receptor by the binding of a benzodiazepine drug increases the degree of inhibition produced by GABA acting at the coupled GABA receptor. Now the latter effect is also produced by the barbiturates, by way of an action at yet a third ('picrotoxinin') receptor that is also closely coupled to the GABA receptor (Fig. 13.11). Thus the two major classes of anti-anxiety drug both increase the inhibitory effects of GABA, as does (though not so clearly) alcohol. It seems possible, therefore, that facilitation of GABAergic inhibition is the key step in anxiolytic drug action. This possibility is reinforced by the discovery of other kinds of drugs (especially the

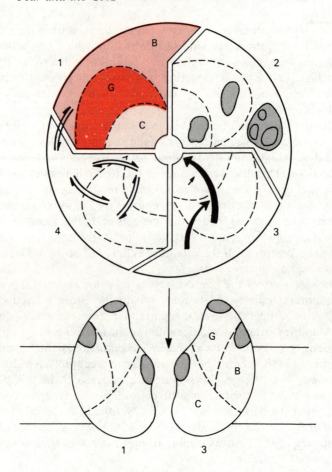

Fig. 13.11. Hypothetical anatomy and functions of the receptor–channel complex proposed by Haefely *et al.* (1985). A view from the extracellular space (above) shows the four monomers (1, 2, 3, 4). Four different functions and interactions are depicted. (1) The three domains: Cl⁻ channel part (c), GABA binding domain or GABA-R (G), and the regulatory domain with BZ (benzodiazepine) binding site, the BZR (B). (2) The ligand binding sites on the channel domain ('picrotoxin binding site', 'barbiturate receptor', and 'channel ligand site'), on the GABA-R (GABA binding site), and on the BZR (binding site for agonists, antagonists, and inverse agonists, composed of not entirely identical, but adjacent and probably partially overlapping subsites). (3) The main function of the complex, namely the GABA-induced gating (opening) of the Cl⁻ channel (large arrow), with the regulation of this gating process by BZ agonists and inverse agonists (medium arrow), and the regulation of Cl⁻ channel properties by ligands of the picrotoxinin binding sites. (4) The four bidirectional coupling functions or domain–domain interactions (between GABA-R and Cl⁻ channel domain, between BZR and GABA-R, between BZR and channel ligands, and between the BZ binding sites of adjacent subunits). On a cross-section through the membrane (see below) are shown the three domains of subunits 1 and 3.

so-called beta-carbolines) which have effects at the benzodiazepine receptor that are exactly opposed to those of the benzodiazepines. These drugs reduce the inhibitory effects of GABA, cause feelings of anxiety in Man, and change animal behaviour in an anxiogenic direction.

The trouble with this hypothesis is that GABA and GABA receptors, like the benzodiazepine receptor, are widely distributed in the brain and spinal cord. There is little reason to suppose that a general facilitation of GABAergic inhibition would produce the highly specific behavioural effects observed after administration of anti-anxiety drugs (see Fig. 12.4). On the contrary, such facilitation might be expected to produce sedative or anti-convulsant effects, both of which are observed after treatment with anti-anxiety drugs. Indeed, much other evidence implicates the GABA/benzodiazepine/picrotoxinin receptor complex in convulsant phenomena. For example, both the beta-carbolines and drugs (such as picrotoxin) which antagonise the effects of the barbiturates at the picrotoxinin receptor facilitate and/or themselves produce epileptic seizures.

Direct experimental tests of the GABAergic hypothesis of anxiolytic drug action have also produced only equivocal support for it. In a series of experiments from my own laboratory, for example, we have sometimes been able to reproduce the anxiolytic effects of the benzodiazepines and barbiturates with other drugs that increase GABAergic transmission by a direct action on the GABA receptor; and we have consistently been able to block the behavioural effects of the benzodiazepines with drugs that antagonise GABAergic transmission by a direct action either on the GABA receptor (bicuculline) or on the picrotoxinin receptor (picrotoxin); but we have never been able to block the anxiolytic action of the barbiturate sodium amytal with these drugs. Thus, for the moment, one can conclude only that facilitated GABAergic transmission may underlie the anxiolytic action of some drugs, but it does not necessarily underlie all anxiolytic drug action.

Even if the GABAergic hypothesis of anxiolytic drug action were entirely correct, we would still need to determine by independent means which of the millions of GABAergic synapses in the brain are the ones that are critically involved in anxiety and its reduction (as distinct from convulsions, sedation, etc.). Thus the indirect approach to the problem remains important, whatever the outcome of tests of the GABAergic hypothesis. Indeed, an important feature of the type of function subserved by GABAergic neurons suggests that it is *only* by the indirect approach that we shall be able to determine *which* GABAergic synapses are involved in anxiety.

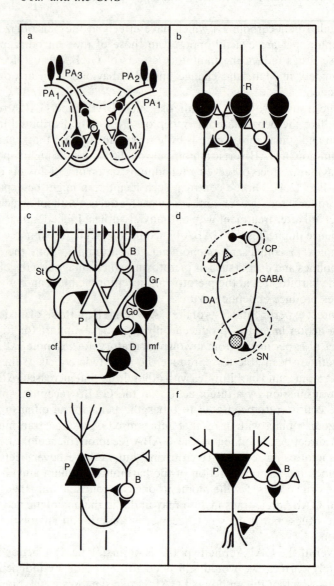

Fig. 13.12. Schematic diagrams of the main neuronal circuits containing GABAergic synapses, i.e. synapses at which transmission is mediated by γ-aminobutyric acid (GABA). At such synapses, benzodiazepines have been found to enhance transmission. GABAergic neurons are shown in white, excitatory neurons in black. (a) Spinal cord (M, motor neuron; PA, primary afferents); (b) dorsal column nuclei (R, relay cell; I, interneuron); (c) cerebellar cortex (P, Purkinje cell; Gr, granule cell; Go, Golgi cell; B, basket cell; St, stellate cell; D, output neuron of Deiters' nucleus; cf, climbing fibre;

There are a few long-axoned GABAergic neurons in the brain. But the great majority of GABAergic cells are interneurons with short axons which participate in the type of circuit depicted in Fig. 13.12. As shown in this figure, GABAergic cells frequently play a critical role in negative feedback loops. These cells are activated by firing in long-axoned neuronal pathways and provide recurrent inhibition to keep that firing within bounds. In such a feedback loop the GABAergic neurons will fire only if the principal cells to which they provide feedback are themselves excited. Now, let us assume (as held by the GABAergic hypothesis of anxiolytic drug action) that the anxiolytic effects of the benzodiazepines and barbiturates are due to an enhancement of the inhibitory action of GABA. This enhancement can take place only at synapses where GABAergic transmission is in fact taking place; and this transmission can take place (on the argument advanced above) only in systems where the related principal cells are themselves firing. In this way, benzodiazepines would selectively dampen activity in anxiety-related brain systems, not because these drugs have a special affinity for such systems, but because it is in these systems that (given appropriate anxiogenic stimulation) neuronal firing of a kind with which they interact is currently taking place.

Brain structures that mediate anxiety

The foregoing arguments suggest that even if anxiolytic drug action indeed turns out to be mediated by facilitated GABAergic inhibition, the direct approach to the study of these drugs will be insufficient to isolate the particular brain regions in which the critical changes are produced. We need, therefore, to turn to the indirect approach favoured by psychologists. Are there any interventions in the brain which produce the kind of behavioural effect (summarised in Fig. 12.4 in the form of the behavioural inhibition system) that is seen after administration of the anti-anxiety drugs?

We have already had forewarning that the answer to this question is 'yes'; as summarised in Table 12.2, there is an extraordinary degree of overlap between the behavioural effects of administration of the anti-anxiety drugs, on the one hand, and lesions to two brain regions,

Caption for Fig. 13.2 (*cont.*).

mf, mossy fibre); (d) neostriatum and substantia nigra (CP, caudate-putamen; SN, substantia nigra; DA, nigrostriatal dopamine pathway; GABA, GABAergic striatonigral pathway); (e) cerebral cortex (P, pyramidal cell; B, basket cell); (f) hippocampus (P, pyramidal cell; B, basket cell). From Haefely (1984).

the hippocampus and the septal area (Figs. 13.9 and 13.10), on the other. Furthermore, this overlap includes many types of behaviour that have been central to the analysis of fearful behaviour that has occupied us in earlier chapters. Thus, like the anxiolytic drugs, septal or hippocampal lesions reduce passive avoidance, retard extinction, improve shuttlebox avoidance, improve bar-press avoidance, impair exploratory behaviour, and eliminate the partial reinforcement extinction effect. In addition, a number of effects that are *not* produced by the anti-anxiety drugs, and which have figured prominently in our analysis of the nature of the psychological action of these agents, are similarly absent from the 'septo–hippocampal syndrome', as we may call the profile of change set out in Table 12.2. Thus, neither septal nor hippocampal lesions alter the double-runway frustration effect, escape behaviour, one-way active avoidance, simple rewarded behaviour, or the threshold for detection of shock.

Note that both in their positive and in their negative effects, septal and hippocampal lesions resemble each other very closely. This is not surprising, since these two regions are interrelated both anatomically and physiologically. The medial septal area sends a cholinergic projection to the hippocampus, and the hippocampus reciprocates with a projection to the lateral septal area; it is the cholinergic septo–hippocampal projection which controls the hippocampal theta rhythm (Fig. 7.3). Particularly when septal and hippocampal lesions have the same behavioural effect as each other, the anti-anxiety drugs have the same effect yet again. Thus it is reasonable to propose the hypothesis (as I did in 1970) that the anti-anxiety drugs reduce anxiety by way of an action, direct or indirect, upon an integrated septo–hippocampal system. And this hypothesis in turn implies that the septo–hippocampal system in part mediates the state of anxiety.

Functions of the septo–hippocampal system

If this is so, we should be able to learn a great deal about anxiety by enquiring what functions are discharged by the septo–hippocampal system. These functions have been the subject of many and varied speculations. I have recently reviewed the different theories that have been advanced and the evidential bases upon which they rest in an attempt to distil a common core of agreement that could find support from many sources of data. The concept that comes closest to this ideal is one that we met earlier in this chapter, in our discussion of Sokolov's work on the perception of novelty: it is the idea that the hippocampus functions as a comparator.

In developing this idea further, I have supposed that the septo–hippocampal system (together with a number of other closely related structures which we shall consider later) has the central task of comparing, quite generally, actual with expected stimuli. The system functions in two modes. If actual stimuli are successfully matched with expected ones ('match'), it functions in 'checking mode' and behavioural control rests with other (unspecified) brain mechanisms. If there is discordance between actual and expected stimuli (the 'novelty' input to the behavioural inhibition system of Fig. 12.4) or if the predicted stimulus is aversive (the inputs to the behavioural inhibition system termed 'signals of punishment' or 'signals of non-reward') – conditions that are jointly termed 'mismatch' – the septo–hippocampal system takes direct control over behaviour and now functions in 'control mode'. In control mode, the septo–hippocampal system operates the outputs of the behavioural inhibition system (Fig. 12.4), though this is achieved with the participation of other brain structures, as we shall see below. In this way, the septo–hippocampal system acts as the computational heart of the behavioural inhibition system, looking out for and detecting anxiogenic stimuli and activating behavioural routines appropriate to them.

Before proceeding further, it is worth asking in general terms what kinds of information processing are necessary for such a comparator to perform its functions (Fig. 13.13). Clearly, the comparator must have access to information both about current sensory events ('the world') and about expected ('predicted') events. However, these two classes of information must themselves be closely interrelated. Predictions can only be generated (a task for which a 'generator of predictions' is needed) in the light of information about the current state of the world. Thus current sensory events must be transmitted to the generator of predictions. In addition, the latter must have access to

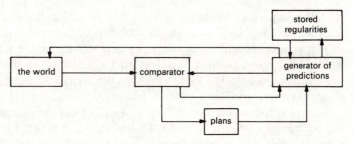

Fig. 13.13. The kinds of information processing required for the successful functioning of the hypothetical comparator. See text for further explanation.

information about past environmental regularities which, in conjunction with the present state of the world, determine the content of the next prediction. But this will be sufficient only if the environment is one over which the subject has no control. If, as in most cases, the subject's behaviour affects the world, the generator of predictions must also have access to information about the next intended set of movements, that is, motor programs or plans. Prediction will now depend on the conjunction of the present state of the world, stored past environmental regularities (learned by way of Pavlovian conditioning), the next intended step in the motor program, and stored past relationships between such steps and changes in the environment (learned by way of instrumental conditioning).

Once made, the prediction must be tested against the world. For this to be possible the right sensory input must be chosen. It follows that the input of sensory information must be selected in the light of what is predicted. Such selection can in principle be accomplished in one of two ways. The subject's motor behaviour can be left under the control of other systems, with selection accomplished by choosing among the sensory events occuring anyway as a result of that motor behaviour; or the selection mechanism can itself take active control over motor behaviour and command appropriate exploratory action. This might range from simple adjustments of sensory organs (e.g. dilating the pupils) to complex patterns of locomotion.

It follows from these arguments that the major computational transactions for which we must seek neural correlates are those set out in Fig. 13.13. To these must be added one further desideratum. For the comparator to work, each prediction must arrive at the same time as the input from the world to which it relates. Thus time has to be quantised: information must flow in small packets or chunks that are synchronised throughout the system.

It is not easy to relate these abstract computational functions to specific circuitry in the brain (though the brain must certainly possess circuits that perform these or similar functions). Nonetheless, I have tried to do so. Fig. 13.14 is a highly condensed summary of the resulting theory. It allots functions, not only to the septo–hippocampal system itself, but also to a number of other structures with which the septo–hippocampal system has close connections. I shall introduce these structures along with the functions I have attributed to them.

Sensory information describing the world enters the system from the entorhinal cortex (Fig. 13.15) located in the temporal lobe. This receives highly processed multimodal information from all the neocortical sensory systems and transmits it to the hippocampal formation

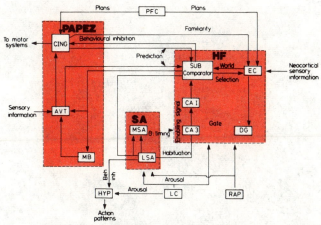

Fig. 13.14. A summary of the theory developed by Gray (1982). The three major building blocks are shown in heavy print: HF, the hippocampal formation, made up of the entorhinal cortex, EC, the dentate gyrus, DG, CA 3, CA 1, and the subiclar area, SUB; SA, the septal area, containing the medial and lateral septal areas, MSA and LSA; and the Papez circuit, which receives projections from and returns them to the subicular area via the mammillary bodies, MB, anteroventral thalamus, AVT, and cingulate cortex, CING. Other structures shown are the hypothalamus, HYP, the locus coeruleus, LC, the raphe muclei, RAP, and the prefrontal cortex, PFC. Arrows show direction of projection; the projection from SUB to MSA lacks anatomical confirmation. Words in lower case show postulated functions; beh. inh., behavioural inhibition. For further explanation, see text.

(i.e. the dentate gyrus plus the hippocampus proper) via the perforant path (Fig. 13.16). One route taken by the nervous impulses coming from the entorhinal cortex goes by way of the basic trisynaptic hippocampal circuit depicted in Fig. 13.16: to the granule cells of the dentate gyrus, thence along the mossy fibres to area CA3 of the hippocampus, thence along the Schaffer collaterals to area CA1 of the hippocampus, and finally to the subicular region (which provides the main output from the hippocampus, just as the entorhinal cortex provides the main input).

This basic hippocampal circuit has some odd features. To begin with, information is also sent directly from the entorhinal cortex to the subiculum. One wonders, therefore, what advantage accrues from sending the same information around the more circuitous hippocampal route. It cannot come from the addition of other precise information, since no other precise information appears to enter the hippocampus. As we shall see, the other major inputs to this structure (from the medial septal area, the locus coeruleus, and the median raphe) are all apparently diffuse, lacking specific sensory or motor features.

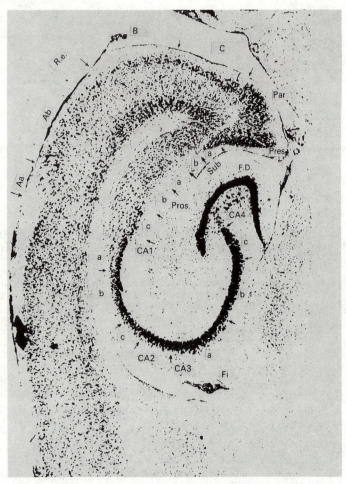

Fig. 13.15. Photograph of a horizontal section of the right side of the adult mouse brain as seen from above. The section passes through the posterior arch of the hippocampus and entorhinal cortex (R.e.). Caudal is up. Note the gradual transition from the six-layered entorhinal cortex through the parasubiculum (Par), pre-subiculum (Pres), subiculum (Sub), prosubiculum (Pros), to the three-layered cortex of the hippocampus proper (CA1–CA4) and the fascia dentata (F.D.). Lorente de No further subdivided some of these areas (a, b, c). Fi is the fimbria. Nissl stain. From O'Keefe & Nadel (1978), after the original work by Lorente de No in 1934.

Furthermore, even the precise information that enters from the entorhinal cortex along the perforant path appears to get scrambled as it passes around the hippocampal circuit. This, at any rate, is the natural implication of the fact that electrical recordings in the entorhi-

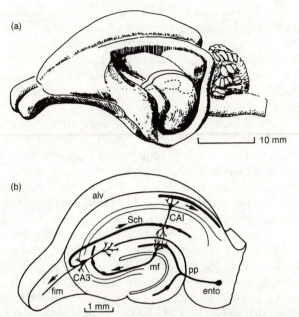

Fig. 13.16. Lamellar organisation of the hippocampus. (a) Lateral view of the rabbit brain with the parietal and temporal neocortex removed to expose the hippocampal formation. The lamellar slice indicated has been presented separately in (b) to show the proposed circuitry; alv, alveus; ento, entorhinal cortex; fim, fimbria; pp, perforant path; Sch, Schaffer collateral.

nal area reveal neurons with specific sensory correlates, whereas equivalent recordings in the dentate gyrus pick up neurons with much less specific or stable sensory correlates.

A final intriguing feature of the hippocampal circuit illustrated in Fig. 13.16 is that there appears to be a 'gate' located between the dentate granule cells and area CA3. Electrical stimulation on the perforant path side of the gate sometimes gives rise to steadily increasing responses in CA3 or CA1 ('potentiation') and sometimes to steadily decreasing response ('habituation'). The important experiments of Menachem Segal in Israel have shown that the dentate/CA3 gate is opened wider if, just before stimulation is applied to the perforant path, the animal is presented with a biologically important stimulus (e.g. a CS for food or for shock), and that this effect is probably due to an input to the septo–hippocampal system from noradrenergic fibres travelling in the dorsal bundle from the locus coeruleus (Figs. 13.2 and 13.3). This is a good example of the capacity of these fibres to increase the signal-to-noise ratio in the target organs that they innervate: that is, they increase the capacity of the hippocampus to handle inputs originating in the entorhinal cortex. As we shall see, the anti-anxiety drugs have an important influence upon this noradrenergic input to the septo–hippocampal system.

To account for these odd features of the hippocampal circuit I have made the following assumptions. First, it is the subiculum (Fig. 13.15) which provides the main comparator function and which initiates activity in the circuits responsible for the generation of predictions (see below). Second, the precise description of the current state of the world which the subiculum requires for these purposes is sent direct from the entorhinal area. However, the signal sent from the entorhinal area to the subiculum is acted upon only if the subiculum also receives an 'enabling signal' (Fig. 13.14) sent via the hippocampal circuit. This circuit therefore functions as a selection mechanism, a function that is served by the dentate/CA3 gate. Finally, selection is operated in favour of biologically important stimuli by way of the noradrenergic input to the dentate/CA3 gate.

The chief reason for picking upon the subiculum as a likely site for the main comparator function lies in the nodal position occupied by this structure within the loops that make up the septo–hippocampal system as a whole (see Fig. 13.14). In particular, the subiculum lies at the origin of a major loop via the Papez circuit, and receives the output from this loop via both the anteroventral thalamus and the cingulate cortex (Fig. 13.14). It was these loops that led Papez in 1937 to make his inspired guess that the limbic structures related to the hippocampus are involved in the elaboration of emotional experience. In a particular version of Papez's hypothesis I have proposed that the circuit that goes from the subiculum via the anteroventral thalamus, the mamillary bodies of the hypothalamus, and the cingulate cortex back to the subiculum is responsible for the generation of predictions, in the sense outlined in Fig. 13.13. According to this hypothesis, neurons in the subiculum first check that a current prediction matches against the input from the world defined by the current input from the entorhinal cortex together with an enabling signal from CA1. This verified prediction is then used to initiate the generation of the next prediction via the Papez loop. This loop has access to stored stimulus–stimulus regularities via the thalamus, to stored response–stimulus regularities via the cingulate cortex, and to the animal's current motor plan also via the cingulate cortex (to which information about motor plans is relayed from the prefrontal cortex). (The phrase 'has access to' in the previous sentence is deliberately vague; storage of past experience is probably a neocortical function, the temporal lobe being a particularly likely site.) The arrival in the subiculum of neuronal messages from the thalamus and the cingulate cortex constitutes the next prediction, which is again matched with the next input from the entorhinal area.

For this flow of information to work properly, time must be quantised, as we have noted above. The hippocampal theta rhythm is ideally suited to perform this function. This rhythm can be recorded, not only in the hippocampal formation itself, but also in the subiculum, the entorhinal area, and the region of the mamillary bodies. To date, wherever theta has been recorded it has turned out to depend upon the same pacemaker cells in the medial septal area. Thus this region is able to pace the passage of information around all the complex circuits depicted in Fig. 13.14. Estimates of the conduction time for messages that traverse the whole Papez circuit give values around 60 milliseconds (msec). This is the same order of magnitude as the intervals between successive peaks of theta waves (about 80–160 msec). Furthermore, there is evidence that when the hippocampus is directly stimulated electrically, the passage of the resultant response around the hippocampal circuit is relatively facilitated at certain points in the phase of the concomitant theta rhythm, suggesting in agreement with the present argument that this rhythm performs a gating function.

A group in Canada under the direction of Case Vanderwolf has produced strong evidence that there are two functionally distinct types of theta rhythm. One of these is generally of relatively high frequency (above 7 Hz in the rat) and strongly related to movement, higher speeds of movement being accompanied by higher frequencies of theta. The other is generally of lower frequency (typically, less than about 8 Hz) and not related to movement; indeed, this rhythm has been observed by Graeff & Quintero to accompany freezing behaviour elicited by electrical stimulation of the median raphe (a serotonergic nucleus in the brainstem). A further distinction between movement-related and immobility-related theta is that the latter, but not the former, may be abolished by systemic injection of anti-cholinergic drugs, such as atropine or scopolamine. This observation is well-established but curious, since most evidence suggests that both types of theta rhythm depend upon the same cholinergic pathway from the septal area to the hippocampus. It is likely, therefore, that immobility-related theta depends in addition upon a second cholinergic pathway that is afferent to the medial septal area and that is more sensitive to systemic treatment with anti-cholinergic drugs than is the septo–hippocampal cholinergic pathway.

I have linked Vanderwolf's two types of theta with the model of hippocampal function developed above by supposing that movement-related theta is seen when the septo–hippocampal system (a phrase

I am now using as shorthand for the whole system depicted in Fig. 13.14) is functioning in checking mode. Theta frequency is higher the faster the speed of movement, because the rate of generation and checking of predictions must also be higher the faster the animal is moving. Immobility-related theta, conversely, is seen when the septo–hippocampal system functions in control mode, operating the outputs of the behavioural inhibition system (Fig. 12.4). It is supposed that in this mode the septo–hippocampal system is actively searching for a way out of the problem the animal is facing (having just encountered a mismatch) and that this involves more information processing (and so a slower rate of information processing) than does the checking mode of hippocampal function. It is for this reason that frequencies are generally slower for immobility-related than for movement-related theta.

Anxiety and the dorsal noradrenergic bundle

The functions attributed to the septo–hippocampal system in the preceding section are highly cognitive. This is by no means at variance with the notion that this system plays a central role in emotional behaviour. Anxiety gives rise to characteristic thought processes (as we shall see in the next chapter) as much as to equally characteristic behaviour patterns; and the two are of course connected. Nonetheless, it would be wrong to suppose that the septo–hippocampal system participates only in anxiety. On the contrary, the model I have proposed treats this system as constituting a kind of interface between emotional and cognitive processes. The computational powers possessed by the septo–hippocampal system are put sometimes at the disposal of emotional behaviour, but sometimes at the disposal of certain specialised and non-emotional forms of problem solving (analogous to the use of a list by a human being on a shopping expedition) into which I shall not enter here. Similarly, the emotion of anxiety makes use of the computational circuits of the septo–hippocampal system, but also (as we shall now see) many other brain regions and pathways. There is, in other words, only partial overlap between the state of anxiety and activity in the septo–hippocampal system: the septo–hippocampal system does more than just participate in anxiety, and anxiety includes more of the brain than just the septo–hippocampal system. This is hardly surprising. It would be little short of miraculous if categories of emotional life derived in the first instance from common sense and introspection turned out, even after the kind of theoretical massage they have undergone in this book, to correspond to the functioning of neatly delimited sub-units of the brain.

I pointed out earlier that there is very great similarity between the behavioural effects of anti-anxiety drugs and damage to the septo–hippocampal system. However, the similarity is not total. One important exception concerns the kind of non-emotional task alluded to above. Consider the radial-arm maze introduced by David Olton. This is shaped like a bicycle wheel, with a central hub from which up to 16 arms radiate out. At the start of a trial, a pellet of food is placed at the end of each arm. The rat's task is to retrieve each pellet, which it does in an extemely efficient manner, hardly ever re-visiting an arm from which it has already extracted the pellet, even though it visits arms in an apparently random manner, and even if it is detained for several minutes or more in the central hub between choices. It is this kind of task which one may regard as analogous to the use of a list by a human being: the rat behaves as though it is able to check off in its head the arms it has already visited (a capacity that Olton has termed 'working memory'). This capacity is severely disrupted by damage to the septo–hippocampal system. But it is affected by anxiolytic drugs only if these are given at doses clearly in excess of the minimal dose required to reduce anxiety in the kind of task from which the concept of the behavioural inhibition system (Fig. 12.4) has been derived.

A second exception to the rule that damage to the septo–hippocampal system affects behaviour in the same manner as the anxiolytic drugs concerns arousal. As we saw in Chapter 10, anxiolytic drugs clearly antagonise a number of behavioural phenomena that are plausibly attributed to an increase in the level of arousal elicited by stimuli associated with punishment or with non-reward. A good example is the partial reinforcement acquisition effect (Fig. 10.10). But there is no clear evidence of similar effects after septal or hippocampal lesions.

We may gain some insight into these exceptions if we consider how the anti-anxiety drugs affect the functioning of the septo–hippocampal system. One possibility is that they act directly within the hippocampus itself. This structure contains a large number of GABAergic neurons, hooked up in the kind of recurrent inhibitory circuits illustrated in Fig. 13.12; and the Swiss physiologist Haefely has shown that benzodiazepines do indeed increase recurrent inhibition in the hippocampus. There is, however, a second possibility. As we have seen, ascending noradrenergic fibres, originating in the locus coeruleus, play an important role in opening the dentate/CA3 gate and biassing the activity of the subicular comparator in favour of biologically significant stimuli (Segal's experiments). One may interpret this phenomenon as indicating the action of a switching mechanism: one that diverts the septo–hippocampal system from playing a role in relatively non-emotional activities (such as solving Olton's radial-arm maze problem), these presumably being controlled by influences that descend to the septo–hippocampal system from neocortical sources, to the more pressing and emotionally toned messages that are coming up from lower brain centres. On this view, the noradrenergic input to the septo–hippocampal system acts like an alarm bell, putting this system to the job of dealing with environmental threats. Where better for anxiolytic drugs to act than here?

And, indeed, there is evidence that (1) stress activates the dorsal ascending noradrenergic bundle, and (2) this activation is counteracted by anxiolytic drugs. Effects of this kind have been demonstrated by Fuxe's group in Stockholm, measuring the rate of utilisation of noradrenaline ('noradrenaline turnover', an indirect but reliable index of the rate of firing in noradrenergic neurons) in terminal regions served by the dorsal bundle. Anxiolytic drugs of several different classes (benzodiazepines, barbiturates, alcohol) all reversed the stress-induced increase in noradrenaline turnover. Evidence from my own laboratory suggests that this antagonism between anxiolytic drugs and the dorsal noradrenergic bundle extends to the fibres destined for the septo–hippocampal system.

We have discussed these experiments before, in Chapter 7 (Figs. 7.2–7.5). They depend upon the fact that via electrodes chronically implanted in the septal area it is possible to drive the hippocampal theta rhythm by applying stimulation (pulses half-a-millisecond in duration) to the septal pacemaker cells at frequencies lying within the natural theta range. The threshold current able to drive theta in this way in the free-moving rat bears a characteristic relation to stimulation frequency, a minimum threshold being found at precisely 7.7 Hz (inter-pulse interval = 130 msec). This minimum in the 'theta-driving curve' is abolished by any of a range of anti-anxiety drugs, which increase the driving threshold selectively at 7.7 Hz (Fig. 13.17), suggesting a relationship between theta at this frequency and anxiety. As we saw in Chapter 7, this hypothesis is supported by both strain and sex differences (the Maudsley Nonreactive strain and females of other strains, both low in fearfulness, lack the 7.7-Hz minimum in the theta-driving curve).

Experiments in which McNaughton, Kelly, and I destroyed the dorsal ascending noradrenergic bundle (Fig. 13.3) by injecting it with the selective neurotoxin 6-hydroxydopamine have demonstrated the probable site at which the anti-anxiety drugs exert their effects on the theta-driving curve. Animals injected with the toxin had less than 10% of normal noradrenaline levels in the forebrain, including the

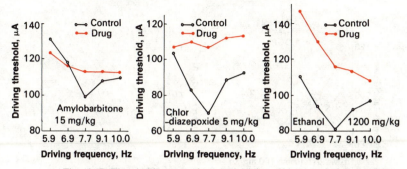

Fig. 13.17. Threshold currents for septal driving of hippocampal theta rhythm (see Figs. 7.3–7.5) as a function of stimulation frequency in the free-moving male rat before (control) and after (drug) injections of three anti-anxiety drugs. Data from McNaughton *et al.* (1977).

hippocampus. The loss of forebrain noradrenaline was accompanied by loss of the 7.7-Hz minimum in the theta-driving curve (Fig. 13.18). Pharmacological experiments in which animals were injected with drugs that selectively impair noradrenergic transmission produced similar results (Fig. 13.18). Thus the most likely site of action of the anti-anxiety drugs, insofar as their effects on the theta-driving curve are concerned, is on neurons that ascend from the locus coeruleus to innervate the septo–hippocampal system. There are GABA receptors both on the cell-bodies of these neurons and on their terminals in the hippocampus. The anti-anxiety drugs could therefore reduce the release of noradrenaline within the hippocampus by an action at either of these points or both. Consistent with this hypothesis, Jane Mellanby and I have shown that muscimol, a GABA receptor agonist, acts on the theta-driving curve like an anxiolytic.

These experiments suggest, then, that at moments of anxiety there is a noradrenergic input to the septo–hippocampal system which has two, perhaps related, effects. First, there is an opening of the dentate/CA3 gate, biassing the subicular comparator so that it preferentially deals with stimuli (arriving in the hippocampal formation from the entorhinal cortex) which bear a relation to biologically important stimuli (food, shock, etc.). Second, the hippocampal theta rhythm is biassed towards frequencies at or close to 7.7 Hz. The relation between this effect and the two types of theta distinguished by Vanderwolf and discussed above remains obscure. Given the general effects of the anxiolytic drugs, it seems likely that they would antagonise immobility-related rather than movement-related theta. Thus it is possible that 7.7 Hz represents a frequency of theta that is close to the upper limit reached by this type of theta. Quintero, Graeff, and I have observed,

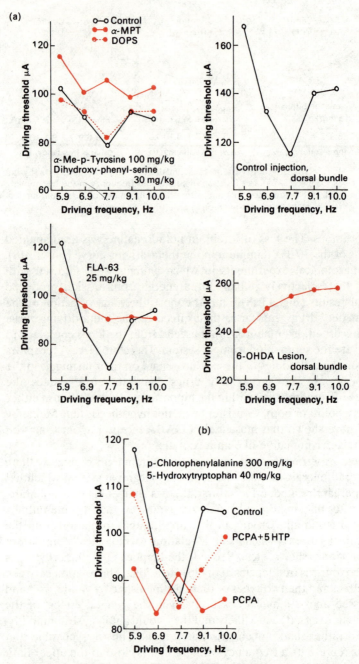

Fig. 13.18. Threshold currents for septal driving of hippocampal theta rhythm (see Figs. 7.3–7.5) as a function of stimulation frequency in the free-moving

in agreement with this suggestion, that the highest frequency of theta which we were able to drive by stimulation of the median raphe in the brainstem (a treatment which elicited strong freezing behaviour) lay very close to 7.7 Hz. If this view is correct, it implies that in states of anxiety the septo–hippocampal system processes information at the fastest rate possible while remaining in control mode.

The importance of the locus coeruleus in states of anxiety has also been emphasised by Gene Redmond and his colleagues at Yale. They have shown that electrical stimulation of this nucleus in a primate, the stump-tailed monkey, leads to frank signs of fear. The results of pharmacological experiments are consistent with this observation. The experiments have used the drugs clonidine and yohimbine. To understand their significance, we must first introduce the concept of the 'auto-receptor'.

As we know, transmitters act by binding to a receptor and thereby altering the function of the cell on which the receptor is located (Fig. 5.6). In some cases, however, the cell upon which the receptor is located is the same cell that releases the transmitter; such receptors are called 'auto-receptors'. Typically, auto-receptors act so as to restrain further activity in the cell concerned; that is, they form part of a homeostatic negative feedback loop. The noradrenergic cells of the locus coeruleus apparently possess such auto-receptors, classified (on pharmacological grounds) as α_2-noradrenergic receptors. It is on these receptors that the drugs clonidine and yohimbine act. Clonidine is an α_2-agonist: it acts via auto-receptors on cells in the locus coeruleus to decrease the firing rate of these neurons. Conversely, yohimbine is an antagonist at the α_2-receptor, reducing negative feedback and so increasing the firing rate of coerulear neurons. It follows from these modes of action that if activity in the locus coeruleus mediates

Caption for Fig. 13.18 (*cont.*)

male rat. In the undrugged animal (open circles) the minimum threshold is at 7.7 Hz (130 msec inter-pulse interval). This minimum is eliminated by blockade of either noradrenaline (a) or serotonin (b). (a) shows the effects of systemic injection of alpha-methyl-p-tyrosine (α-MPT), which blocks the synthesis of both dopamine and noradrenaline, together with the restoration of the curve to normal by subsequent administration of dihy-droxyphenylserine (DOPS), which acts as a substrate for the synthesis of noradrenaline only; of systemic FLA-63, which blocks the synthesis of norad-renaline only; and of destruction of the dorsal ascending noradrenergic bun-dle by local injection of the selective neurotoxin 6-hydroxydopamine (6-OHDA). (b) shows the effects of systemic *p*-chlorophenylalanine (PCPA), 300 mg/kg, which blocks the synthesis of serotonin, followed by 5-hydroxy-tryptophan (5-HTP), 40 mg/kg, to act as a substrate for renewed synthesis of this transmitter. Data from McNaughton *et al.* (1977).

anxiety, clonidine should reduce and yohimbine increase anxiety. Results consistent with these predictions have been reported in both animals and Man.

A further way to test the hypothesis that ascending noradrenergic systems are involved in anxiety is to destroy the dorsal noradrenergic bundle by local injection of 6-hydroxydopamine. As this hypothesis predicts, Lila Tsaltas and I found that such lesions alleviate punishment-induced suppression of bar-pressing (which the reader will recognise as a characteristic anxiolytic effect); while Susan Owen and I have demonstrated abolition of the partial reinforcement extinction effect (cf. Fig. 10.7 and 10.8) after the same lesion. These and other findings, then, attest to the role of the dorsal noradrenergic bundle in anxiety and in the mediation of anti-anxiety drug effects.

Redmond's group, however, has stressed not just the projection of the dorsal bundle to the septo–hippocampal system, as I have so far done, but rather the manner in which the locus coeruleus projects throughout the entire brain and spinal cord (Fig. 13.3). In effect, Redmond proposes that the locus coeruleus acts as a general alarm bell, set swinging by any kind of threat and rousing widely distributed brain systems to emergency action. This view is essentially the same as the one I adopted earlier in the chapter, according to which the locus coeruleus is an arousal mechanism; and it is consistent with the evidence that noradrenergic afferents boost the efficiency (increase the signal-to-noise ratio) of the structures that they innervate. More specific behavioural observations are also consistent with a role for locus coeruleus neurons in arousal. Thus Susan Owen and I found that the partial reinforcement acquisition effect, a phenomenon that is best explained in terms of arousal and which is abolished by sodium amytal (Fig. 10.10), is absent in animals in which the dorsal bundle has been destroyed.

It is in the locus coeruleus and the dorsal bundle, then, that we find the output of the behavioural inhibition system (Fig. 12.4) that we failed to find within the septo–hippocampal system, namely, that of increased arousal. This output is effected by way of many of the different efferents from the locus coeruleus. Thus the enhancement of fixed action patterns (e.g. sexual behaviour, as in Beach & Fowler's experiment, described in Chapter 10) or reflex behaviour (as in the potentiated startle reflex) are likely to be mediated by noradrenergic afferents to the hypothalamus (Fig. 13.14) or brainstem, respectively. It remains unclear, however, how septo–hippocampal and coerulear influences on other brain structures are co-ordinated with each other, as they must be if they are jointly to subserve the functions I have attributed to them.

The opiate withdrawal syndrome

Redmond's group has emphasised a further important aspect of the functions of the locus coeruleus. This nucleus is rich in opiate receptors, and (like anti-anxiety drugs) opiates depress the firing of coerulear neurons. These observations have formed the basis of an interesting account of the distressing symptoms seen during opiate withdrawal, proposed by Gold, Redmond & Kleber. According to these authors, exogenous opiates initially depress firing in coerulear neurons but with continued administration of the drug there is a return to near-normal firing rates (that is, tolerance to the drug develops). Then, when the drug is withdrawn from an addicted individual, there ensues a rebound hyperactivity in coerulear neurons, or in the receptors that are post-synaptic to terminals from these neurons: this hyperactivity gives rise to the withdrawal syndrome. In support of this hypothesis it has been demonstrated that clonidine, which inhibits firing in coerulear neurons but by way of an α_2-noradrenergic receptor (see above) not an opiate receptor, is an effective treatment for withdrawal symptoms in Man.

If opiate withdrawal symptoms are due to hyperactivity in the locus coeruleus system, and if (as proposed here) activity in this system plays a key role in anxiety, it follows that the opiate withdrawal syndrome is a form of anxiety, perhaps a particularly intense form. This inference is supported by a consideration of the actual symptoms that make up the syndrome. Besides actual reports of anxiety, these include yawning, perspiration, lacrimation, goose flesh, tremors, hot and cold flushes, increased blood pressure, insomnia, increased rate and depth of breathing, increased pulse rate, and restlessness. Many of these changes also occur in states of anxiety. They may well reflect the activity of coerulear (or other noradrenergic) fibres that descend into the spinal cord, since they are suppressed by clonidine. It is possible, therefore, that the autonomic symptoms of anxiety are mediated, at least in part, by these descending coeruleo-spinal projections.

There is one difficulty with this account. In experiments with animals, exogenous opiate drugs, such as morphine, fail to alter behaviour in the manner of the anti-anxiety drugs. Indeed, the double dissociation between the effects of anti-anxiety drugs (which block responses to stimuli associated with pain but not responses to pain itself) and those of the opiates (which do the opposite) has been an important strand in the chain of evidence suggesting that separate systems mediate the behavioural effects of conditioned and unconditioned aversive stimuli respectively. Yet Redmond & Gold's work implies that opiates and anti-anxiety drugs should both reduce anxiety, since they both reduce the firing of neurons in the locus coeruleus.

In an attempt to resolve this dilemma I have suggested that there are different populations of coerulear neurons. One population has GABA/benzodiazepine/picrotoxinin receptor complexes and projects to the forebrain, where it facilitates the cognitive and behavioural aspects of anxiety; anti-anxiety drugs produce many of their effects by way of an action at these receptors. A second population of neurons has opiate receptors and projects into the spinal cord, where it mediates the autonomic symptoms of anxiety; here is where the opiates act. Both populations possess in addition α_2-noradrenergic receptors; in this way, clonidine is able to suppress all three types of symptom – behavioural, cognitive, and autonomic. This hypothesis has not yet been subjected to experimental test.

Behavioural inhibition and the ascending serotonergic pathways

We have so far located many of the functions of the behavioural inhibition system (Fig. 12.4) in either the septo–hippocampal system or the locus coeruleus system without yet saying anything about behavioural inhibition itself, that is, the suppression of ongoing behaviour that is such a cardinal feature of the operation of the behavioural inhibition system. To some extent this omission is a reflection of ignorance. Remarkably little is known about the way in which the septo–hippocampal system makes contact with motor systems (as it must, if it is to influence actual behaviour). Nonetheless, certain speculations are possible.

These are in part encapsulated in Fig. 13.14, which shows two possible routes: inhibition of motor plans by way of the projection from the subiculum to the cingulate cortex (a region which, together with the pre-frontal cortex, appears to be involved in the running of motor programs); and inhibition of fixed action patterns (eating, drinking, sexual behaviour, etc.) by way of a pathway shown by Albert and his collaborators to descend to the hypothalamus from the lateral septal area. In addition, recent anatomical work has demonstrated a large projection from the subiculum to the nucleus accumbens and other regions of the ventral striatum. This part of the brain appears to play a key role in transmitting incentive motivation to motor systems; that is, it discharges the functions attributed to the reward mechanism in Fig. 12.1 and 12.2. The input to the ventral striatum via the mesolimbic dopaminergic pathway from neurons located in the ventral tegmental area of the brainstem (Fig. 13.19) is apparently responsible for conveying information concerning the availability of

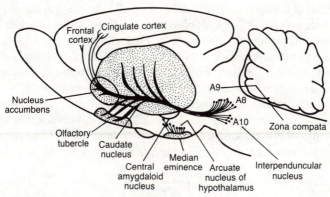

Fig. 13.19. The major dopaminergic pathways of the rat forebrain. From Cooper et al. (1982).

reward; striatal neurons are then responsible for the initiation of appropriate motor behaviour. It is consistent with this view that the mesolimbic dopaminergic pathway is a major substrate for electrical self-stimulation of the brain. These same neurons appear also to mediate the powerfully rewarding effects of drugs of abuse such as cocaine and heroin. Thus the subicular projection to the ventral striatum is well situated to provide the route by which the behavioural inhibition system inhibits rewarded instrumental behaviour.

Interestingly, there is also an inhibitory connection from the mesolimbic dopaminergic pathway to the septo–hippocampal system. Galey *et al.* have shown that the terminals of this projection in the lateral septal area inhibit the behavioural effects of activating the cholinergic input from the medial septal area to the hippocampus. Destruction of these terminals by local injection of 6-hydroxy-dopamine caused a number of behavioural changes that were opposite in sign to those produced by septal or hippocampal damage, as well as increasing cholinergic activity in the hippocampus. These data are consistent with the existence, as postulated in Figs. 12.1 and 12.5, of reciprocally inhibitory links between a reward mechanism (the mesolimbic dopaminergic system) and a mechanism for behavioural inhibition (the septo–hippocampal system).

In discharging, by whatever route, its function of behavioural inhibition (in the narrow sense) the septo–hippocampal system appears to act in close partnership with a second ascending monoaminergic pathway, one that uses serotonin as its transmitter.

The serotonergic pathways that innervate the septo–hippocampal system (along with many other forebrain structures) originate in the median and dorsal raphe nuclei (especially the median raphe) in the brainstem (Fig. 13.20) and follow similar routes to those taken by the noradrenergic fibres illustrated in Fig. 13.3. Like these fibres, the serotonergic afferents innervate the septal area and hippocampus extensively but diffusely. Further similarities between the two systems appear when we consider the way in which they respond to stress. Stressors of many kinds increase serotonin turnover (an indirect but reliable measure of firing rate), and this increase is reversed by anxiolytic drugs of diverse kinds. Some evidence suggests that like the noradrenergic input to the septo–hippocampal system, the serotonergic

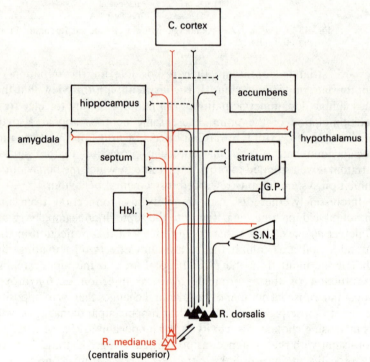

Fig. 13.20. Schematic representation of the main ascending projections from serotonin-containing cells of the dorsal (R. dorsalis) and the median (R. medianus, shown in red) raphe nuclei. These nuclei are the common source of overlapping and diffuse serotonergic input to various brain structures although some areas (striatum, nucleus accumbens or lateral septum) are principally innervated by fibres arising from the dorsal raphe whereas other regions (hippocampus, medial septum) mainly receive input from the median raphe. The arrows located between the dorsal and the median raphe indicate that these nuclei are functionally interconnected. C. cortex, cerebral cortex; G.P., globus pallidus; Hbl., habenular nuclei, S.N., substantia nigra. From Soubrié (1986).

input increases the ease with which a stimulus crosses the dentate/CA3 gate. Thus many of the reasons that led to the proposal of a role in anxiety for the noradrenergic fibres innervating the septo–hippocampal system apply with equal force to its serotonergic afferents. And, indeed, the hypothesis that anti-anxiety drugs alleviate anxiety by reducing activity in serotonergic neurons was advanced earlier than the corresponding noradrenergic hypothesis, by Graeff & Schoenfeld and by Larry Stein. Note that the two hypotheses are not mutually exclusive, although they have often been treated as though they were.

The serotonergic hypothesis of anxiolytic drug action is supported by several lines of evidence. When tests in which behaviour is suppressed by punishment are used, drugs that interfere with serotonergic transmission reliably reverse this behavioural suppression. Conversely, serotonergic agonists produce behavioural suppression. These effects have been related to the serotonergic pathways that innervate the forebrain in experiments in which the dorsal and median raphe nuclei have been stimulated directly. This treatment has been shown by Stein to suppress both punished and unpunished behaviour in the Skinner box, and these effects were reversed by injections of a benzodiazepine drug. Similarly, Graeff & Silveira Filho stimulated the median raphe electrically and observed suppression of bar-pressing accompanied by obvious signs of fear (crouching, defecation, urination, and piloerection). The reverse type of effect was demonstrated by Thiébot, Jobert & Soubrié: the behavioural suppression elicited by a signal associated with shock was reversed by injection of either chlordiazepoxide (Librium) or GABA directly into the dorsal raphe. These experiments are all consistent with the hypothesis that the behavioural inhibition produced by warnings of punishment is mediated by activity in ascending serotonergic fibres and that the anxiolytic drugs alleviate such behavioural inhibition by blocking activity in these pathways. Equally consistent results have been obtained in further experiments in which the ascending serotonergic fibres were destroyed by injecting them with a selective neurotoxin (5,6- or 5,7-dihydroxytryptamine).

It is clear that the reduction in response suppression seen in animals with impaired forebrain serotonergic transmission is not due to a loss of sensitivity to the primary aversive reinforcer. On the contrary, sensitivity to pain is, if anything, increased in such animals (later on, we shall see why this might be so). As in the comparable case of septal or hippocampal lesions, this change seems to be one of motor disinhibition. Thus the threshold for jumping in response to shock is lowered by injection of a drug which blocks the synthesis of serotonin but the threshold for the detection of shock is unchanged. There is other evidence for a more general loss of motor inhibition in animals with impaired serotonergic transmission. In the Open Field there is an increase in ambulation, provided the test is conducted with bright lights and noise (presumably increasing fear and so reducing movement in the untreated animals; see Chapter 4). In other tests, increased movement has been seen only if the environment is novel, again suggesting that fear-induced inhibition of movement is particularly susceptible to the effects of reduced serotonergic transmission. There is evidence that loss of motor inhibition is strongly related to damage to the serotonergic input specifically to the septo–hippocampal system. This effect is seen more clearly after damage to the median than to the dorsal raphe (and the median raphe is more closely related to the hippocampus). Furthermore, the degree of motor disinhibition seen after lesions of the median raphe is well correlated with the extent of loss of serotonin in the hippocampus. In a particularly neat experiment Williams & Azmitia injected 5,7-dihydroxytryptamine directly into the fornix-fimbria (Fig. 13.10), destroying only the serotonergic afferents to the hippocampus, and increased motor activity to a degree proportional to the ensuing loss of hippocampal serotonin.

In contrast to experiments on motor inhibition induced by stimuli associated with punishment or by novelty, tests involving non-reward have been much less sensitive to the effects of interference with serotonergic transmission. In my own laboratory, for example, Nicola Davis was unable to affect either resistance to extinction or the partial reinforcement extinction effect by destruction of forebrain serotonergic fibres using 5,7-dihydroxytryptamine. Similarly, Tye, Everitt & Iversen found no effect of this treatment upon responding during extinction of a bar-pressing response, even though the same animals showed release of bar-pressing suppressed by punishment. In some ways, these results neatly complement those obtained after destruction of the dorsal ascending noradrenergic bundle. For animals treated in this way show clear changes in non-rewarded behaviour, but much slighter changes in punished responding. This pattern of results might encourage the hypothesis that serotonergic fibres are specialised to deal with punished behaviour, and noradrenergic fibres to deal with non-rewarded behaviour. However, Tsaltas and I have shown reliable increases in punished responding after dorsal noradrenergic bundle lesions (though these increases are smaller than those seen after anxiolytic drugs or lesions of serotonergic systems); and there have been several reports of slight impairments in extinction after drugs that interfere with serotonergic transmission. Thus this hypothesis would be too simplistic.

A more fruitful place to look for the distinction between the functions of noradrenergic and serotonergic pathways is on the output side of the behavioural inhibition system. Stimuli associated with punishment and non-reward, respectively, both elicit all the outputs of the behavioural inhibition system (Fig. 12.4). However, it is usually the case that response suppression (becoming in the extreme freezing) is a more prominent feature of the behaviour of an animal threatened with, say, shock than of an animal faced merely with the loss of a reward. This distinction does not contradict the close equation between fear and frustration emphasised throughout this book. This equation allows for differences in the intensity in the unitary state set up by either conditioned fear or conditioned frustrative stimuli (see the start of Chapter 10). We may therefore suppose that behavioural inhibition, in the narrow sense, occurs at relatively higher levels of intensity of anxiety; that stimuli associated with punishment are on average more likely to elicit high-intensity anxiety than stimuli associated with non-reward; and that ascending serotonergic projections mediate (in conjunction with the septo–hippocampal system) behavioural inhibition, whereas ascending noradrenergic fibres are more concerned with increments in arousal and attention.

It is clear that according to this formulation there is no opposition between the hypothesis of a serotonergic role in anxiety and the hypothesis of a noradrenergic role. On the contrary, the two monoaminergic pathways appear to play complementary parts. Under appropriate conditions, the ascending noradrenergic input to the septo–hippocampal system prompts this into applying its computational powers to the analysis of anxiogenic stimuli; while the serotonergic input prompts it to inhibit motor behaviour (via septal projections to the hypothalamus, and hippocampal projections to the cingulate cortex and ventral striatum; see above). At the same time, the locus coeruleus system increases the activity of a number of other regions widely distributed in the brain (so exercising its general arousal function) and operates a number of changes in the autonomic nervous system via its descending output to the spinal cord. Co-ordination between the noradrenergic and serotonergic pathways may take place in part in the septo–hippocampal system itself and in part by way of the reciprocal projections that are known to connect the locus coeruleus and the raphe nuclei. Finally, as we have seen, the anti-anxiety drugs apparently produce their effects by acting on both the noradrenergic and the serotonergic systems.

Counter-conditioning

The anti-anxiety drugs block the acute behavioural effects of anxiogenic stimuli; but, in addition, they impair the capacity of the animal to develop tolerance for such stimuli. Is the latter type of effect also mediated by the septo–hippocampal system?

One series of experiments investigating this possibility has started from the observation that the anti-anxiety drugs abolish the partial reinforcement extinction effect (Fig. 10.8). Similar results are obtained after lesions of the septo–hippocampal system. Thus, Henke was able to abolish the partial reinforcement extinction effect by destroying the entire anterior septal area, including both the medial and lateral nuclei; and Rawlins, Feldon and I have found the same result after either destruction of the hippocampal formation or section of the fibres (in the fornix and fimbria) which connect the septal area and hippocampus. The same result again emerged when Owen and I destroyed the dorsal ascending noradrenergic bundle by injecting 6-hydroxydopamine into this pathway. In each case, the changes seen were like those produced by administration of the anti-anxiety drugs (Fig. 10.8): resistance to extinction was increased in the lesioned animals trained under continuous reinforcement and reduced in those trained under partial reinforcement (Fig. 13.21).

Now, as we saw in Chapter 10, the anti-anxiety drugs alter the partial punishment effect in much the same way that they alter the partial reinforcement extinction effect (Figs. 10.8 and 10.9). We therefore expected that damage to the septo–hippocampal system would also have this effect. However, several experiments conducted by Brookes in my laboratory failed to alter the partial punishment effect by lesions to either the septal area or the hippocampus. It is clear, therefore, that, while both septo–hippocampal damage and anti-anxiety drugs impair an animal's ability to develop tolerance for aversive stimulation, they do so by affecting different mechanisms.

The easy way out of Brookes' results is to suppose that the septo–hippocampal system mediates tolerance for non-reward (the partial reinforcement extinction effect) but not punishment (the partial punishment effect). Rawlins, however, suggested an alternative possibility: that the effects of septo–hippocampal damage depend critically upon the interval that separates the aversive and appetitive events in a counter-conditioning paradigm. If this interval is sufficiently short, Rawlins proposed, the animal could do without its hippocampus, the business of associating aversive and appetitive events being carried out elsewhere in the brain. The hippocampus is necessary, in contrast, if the interval between these events exceeds a critical value (of the

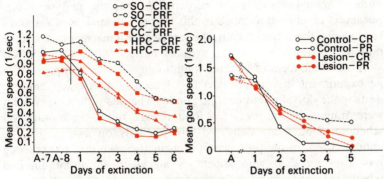

Fig. 13.21.(a) Blockade of the partial reinforcement extinction effect by lesions to the hippocampus. Running speeds on the last 2 days of acquisition (A-7 and A-8) and during extinction (6 trials/day) are shown as a function of continuous reinforcement (CRF) or partial reinforcement (PRF) during acquisition in groups of rats that had had sham operations (SO), removal of cerebral cortical tissue equivalent to that damaged by the hippocampal lesions (CC), or removal of the entire hippocampal formation by aspiration (HPC). From Rawlins *et al.* (1980*a*). (b) Blockade of the partial reinforcement extinction effect by destruction of the dorsal ascending noradrenergic bundle. Speeds in the goal section of the alley are shown for the last day of acquisition (A) and during extinction as a function of continuous reinforcement (CR) or partial reinforcement (PR) during acquisition in rats that had sustained an injection of the catecholamine-selective neurotoxin 6-hydroxydopamine into the dorsal bundle (reducing hippocampal and cerebral cortical levels of noradrenaline by about 90%) or a control operation. From Owen *et al.* (1982).

order of many seconds or a few minutes depending upon the task, as shown in subsequent experiments testing Rawlins' hypothesis).

Applying this analysis to the two phenomena of interest to us here, we note that the critical interval is very short in the partial punishment effect (since the rat typically eats the food reward immediately after being shocked in the goalbox); but in the partial reinforcement extinction effect it is as long as the inter-trial interval (of the order of 5 minutes in the experiments described above). On Rawlins' hypothesis, therefore, we should be able to demonstrate counter-conditioning using non-reward in animals with hippocampal damage, provided we radically reduce the interval between non-reward and the following reward. In agreement with this prediction, Rawlins and Feldon were able to demonstrate an intact partial reinforcement extinction effect in hippocampectomised rats by reducing the inter-trial interval to just a few seconds.

On the basis of these and other results Rawlins has proposed that the hippocampus acts as a high-capacity, intermediate-term memory buffer. This hypothesis is not incompatible with the analysis of hippocampal function developed in this chapter; indeed, Rawlins and I have shown that, theoretically, the same general machinery can serve both as a comparator and as a memory buffer.

Yet another form of counter-conditioning has been implicated in an experiment in which Tsaltas and I used 6-hydroxydopamine to destroy the dorsal noradrenergic bundle. The rats in this experiment were trained to bar-press for food reward on a random-interval schedule, food being delivered unpredictably on average once a minute. Then, in the presence of a distinctive stimulus, bar-pressing was also punished with shock on an identical random-interval schedule. In one condition the two schedules were independent, food and shock delivery being unrelated to each other. In a second condition, a single schedule determined both food and shock delivery, so the animal took the shock and at once ate the food. The latter condition permits Pavlovian counter-conditioning (Chapter 9) to occur, and so it did: in control animals the amount of suppression of bar-pressing was less when food regularly followed shock than when food and shock were unrelated. But the counter-conditioning effect was completely absent in the animals with dorsal bundle lesions.

However, while Tsaltas' experiment implicates ascending noradrenergic fibres in Pavlovian counter-conditioning, in the present context her results are unexpected. For, using a paradigm described in Chapter 9 to produce Pavlovian counter-conditioning, McNaughton and I found no change in this phenomenon when the animals were injected with chlordiazepoxide (Fig. 9.6). Thus, while both the septo–hippocampal system and the dorsal noradrenergic bundle play roles in counter-conditioning, in neither case does this role at present appear to be of a kind that could mediate the effects of the anti-anxiety drugs: these drugs block the partial punishment effect, yet septo–hippocampal damage does not; and dorsal bundle lesions block Pavlovian counter-conditioning, yet the drugs do not.

Toughening up

The experiments described in the previous section were concerned with associative mechanisms for the development of behavioural tolerance for stress. But, as we saw in Chapter 9, tolerance for stress may also be increased non-associatively. If we enquire about the neural basis for such non-associative effects, we are once again led to the septo–hippocampal system and its noradrenergic afferents.

Consider Weiss's experiments on toughening up. These experiments showed that one session of inescapable shock is able to interfere with subsequent escape and avoidance learning; but 15 consecutive such sessions no longer produce this interference. In parallel with these behavioural observations Weiss has shown that one session of inescapable shock causes a fall in brain noradrenaline levels, but this fall is not seen after multiple sessions. This appears to be due to an elevation in the capacity of the neuron to synthesise noradrenaline, since Weiss also reports that multiple shock sessions cause an increase in the activity of the rate-limiting enzyme in this synthesis, tryosine hydroxylase (Fig. 5.5), especially in the locus coeruleus. This phenomenon appears to be similar to the trans-synaptic enzyme induction (Fig. 5.9) that is seen in the adrenal medulla and sympathetic nervous system when animals are stressed, although the detailed mechanism underlying Weiss's observations has not yet been elucidated. (Induction of tyrosine hydroxylase in the brain has also been seen by other investigators after either cold stress or drug treatments. The function of the increased tyrosine hydroxylase activity is presumably to allow the neuron to maintain sufficiently high levels of noradrenaline to meet the heavy demands for transmitter release due to stress.)

Weiss's experiments, then, implicate the locus coeruleus system in non-associative toughening up. Experiments in my own laboratory have in addition implicated the septo–hippocampal system.

These experiments start out from two previous findings already known to you: (1) the abolition of the partial reinforcement extinction and partial punishment effects by the anti-anxiety drugs (Figs. 10.8 and 10.9); and (2) the elimination by these drugs of the 7.7-Hz minimum in the theta-driving curve (Fig. 13.17). Now, let us suppose that the latter effect plays a causal role in the former. One might then be able to produce effects opposite in sign to those of the anti-anxiety drugs by 'injecting' 7.7-Hz theta into the rat's brain by way of appropriate electrical stimulation of the septal pacemaker cells that control theta. Let us suppose further that the process which gives rise to the partial reinforcement and partial punishment effects, and which is blocked by the anti-anxiety drugs, is essentially non-associative even though it is buried in an associative paradigm (a point to which we return below). It then follows that it should be possible to produce toughening up by a course of theta-driving stimulation applied prior to any behavioural training whatsoever.

In accordance with this line of reasoning Lee Holt and I implanted rats with septal stimulating and hippocampal recording electrodes and gave some of them a course of 10 days' theta-driving stimulation (15 6-second trains of pulses per day); other, control animals were treated identically, except that they were not stimulated. Only after this were the rats (without further stimulation) trained to bar-press for food reward. We then, in three separate experiments, attempted to suppress bar-pressing by extinction, punishment with foot-shock, or presentation of a stimulus paired with response-independent foot-shock (on-the-baseline conditioned suppression). In each experiment the previously stimulated rats were more resistant to suppression of bar-pressing than the controls, even though the critical behavioural measurements were made some 3–4 weeks after the end of the period of brain stimulation.

Now, electrical brain stimulation inevitably affects many pathways besides the intended one. Additional evidence is needed, therefore, to show that Holt's effects were indeed due to the driving of hippocampal theta that the septal stimulation produced. We have indeed obtained such evidence. It is possible to block hippocampal theta (substituting a low-voltage, desynchronised, predominantly high-frequency pattern of waves) by high-frequency septal stimulation. The effect of such stimulation upon paths other than the one controlling theta would be expected to be greater than low-frequency, theta-driving stimulation. Thus, if the critical aspect of low-frequency stimulation is that it drives theta, then high-frequency septal stimulation should not increase later tolerance for stress; but, if other pathways are involved, high-frequency stimulation should, if anything, produce a bigger effect than low-frequency stimulation. When Holt and I did this experiment we found that high-frequency, theta-blocking septal stimulation *decreased* the resistance to extinction of a bar-press response acquired subsequent to the period of stimulation – strong evidence that low-frequency stimulation indeed produces its effect because it drives theta.

In other experiments Jonathan Williams and I have investigated the role of the frequency of the septal stimulating current and the theta rhythm it elicits. Increased resistance to extinction was seen after stimulation at 7.7 or 8.4 Hz. But a slight reduction in frequency from 7.7 to 7.5 Hz (that is, a lengthening of the inter-pulse interval from 130 to 133 msec) was sufficient to reverse the effect, subsequent resistance to extinction now being decreased. This exquisite dependence of the behavioural effects of theta-driving upon the exact frequency of the stimulating current corresponds well with the frequency-dependence of the effects of the anti-anxiety drugs upon the theta-driving curve (Fig. 13.17), suggesting that common processes may be involved.

As we have seen, the toughening up seen in Weiss's experiments appears to depend upon changes in noradrenergic neurons. Given the special relationship between 7.7-Hz theta driving and the noradrenergic input to the septo–hippocampal system (Fig. 13.18), it seemed possible that Holt's effect might also involve changes in these neurons. Graham-Jones, Fillenz & Holt therefore measured hippocampal tyrosine hydroxylase activity in the brains of rats that had been given 7.7-Hz theta-driving stimulation, with or without subsequent bar-pressing training and extinction. At periods of 15–33 days after stimulation there was an elevation of tyrosine hydroxylase activity; this effect was not observed after high-frequency, theta-blocking septal stimulation. Thus the behavioural changes produced by theta-driving stimulation may depend upon the same kind of trans-synaptic enzyme induction that is observed after chronic stress in the sympathetic nervous system and adrenal medulla (Chapter 5). As in the sympathetic nervous system, we have a cholinergic pre-synaptic cell (the septal projection to the hippocampus) which synapses upon a noradrenergic cell (but at the terminal of this cell in the hippocampus, not the cell-body in the locus coeruleus). However, even supposing the neurochemical and behavioural effects of theta-driving stimulation are causally related, the mechanism by which the one gives rise to the other is at present unknown.

The arguments we have pursued above assume at several points that it is a non-associative process whose blockade by the anti-anxiety drugs underlies their abolition of the partial reinforcement extinction effect. This assumption is apparently at variance with a conclusion reached in Chapter 10, namely, that the anti-anxiety drugs antagonise Amselian conditioned frustration, an associative process. This contradiction may, however, be more apparent than real.

There are two points at which associative mechanisms enter into Amsel's account of the partial reinforcement extinction effect. First, frustration must become conditioned to stimuli preceding non-reward. Second, the cues provided by conditioned frustration become discriminative stimuli for continued performance of the instrumental response, which is rewarded in the presence of these cues. It is the second of these processes which is usually thought to rob non-reward of its capacity to disrupt the behaviour of the partially reinforced animal. But suppose that conditioned frustration loses its capacity to disrupt behaviour simply in virtue of its repeated elicitation. This, a non-associative mechanism, is not incompatible with the associative mechanism described above; indeed, under normal conditions, both might act to create tolerance for non-reward. Now let us suppose that 7.7-Hz theta driving mimics the central effects of conditioned frustration. The non-associative mechanism of adaptation proposed above could then give rise to increased resistance to extinction as observed in Holt's experiments. Further, since conditioned frustration and fear are identical, the same mechanism would give rise to the increased resistance to punishment and to conditioned suppression that he also observed.

The fight/flight system

It is time to descend from these rarefied heights of the theory of learning to deal with simpler matters: the systems that mediate responses to unconditioned aversive stimuli. These responses include such behaviour patterns as striking, biting, the adoption of an upright defensive posture with forepaws raised (appropriately enought termed a 'boxing' posture), flight, upwards rearing and jumping, and hissing and squealing. It has proved possible either to elicit these behaviour patterns by brain stimulation or to eliminate them by brain lesions, so allowing one to map the brain systems concerned. A number of attempts have been made in the drawing of such maps to distinguish between brain regions mediating subgroups among these behaviour patterns, e.g. between defensive attack and flight, or between the defense used against a predator and the submissive postures shown towards conspecifics. But none of these subdivisions has yet met universal approval; and it is clear that if different brain systems mediate different parts of the totality of defensive behaviour patterns listed above, these systems are closely intermingled and share much the same brain space. Thus, since finer distinctions of this kind are in any case not germane to the arguments pursued in this book, I shall speak here only of a single fight/flight system (behavioural arguments also motivating this decision were presented in Chapter 12).

Fig. 13.22 shows results obtained by De Molina & Hunsperger in Zurich in their work with cats. On the basis of the behaviour elicited by central stimulation, they describe a system with its highest level in the amygdala, which then descends along a tract called the stria terminalis to the ventromedial hypothalamus and thence to the central grey of the midbrain. In the black area in this figure, stimulation produced defensive attack against suitable objects in the environment, and in the pink area surrounding the black area it usually produced flight. In the red area, both kinds of effect were produced. That they were dealing with a single system at different levels of the nervous system was shown by combined stimulation and lesion experiments: when the ventromedial hypothalamus was removed, stimulation of the amygdala no longer had these effects, and removal of the central grey similarly abolished the effects of hypothalamic stimulation.

De Molina & Hunsperger's scheme has required one major modification: Hilton & Zbrozyna have shown that the amygdala influences the medial hypothalamus, not by way of the stria terminalis, but by the more diffuse ventral amygdalofugal pathway. With this exception, many experiments have confirmed the role of the structures picked out in Fig. 13.22. Recordings made from single neurons (via chronically implanted electrodes) have shown that firing in both the central grey and the ventromedial hypothalamus is well correlated with defensive attack or attempts to escape. Other experiments have shown that animals readily learn to switch off electrical stimuli applied to the same hypothalamic and midbrain sites from which fight or flight is

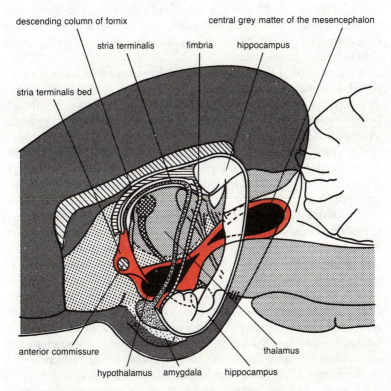

Fig. 13.22. Results of De Molina & Hunsperger's experiments on the effects of electrical stimulation of the brains of cats. In the black area in the central grey matter and the hypothalamus, stimulation produced defensive attack; in the pink area surrounding the black area, it produced flight; and in the red area in the amygdala, the stria terminalis and the bed nucleus of the stria terminalis, it produced growling, miaowing, and flight.

elicited, suggesting that an aversive emotional state is generated by the electrical stimulation.

In the central grey it is likely that electrical stimulation owes its aversive quality to its capacity to elicit pain; as we have seen (Fig. 13.5), this region receives a major input from spinal neurons mediating pain. This inference is confirmed by reports from people who have undergone electrical stimulation of the central grey in preparation for neurosurgery: they speak of poorly localised sensations of pain, accompanied by strong feelings of fear or impending death. Similar reports are given by people who have been stimulated in the amygdala. In addition to behavioural changes, stimulation of the central grey and medial hypothalamus elicits marked autonomic changes, including increases in blood pressure and heart rate, muscular vasodilation, constriction of blood vessels in the viscera and skin, and hyperventilation. Nashold, Wilson & Slaughter, who reported some of these observations, comment that they resemble the changes seen in patients experiencing extreme pain.

We have several times developed the argument that there is a functional equivalence between unconditioned painful stimuli and unconditioned non-reward. If this equivalence is based upon a common brain system, then brain lesions that reduce defensive aggression should also impair behavioural responses to unconditioned frustrative non-reward. The results of experiments in which the amygdala has been destroyed agree with this prediction. Such lesions are an effective means of taming animals. In addition, Henke has reported an interesting pattern of change when rats with amygdalar lesions were tested in the double runway (Fig. 10.3). These animals showed no sign of the normal frustration effect (i.e. faster running in the second alley after non-reward, relative to reward, in the first goalbox). Yet the same animals showed a normal partial reinforcement extinction effect, which is controlled by secondary frustrative stimuli (Chapter 10). This pattern of change is exactly opposite to the one seen after anxiolytic drugs or septo–hippocampal damage. Thus Henke's results fit comfortably with the hypothesis that the amygdala mediates responses to both unconditioned pain and frustration, but not responses to conditioned stimuli associated with these primary aversive events.

The interactions between the three levels of the fight/flight system – amygdala, hypothalamus, and midbrain – have been subjected to further investigation in experiments by Kling and his collaborators, who found that cats rendered docile by lesions to the amygdala were subsequently made extremely ferocious by further lesions to the ventromedial hypothalamus. Cats first made ferocious by lesions in the hypothalamus were, on the contrary, unaffected by subsequent amyg-dalectomy. As Deutsch & Deutsch point out, 'one hypothesis consistent with these findings is that the amygdala exerts an inhibitory influence on structures in the ventromedial nucleus and that this nucleus in turn inhibits other structures responsible for the elicitation of fear and rage' – presumably the central grey of the midbrain.

Other experiments have been concerned with the interactions between lesions to the amygdala and lesions to the septal area. Damage of the latter kind has been frequently observed to give rise to a syndrome known as 'septal hyper-reactivity', which includes an increase in the magnitude of the startle reflex, exaggerated aggressive behaviour in response to handling by the experimenter, increased resistance to capture, and increased squealing. This syndrome was abolished by subsequent lesions to the amygdala, or prevented by prior lesions to the amygdala, in an experiment by King & Meyer. In view of these findings, we may perhaps extend Deutsch & Deutsch's hypothesis into the pattern of interconnections shown in Fig. 13.23.

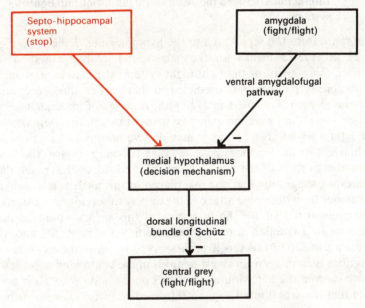

Fig. 13.23. The major components of the proposed fight/flight system. The final common pathway for fight/flight behaviour is in the central grey; this is under inhibitory control from the medial hypothalamus. This inhibitory control may be disinhibited by impulses arriving from the amygdala or intensified by an input from the behavioural inhibition system (probably by way of a pathway descending from or through the lateral septal area).

We suppose that the ventromedial nucleus of the hypothalamus exercises a tonic (i.e. normally present) inhibition over the midbrain central grey and that this latter area is the ultimate executive organ for either flight or defensive attack. We further suppose that this inhibitory path may itself either be inhibited from the amygdala, releasing fight/flight, or be intensified by influences proceeding from the septo–hippocampal system. These latter influences would counterbalance the amygdaloid response to novel or threatening stimuli, leaving the outcome (passive avoidance or fight/flight) to be determined by which of the inputs to the ventromedial hypothalamus is stronger under a particular set of environmental conditions. This scheme, as well as accounting for the pattern of interactions between septal, amygdaloid, and hypothalamic lesions just described, also makes sense out of the common observation that under conditions of danger, both freezing and fight/flight patterns may be intensified at the same time and alternate rapidly with each other.

Interactions between the behavioural inhibition and fight/flight systems

The route from the septal area to the hypothalamus depicted in Fig. 13.23 provides a means for the behavioural inhibition system to restrain the activity of the fight/flight system; the existence of such an inhibitory pathway was predicted in that part of the conceptual nervous system illustrated in Fig. 12.5. A series of experiments by Albert has shown that this pathway passes through or originates in the lateral septal area (where it may receive an input from the subiculum and/or the hippocampus) and then descends towards the ventromedial nucleus of the hypothalamus. David Adams has made the interesting suggestion that one role played by this nucleus is to switch behaviour from defensive attack to the display of submissive postures. The optimal trigger for the operation of this switch would be the presence of a familiar, dominant conspecific (cf. Chapters 2 and 6). Evidence in support of this hypothesis is so far slim, though Adams describes recordings from single neurons in the ventromedial nucleus which showed the behavioural correlates one would expect. It is possible that the septal input to the ventromedial nucleus is responsible for conveying information concerning social interaction. Consistent with this hypothesis, septal lesions increase the time spent in social contact in a variety of species (rats, mice, hamsters, cats, etc.).

Other possible loci of inhibitory control over the fight/flight system lie in the central grey. An important series of experiments by Graeff and his co-workers has analysed the neurochemistry of this region. Neurons in the central grey appear to be under three kinds of inhibitory control. One is GABAergic; this provides the basis for the action of the anti-anxiety drugs, which (as we saw in the last chapter) reduce the aversive effects of electrical stimulation at the midbrain or hypothalamic levels of the fight/flight system. A second is serotonergic. Whereas drugs that antagonise serotonergic transmission reduce behavioural inhibition (e.g. in a punished bar-pressing task), these agents enhance the aversive effects of stimulation in the fight/flight system. Activation of the serotonergic input to the central grey could provide a second route by which the behavioural inhibition system controls the fight/flight system; but, since this route does not go by way of the septal area or hippocampus, it is unclear how its activity could be co-ordinated with activity in the latter structures.

A third type of inhibitory control over central grey neurons is opioid. The central grey contains opioid peptides and receptors. Micro-injection of morphine into this region causes analgesia, as does electrical stimulation of the lateral edge of the central grey. Both these effects are blocked by section of a distinct pathway that descends in the dorsolateral spinal cord. Basbaum & Fields have shown that this pathway has its origin in the raphe magnus, a serotonergic nucleus in the medulla which receives an excitatory input from the central grey. These workers have therefore proposed the model for endogenous pain control illustrated in Fig. 13.24. Essentially, this model is a negative feedback circuit: pain signals ascending in the spinal cord arrive at the central grey and, via the descending pathway just described, restrain further activity in pain transmission cells (cf. Fig. 13.7). However, the model also provides a possible basis for the analgesic effect of conditioned fear. As we saw in Chapter 2, one effect of presenting an animal with a conditioned fear stimulus is a reduction in pain. This effect is blocked by the opiate antagonist naloxone, suggesting that it is mediated by endorphins; these perhaps act by way of the opiate receptors in the central grey. Consistent with this hypothesis, Miczek has shown that naloxone injected directly into this region blocks the analgesia normally induced in a mouse that is defeated in a fight with another mouse.

The decision mechanism

The function attributed to the ventromedial nucleus of the hypothalamus in the preceding two sections is essentially that of deciding between defensive attack and passive avoidance (the latter

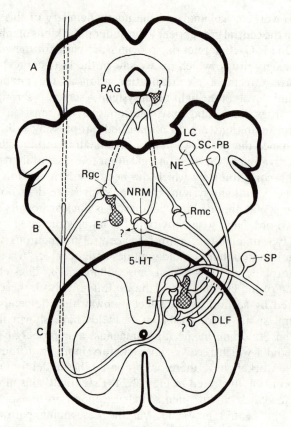

Fig. 13.24. The endogenous pain control system as proposed by Basbaum & Fields (1978). A: Midbrain level. The periaqueductal grey (PAG), an important locus for stimulation-produced analgesia, is rich in enkephalins (E) and opiate receptors, though the anatomical details of the enkephalinergic connections are not known. Micro-injection of small amounts of opiates into PAG also produces analgesia. B: Medullary level. Serotonin (5HT)-containing cells of the nucleus raphe magnus (NRM) and the adjacent nucleus reticularis magnocellularis (Rmc) receive excitatory input from PAG and, in turn, send efferent fibres to the spinal cord. C: Spinal level. Efferent fibres from the NRM and Rmc travel in the dorsolateral funiculus (DLF) to terminate among pain-transmission cells concentrated in laminae 1 and 5 of the dorsal horn. The NRM and Rmc exert an inhibitory effect specifically on pain-transmission neurons. The pain-transmission neurons, which are activated by substance P (SP)-containing small-diameter primary afferents, project to supraspinal sites and indirectly, via the nucleus reticularis gigantocellularis (Rgc), contact the cells of the descending analgesia system in the PAG and NRM, thus establishing a negative feedback loop. Catecholamine-containing neurons of the locus coeruleus (LC) in rat and subcoeruleus–parabrachialis (SC-PB) in cat may also contribute to pain-modulating systems in the DLF. (NE, norepinephrine, i.e. noradrenaline.) From Melzack & Wall (1983).

behaviour perhaps taking the form, if Adams' hypothesis is correct, of submissive postures). We may begin to see this function in a wider context by considering the results of the Olds' self-stimulation studies (Chapter 12) in more detail than we have hitherto. With certain electrode placements Olds and his collaborators obtained some quite remarkable effects:

The animal appeared to respond to turn the stimulus on even though each application of the stimulus seemed aversive. It was as though the animal could not stay away from the stimulus even though it could not stand it. The stimulus appeared to be both rewarding and aversive at the same time . . . . Very energetic efforts were made to escape from the test chamber, and when these were successful the animal would not come back of its own accord. But when the animal was brought back and forced to remain in the test chamber, it did maintain a substantial rate of 'positive reinforcement' behaviour (still struggling futilely to escape). (Olds & Olds, 1965)

In a major review of the experiments done in their own laboratory, James and Marianne Olds come to the not unnatural conclusion that these electrodes are simultaneously stimulating two incompatible systems. Having further studied the location of the points from which these effects were obtained, they describe their results as follows:

To understand the organisation discovered, one may imagine a black box with two large multichannelled cables entering from different sides, along with innumerable smaller cables. The black box appeared to be mainly a junction point between the two very large cable systems. The finding was:
1. Stimulation of one of the cables yielded pure positive reinforcement;
2. Stimulation of the other cable yielded pure negative reinforcement; and
3. Stimulation of the black box itself yielded the mixed responses. The black box was the midline system of hypothalamic nuclei, all of which appeared to be involved. The two large cables were the medial forebrain bundle and the periventricular system of fibres. (Olds & Olds, 1965)

These results are pictured in Fig. 13.25.

Now we have already met one of these midline hypothalamic nuclei: the ventromedial nucleus, to which we allotted a role in the selection of which form of aversive behaviour to display as a result of inputs from the septo–hippocampal system and the amygdala. We are now in a position to expand this suggestion into the hypothesis that the midline system of hypothalamic nuclei has a more general role in the selection between approach behaviour and aversive behaviour. In terms of the conceptual nervous system set out in Chapter 12, this amounts to the hypothesis that the midline hypothalamic nuclei constitute the 'decision mechanism'.

The midline nuclei are well qualified for this role. There are several such nuclei, related to different forms of motivated behaviour – eating, control of water balance, temperature regulation, etc. They all appear to be intimately related both to a final common pathway for fight/flight (the periventricular system of fibres, also known as the dorsal longitudinal bundle of Schütz, and to a final common pathway for approach (the medial forebrain bundle), both of which descend into the mid-brain (see Fig. 13.25). They also receive inputs from the amygdala, able to promote fight/flight; from the septo–hippocampal system, promoting passive avoidance; and from the medial forebrain bundle and the lateral hypothalamus, promoting consummatory behaviour. Between them, these connections appear capable of providing the pattern of inhibitory links postulated in Fig. 12.5 as relating (1) the behavioural inhibition system (mediating the effects of conditioned aversive stimuli), (2) mechanisms for unconditioned consummatory behaviour, and (3) mechanisms for unconditioned fight/flight behaviour.

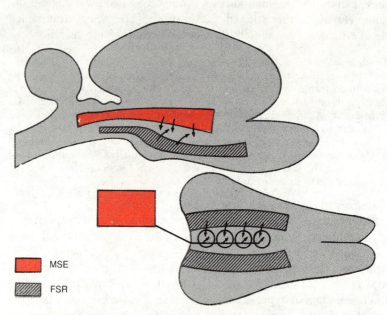

MSE

FSR

Fig. 13.25. Olds' map of rewarding (the 'forebrain substrate of reward' or FSR) and punishing (the 'midbrain substrate of escape' or MSE) points in the brain on schematic saggital and horizontal sections of a primitive mammalian brain. Pure rewarding effects of electrical stimulation were found in the FSR, pure punishing effects in the MSE, and ambivalent or mixed effects in the nuclei (circled) into which both systems project.

Classical conditioning

One function of aversive stimuli is to elicit fight/flight behaviour; and this, as we have seen, is mediated by a system of which the forebrain component is the amygdala (Figs. 13.22 and 13.23). A second function of such stimuli is to act as the UCS for the Pavlovian conditioning of fear. There is evidence that this function, too, is discharged by the amygdala.

A half a century ago Klüver & Bucy reported some remarkable phenomena following lesions to the temporal lobes of the cerebral hemispheres in monkeys. Their lesions involved, besides the temporal cortex, the hippocampus and the amygdala, but subsequent research has shown that the most important structure in the effects they obtained is probably the amygdala. The most striking aspects of the 'Klüver–Bucy syndrome' can be summarised by saying that the lesioned animals appear no longer to know nor to be able to learn what sort of reinforcing event follows upon which stimuli. They try to eat inedible objects or to copulate with sexually inappropriate objects (although experiments fail to disclose any sensory loss *per se*). At the same time, they lose their fear of other animals, of Man, and of normally dangerous objects such as a hissing snake or a burning match. Findings of this kind have been made, not only in monkeys, but in many other species. Increased tameness as a result of lesions to the amygdala has been seen in rats, cats, lynxes, and agoutis, besides monkeys, although in a few cases the opposite result – increased aggressiveness – has been observed. Direct studies of classical conditioning of fear to an initially neutral stimulus have shown this to be impaired in both cats and monkeys.

We may propose, then, that the amygdala is critically involved in the classical conditioning of rewarding and punishing significance to initially neutral stimuli. The consequence of lesions in this area is that the animal mixes up rewards and punishments, and rewards of different kinds, approaching stimuli normally shunned, or misdirecting sexual behaviour to edible objects and vice versa.

This is as far as we can take our mapping of the conceptual nervous system for avoidance behaviour on to the real brain. We have been able to make a number of tentative identifications of components in the one with structures in the other. These identifications are listed in Table 13.1. Armed with this table we can imagine the following sequence of events when an animal is faced with a stimulus which is at first novel, but which acquires some sort of significance as a result of the environmental regularities in which it participates.

Table 13.1 *An attempted equivalence between the components of the conceptual nervous system discussed in Chapter 12 and the brain structures described in Chapter 13*

Component of the conceptual nervous system	Structure in the real nervous system
Arousal mechanism	Ascending noradrenergic fibres (Fig. 13.3)
Control of unconditioned fight or flight	Central grey of the midbrain, medial hypothalamus, amygdala (Fig. 13.22)
Behavioural inhibition system (BIS)	Septo–hippocampal system and associated structures (Figs. 13.9 and 13.14)
Reward mechanism for approach behaviour	Ascending dopaminergic fibres (Fig. 13.19), dorsal and ventral striatum
Decision mechanism	Midline hypothalamic nuclei (Fig. 13.25)
Reciprocal inhibitory links between BIS and reward mechanism	Subiculo–striatal projection; dopaminergic projection to lateral septal area
Comparator for novelty/familiarity	Hippocampus (Fig. 13.9)
Classical conditioning of reward or punishment	Amygdala

On the first few occasions when the animal is exposed to the novel stimulus, the septo–hippocampal 'stop' system dominates, and inhibition of both approach and fight/flight is maintained. As the meaning of the stimulus changes with repetition (through the operation of the hippocampal analysis of familiarity), or with the attachment of rewarding or punishing significance by following unconditioned stimuli (acting via the amygdala), the relative strengths of the different inputs to the medial hypothalamus will change correspondingly, until either approach, escape, or attack behaviour dominates. Skilled escape behaviour and active avoidance behaviour, in accordance with the conclusions we reached in Chapter 11, would be learnt on the basis of inputs arriving from the reward system. Frustrative non-reward, on the other hand, in accordance with the views we advanced in Chapter 10, would act via the septo–hippocampal system.

Theories are not facts, but guides to useful ways of looking at the facts. As a blueprint of reality, the mapping of the conceptual nervous system for avoidance behaviour on to the real nervous system which has been proposed in this chapter is bound to be wrong. We must hope that it is not so wrong that it cannot serve as a useful guide to the physiological psychologist in his design of informative experiments, and to the neurologist and the psychiatrist in their interpretations of the physiological basis of disorders of behaviour. There is also one further way in which this kind of speculation may be of use: it can help us in the construction of a likely theory of the physiological basis of the organisation (and disorganisation) of human personality. But this is part of the business of our final chapter.

14   Man : neurosis, neuroticism, therapy

Can the principles we have been able to abstract from the study of fear and avoidance learning in animals help us to understand similar forms of behaviour in Man? Throughout the book I have tried to draw the implications of particular experiments for particular aspects of human behaviour. In this, our final chapter, I shall try to consider in a more general way the organisation of fearful behaviour in Man and the light that is thrown on this organisation by the experimental study of fear in animals.

In this, as in so many other areas of biology, much can be learned about normal functioning from a study of disorders of function. For this reason, and because of the intrinsic interest of the subject, we shall confine our attention to those kinds of behaviour which are called 'neurotic' in the psychiatric hospital. For the most part, these kinds of behaviour can be regarded as fear gone wrong, either because it is excessive, or because it is inappropriate, or because it has no apparent object. This is not a description of the neuroses which will by any means meet with universal agreement from those concerned with the treatment of these disorders. But I shall try to justify its aptness later in the chapter.

Anxiety and the human brain

Our confidence in the applicability to Man of the ideas developed in this book would be greater if there were reason to believe that the same systems participate in anxiety in animals and in the human experience of anxiety. There is indeed evidence to this effect.

The centre-piece in the edifice constructed in the previous chapter was the septo–hippocampal system. As we saw, lesions in this system reproduce many of the behavioural effects of the anti-anxiety drugs (Table 12.2). Furthermore, these substances produce characteristic changes in the theta-driving curve (Fig. 13.17) that are paralleled by differences between (1) strains of rats selectively bred for high and low fearfulness, and (2) the sexes (which have naturally differing levels of fearfulness; Chapter 7). The septo–hippocampal system, therefore, is the first place we would look for differences between people with characteristically high or low levels of susceptibility to anxiety. Recently, in a study using the non-invasive technique of positron-emission tomography (PET scanning) to visualise the levels of activity of different brain regions, Reiman *et al.* have reported results in line with this expectation. Patients suffering from spontaneous attacks of panic differed from controls in only one brain region: namely, the one containing the major input to the septo–hippocampal system (the entorhinal cortex) and the origin of its major output (the subicular region). The nature of the observed difference, however, was unexpected: controls showed equal activity in this region on both sides of the brain, whereas the patient sample showed relatively more activity in the right hemisphere.

An almost equally important role in the neural system described in the last chapter is played by the locus coeruleus and the noradrenergic fibres to which this nucleus gives origin. Much of the noradrenaline released by these figures undergoes metabolic conversion to 3-methoxy-4-hydroxyphenylglycol (MHPG). About half the MHPG found in human plasma probably comes from this source, thus providing an indirect measure of the activity of locus coeruleus neurons. The evidence reviewed in the previous chapter suggests that coerulear neurons should be particularly active during states of anxiety, and conversely that one way to induce anxiety would be to boost the activity of these neurons. Both these predictions were substantiated in a study by Charney and his co-workers using yohimbine. As we saw in the last chapter, this drug is an antagonist at α_2-noradrenergic receptors and increases the firing rate of locus coeruleus neurons by blocking auto-receptor mediated inhibition. In Charney's experiment, yohimbine administration increased the levels of both self-reported anxiety and plasma MHPG and these two effects were well correlated with each other. Furthermore, both effects were greater in patients who suffered from panic attacks than in controls.

There is also evidence that the other major monoamine transmitter discussed in the last chapter – serotonin – discharges the same functions in Man as in animals. Recall that pharmacological blockade of serotonergic transmission alleviates punishment-induced suppression of instrumental behaviour. This is also a classic effect of the anti-anxiety drugs. Many theorists have therefore supposed that blockade of serotonergic transmission should reduce anxiety in Man. A closer look at the role played by serotonergic pathways in response suppression, however, shows the dangers of arguing in this way directly from a particular drug effect in animals to a global concept (like anxiety) in Man without consideration of the mechanism by which the drug effect is produced.

As we saw in Chapter 13, the major role of serotonergic systems appears to be that of motor inhibition. This role is mediated in part by the pathway from the median raphe to the septo–hippocampal system. However, several other target organs for the serotonergic projections (Fig. 13.20) appear also to be involved. Soubrié's research has implicated in particular the substantia nigra, itself the origin of an ascending dopaminergic projection to the corpus striatum whose motor functions are well known. Now, it is by impairing motor inhibition that anti-serotonergic drugs alleviate punishment-induced response suppression. Consider, then, what the effects would be if environmental anxiogenic stimuli had triggered a state of anxiety which would normally eventuate in the behavioural inhibition output of the behavioural inhibition system (Fig. 12.4), but this output was blocked because of impaired serotonergic transmission. Would such a blockade of motor inhibition reduce anxiety? Surely not: the person concerned would continue to experience anxiogenic stimulation but be unable to cope with it in the normal manner. Under some conditions, such a predicament might even give rise to an increase in anxiety. Just such an increase has been reported by Graeff and his co-workers when a serotonin receptor blocker was administered to human subjects during a simulated public speaking task.

More direct evidence for serotonin-mediated response inhibition in human beings comes from a series of studies in which the levels of 5-hydroxyindoleacetic acid (5-HIAA), the principal metabolite of serotonin, have been measured in cerebro-spinal fluid (CSF). Data from many animal experiments indicate that serotonergic mechanisms inhibit aggressive behaviour, perhaps by way of the serotonergic input to the central grey described by Graeff (Chapter 13). It is consistent with these data that 5-HIAA levels in human CSF correlate negatively with a history of aggressive behaviour directed towards either others or oneself (suicide attempts). Low 5-HIAA levels have also been found among criminals, especially those who have been incarcerated for violent and aggressive behaviour. In agreement with Graeff's findings (above) showing that when serotonergic transmission was blocked pharmacologically there was a rise in self-reported anxiety, patients with low levels of CSF 5-HIAA have also been found to be high on measures of anxiety.

One difficulty with these studies is that the 5-HIAA in CSF comes mainly from the spinal cord, not the brain. But, in answer to this objection, Stanley, Virgilio & Gershon have investigated post-mortem brain material from suicides and report reduced binding of imipramine in the frontal cortex. Since imipramine binding is associated with the neural uptake mechanism for serotonin, this observation suggests a reduced serotonergic input to the frontal cortex in people who commit suicide, in agreement with the lowered CSF 5-HIAA in people who merely attempt it.

Further evidence that the model developed in Chapter 13 is on the right lines comes from patients who, as a last resort, undergo 'psychosurgery' for the treatment of severe and chronic anxiety or depression. One operation used in this type of surgery is to damage the cingulate cortex. This treatment has served successfully to improve patients who have suffered for years from states of anxiety, especially obsessive-compulsive neurosis and anxiety state (see below). Given the role allotted to the cingulate cortex in the functioning of the comparator circuits depicted in Fig. 13.14, these therapeutic effects of cingulate lesions are not unexpected. However, they pose a problem. As you might suppose, psychosurgery is only carried out if the patient has first proved resistant to other, less drastic forms of treatment, including drugs. Thus the patients who benefitted from cingulate lesions had first failed to benefit from medication with anxiolytic drugs. Yet the model illustrated in Fig. 13.14, and from which one can predict (in general terms) that cingulate lesions might reduce anxiety, is based in the first instance upon the behavioural effects of the anxiolytic drugs.

As a way out of this apparent contradiction I have proposed that the cingulate cortex lies on a descending pathway (see Fig. 13.14) by which the neocortex is able to control the activities of the septo–hippocampal system and its associated comparator circuits in a manner that is more or less independent of the ascending monoaminergic control (the noradrenergic and serotonergic inputs to the septo–hippocampal system) which concerned us in the last chapter. This descending pathway also involves the prefrontal cortex, a neocortical region that appears to have achieved particular importance in the human species (Fig. 14.1). Therapeutically inflicted damage to this region (prefrontal lobotomy, or the fibre-cutting operation known as 'leucotomy') has a longer, richer, and far more alarming history than that of cingulectomy. It was originally introduced as a treatment for schizophrenia, for which, many thousands of gratuitous operations later, it was found to be of no use whatsoever. But it turns out to alleviate those same symptoms of anxiety, obsessions, and depression upon which cingulate lesions act.

The involvement of the prefrontal cortex in emotional behaviour appears to be phylogenetically old. In monkeys, stimulation of the medial orbital cortex (that is, the region located just behind the orbits

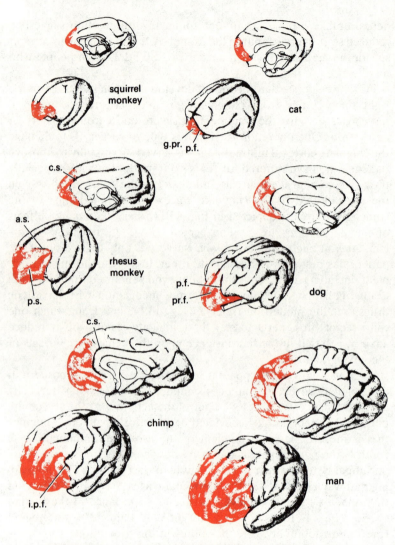

Fig. 14.1. The prefrontal cortex (marked by shading) in six different species. Abbreviations: a.s., arcuate sulcus; c.s., cingulate sulcus; g.pr., gyrus proreus; i.p.f., inferior precentral fissure; p.f. presylvian fissure; p.s., principal sulcus; pr.f., proreal fissure. From Fuster (1980).

of the eyes) produces a variety of changes in vegetative and autonomic responses, including slowing or arrest of respiration, changes in blood pressure, pupillary dilation, increased plasma cortisol, and decreased levels of circultting eosinophils (Chapter 5); these effects are probably mediated by connections linking the prefrontal cortex to the hypothalamus. Lesions of the orbitofrontal cortex in monkeys and of the corresponding regions in the rat (but located in this species in the lateral sector of the frontal cortex) produce some of the same effects (e.g. increased resistance to extinction, impaired reversal learning) that are seen after damage to the septo–hippocampal system or administration of anxiolytic drugs (Table 12.2). In the rat, stress of various kinds has been shown to increase the turnover of dopamine (which, like noradrenaline and serotonin, is a monoamine transmitter) in the prefrontal and cingulate cortices, indicating increased activity in dopaminergic terminals in these regions. This stress-induced increase in dopaminergic activity is blocked by the anti-anxiety drugs, as is the corresponding increase in noradrenergic and serotonergic neuronal activity (Chapter 13). These changes in dopaminergic firing rate may be secondary to an input from the locus coeruleus to the relevant cell bodies, which are located in a nucleus in the ventral tegmental area of the brainstem denominated 'A 10'. In line with this possibility, Glowinski has observed decreased dopamine turnover in prefrontal cortex after destruction of noradrenergic terminals in the A 10 region. A role in anxiety for the dopaminergic projection to the prefrontal cortex is indicated by reports that in rats the response-suppressant effects of punishment are reduced after either ablation of the prefrontal cortex or destruction (by local injection of 6-hydroxydopamine) of the dopaminergic terminals in this area.

Anatomically, there are close connections between the prefrontal cortex and the septo–hippocampal system; these appear to be stronger in primates than in sub-primates. Thus, in monkeys there are projections from the prefrontal cortex to the entorhinal cortex, the hippocampus, and the septal area, as well as to the cingulate cortex and the mammillary bodies of the hypothalamus (to which, as we saw in the last chapter, the subiculum also projects). As well as being involved in emotional behaviour, the prefrontal cortex appears to have a role in the organisation and execution of motor programs. Such a role is consistent with the hypothesis, proposed in Chapter 13, that the prefronto-cingulate projection is the route by which information about motor plans is made available to the Papez loop (Fig. 13.14) for use in the generation of predictions to be sent on to the subiculum. Thus the therapeutic effects of prefrontal damage in human patients may be due to an impairment in the direct control over emotional behaviour apparently exercised by this region (see above), or a consequent alteration in the functioning of the septo–hippocampal comparator circuits, or both.

A further possibility is that the prefrontal cortex has acquired in Man the new function of mediating control over the septo–hippocampal system by neocortically located language systems. This possibility is supported by a number of lines of evidence. People who have suffered lesions in the prefrontal region from tumours or gun-shot wounds display a curious dissociation between their capacity to describe correctly the contingencies or rules that ought to govern their behaviour and their inability actually to follow these rules. In a discrimination task, for example, in which the correct rule is to make an active motor response upon presentation of one stimulus and to refrain from responding upon presentation of a second, such a patient will understand and correctly describe the task he has been set, but he will be unable to refrain from responding on the 'no-go' as well as the 'go' trials. (The same kind of behavioural impairment is seen in rats with damage to the septal area or hippocampus.) It is as though damage to the prefrontal cortex interrupts the circuits that normally allow a verbally coded formulation of a rule to influence the behaviour prescribed by the rule. This inference is supported by the studies of Homskaya. In normal subjects or patients with brain lesions to regions other than the prefrontal cortex, the galvanic skin and vasomotor orienting responses (Chapter 13) appear and disappear rapidly in response to verbal instructions (e.g. the information that a stimulus will or will no longer be followed by a painful electric shock). But Homskaya showed that such responses to verbal instructions are lacking in patients with prefrontal damage (who nonetheless show more or less normal Pavlovian conditioning and extinction of these responses in the absence of verbal instruction). Prefrontal damage, then, appears to produce a relatively selective impairment in the verbal control of orienting responses.

These observations, and the interpretation I have placed upon them, suggest an explanation of the puzzle, alluded to above, that is posed by the ability of cingulate and prefrontal damage to alleviate symptoms of anxiety that are resistant to the anti-anxiety drugs. We need to suppose that a descending pathway from the prefrontal cortex via the cingulate cortex (and perhaps also via the entorhinal cortex) constitutes an alternative alarm bell acting instead of, or in tandem with, the phylogenetically older monoaminergic afferents (noradrenergic and serotonergic) to activate the septo–hippocampal system. This descending pathway is likely to be of particular importance in handling the kind of threat which depends for its formulation upon the semantic capacities of a language system (e.g. the threat that one might contract cancer). These considerations are consistent with the type of patient who has been found to benefit from psychosurgery: namely, patients with obsessional symptoms (see below) or phobic behaviour that has become widely generalised by way of essentially linguistic mechanisms such as semantic association.

Evidence from studies of the human brain, then, is more or less directly in agreement with the conclusions we reached in the last chapter: Reiman *et al.*'s PET scanning study picked out the inputs and outputs of the septo–hippocampal system as the structures that differentiate patients with panic attacks from normal people; Charney's work measuring plasma MHPG implicates the locus coeruleus in anxiety and the different susceptibility to anxiety of patients with panic attacks and controls; and the success of psychosurgery implicates the cingulate cortex (part of the Papez loop) and the prefrontal cortex (which has close association with both the cingulate cortex and the septo–hippocampal system) in the obsessive–compulsive syndrome and widely generalised phobic behaviour. Armed with this evidence, let us press on further with the examination of anxiety in our own species.

Neurosis and psychosis

We began this book by distinguishing two different sorts of question, the core questions of motivation and personality; for our present concerns they become: (1) what causes neurotic behaviour? and (2) which individuals are predisposed towards neurotic behaviour (have a high degree of 'neuroticism')? But we also pointed out that the two sorts of question are so closely linked that an answer to one of them may often give the key for the answer to the other. Nowhere is this more true than in the study of human neurosis. The study of personality has been of central importance in answering several questions about the nature of the neurotic process, to the extent that the best definition of neurotic behaviour we can now give is probably 'the type of behaviour to which neurotic people are especially prone'.

Psychiatric disorders in general may be defined as disorders of behaviour or feeling. This includes such extremely diverse problems as anxiety, depression, obsessional rumination, persistent delin-quency, mental retardation, hallucinations, delusions, disorders of thinking, hysterical paralysis, psychogenic ulcers, and premenstrual tension. The difficulty is to group these symptoms together into recog-nisable disease entities of the kind familiar in physical medicine. The classical way to do this is to find a single causal agent (e.g. a particular bacillus or virus) which produces a multiplicity of symptoms; then this cluster of symptoms occurring together is a single disease. This type of classification is not yet commonly possible in psychiatric medicine, and for many problems may never be. The most useful classificatory principles so far available have emerged from investigations which have sought for clusters of symptoms going together in different groups of people.

Consider the question of the relation between neurosis and psychosis. Psychosis includes principally schizophrenia and certain kinds of depression (though, as we shall see, some other kinds of depression are more accurately classified as neurotic). Schizophrenia is itself a rag-bag of symptoms, including thought disorder, loss of emotional relationships with others, incongruity of emotion, paranoid suspicion, delusions, and hallucinations. In its classic form, the psychotic kind of depression consists in a more or less regular alternation, over periods of days, months, or years, between states of deep depression and states of extreme euphoria and excitement (known as 'mania'), with no apparent precipitating factors in the patient's life history; but it is far more common to see repeated periods of depression without the intervening manic episodes. Now one of the few testable statements it is possible to extract from psychoanalytic theory is that psychosis is a worse form of the same illness which is neurosis. This is also one of the few statements in psychiatry to have achieved the distinction of definite disproof.

The argument leading to this disproof is a simple one. If psychosis is merely a worse form of neurosis, the people who suffer from the one should be no different from those who suffer from the other; it should be possible to describe the differences between neurotic and normal people in such a way that psychotics then differ from normals in the same way, only more so. On the alternative view, psychosis is an altogether different form of disorder from neurosis; in that case, psychotics should differ from normals in ways which are distinct from the differences between normals and neurotics. These two hypotheses have been illustrated graphically by Eysenck (Fig. 14.2).

The way to test the two hypotheses is equally simple. It is to subject groups of known neurotics, psychotics, and normals to a series of behavioural and physiological tests; to look for the tests which best differentiate normals from neurotics, on the one hand, and from psychotics, on the other; and to see how psychotics compare with normals on the best tests of neuroticism and also how neurotics compare to normals on the best tests of psychoticism. The results of investigations of this nature, largely carried out by Eysenck's group in London and by Cattell's in Illinois, are quite definite: by and large,

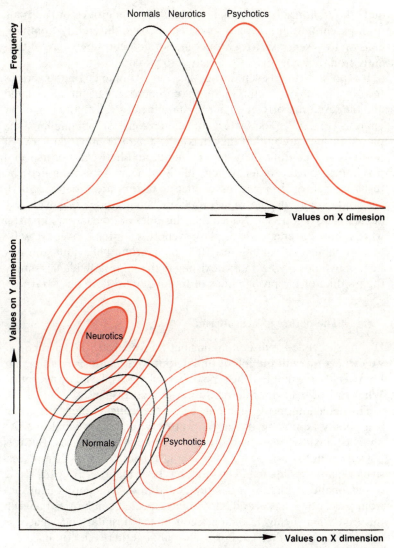

Fig. 14.2. Diagrammatic representation of the one-dimensional and the two-dimensional hypotheses of mental abnormality. On the one-dimensional hypothesis (*above*) psychotics and neurotics both differ from normals in the same way (along the X-dimension), but psychotics more so. On the two-dimensional hypothesis (*below*), neurotics differ from normals along one dimension (the Y-dimension), and psychotics differ from normals along another dimension (the X-dimension).

quite different tests differentiate normals from neurotics and normals from psychotics; and psychotics do not behave differently from normals on tests sensitive to neuroticism, nor do neurotics behave differently from normals on tests sensitive to psychoticism.

It is possible to go even further. The scores obtained on the various tests by the subjects in this kind of experiment may be intercorrelated and the resulting correlation matrix simplified by the method of factor analysis (Fig. 4.2). The factors, or 'dimensions', so obtained may be placed in such a way that they maximise the discrimination between normals and neurotics, on the one hand, and between normals and psychotics, on the other. When this is done, it transpires that the resulting neuroticism and psychoticism factors are independent of each other, any degree of neuroticism being compatible with any degree of psychoticism. Notice that this does *not* mean that a person who is neurotic cannot also display psychotic symptoms, or conversely: the probability of finding any item of neurotic behaviour and any item of psychotic behaviour coexisting in the same individual is simply the product of the probabilities of finding either item on its own.

The nature of neuroticism

Having succeeded in discriminating between neurotics and psychotics, we can get on with the job of finding out what neurosis *is*; for so far I have said only that it is the type of behaviour displayed by neurotics. What sort of behaviour do neurotics, in fact, display?

The behaviour most commonly called 'neurotic' in the clinic has in it a goodly admixture of fear and anxiety; examples would be the phobias, anxiety states, and obsessive–compulsive neurosis. But it is not sufficient to describe neurosis simply as behaviour involving excessive or inappropriate fear. This, too, has been made clear by the study of personality, which, if it has succeeded in discriminating neurotics from psychotics, has also disclosed unsuspected similarities between the traditional neurotic patient seen in the hospital and what was thought to be another kind of person althogether: the criminal.

Let us consider the extensive work of Eysenck's group both on the features of personality which discriminate neurotics as a whole from normals and on those which differentiate the various neurotic groups from one another. Their basic findings are set out in Fig. 14.3. The axes of this figure consist of scores on two independent dimensions of personality which have emerged from a large number of factor-analytic studies of the kind we have described above. The vertical axis is the neuroticism dimension which we have already met; the

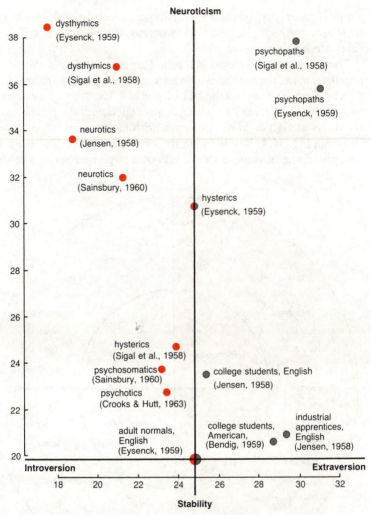

Fig. 14.3. Scores of various neurotic, criminal, and normal groups on the two factors of neuroticism and introversion–extraversion obtained from the Maudsley Personality Inventory. The references indicated by the names and dates may be found by consulting Eysenck (1967), from whom the figure is taken.

horizontal axis represents the important personality dimension known as 'introversion–extraversion'. All the traditional neurotic groups have high scores on neuroticism; but so do such groups as juvenile delin-quents, unmarried mothers, and psychopaths. The thing which dis-criminates between the latter groups and, say, the phobics and obsessionals is their score on the introversion–extraversion scale: traditional neurotics are highly introverted, whereas those who break society's rules are highly extraverted.

The extraversion dimension is found equally strongly in the normal population, i.e. among those who have neither come into the psychiatric clinic nor fallen into the net of the law. The type of personality trait or mode of behaviour which is associated with the possible combinations of high or low neuroticism with high or low extraversion is shown in Fig. 14.4. In this figure the trait names printed in the outer ring give an indication of the results of a large number of statistical

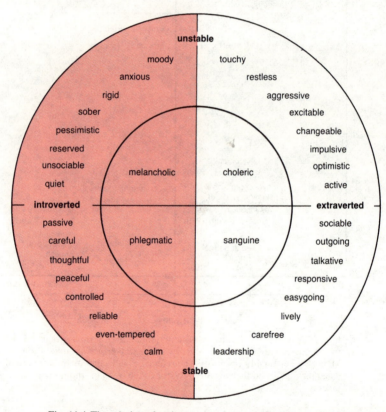

Fig. 14.4. The relation of various traits of personality to the major dimensions of neuroticism (the vertical axis) and introversion–extraversion (the horizontal axis), as determined by numerous factor-analytic investigations. The inner ring shows how the ancient 'four temperaments' of Hippocrates and Galen may relate to the findings of modern experimental studies. In the outer ring, the closer the trait names are together, the higher is the empirically observed correlation between them, according to the conventions set out in Figure 4.3.

investigations of the patterns of individual differences found in such devices as questionnaires and rating scales, as well as a number of direct behavioural tests. The closer the trait names are together, the higher is the empirically observed correlation between them; when the angle between them is about 90 degrees, there is no relation at all, and between 90 and 180 degrees the correlation is negative. The relation of these traits to the dimensions of neuroticism and extraversion is shown by the appropriate axes crossing the figure.

In Figs. 14.3 and 14.4, then, we see both which kinds of disorder are classified as neurotic and what kind of personality is associated with the two main groups of neurotic disorder, called by Eysenck 'dysthymic' (the introverted neuroses) and 'hysterico-psychopathic' (the extraverted neuroses). Let us fill this picture in with a description of the typical forms of behaviour displayed by the individuals at the extreme ends of the two dimensions of personality which concern us. For neuroticism, Eysenck & Rachman, in their book *The Causes and Cures of Neurosis*, have this to say:

At the one end we have people whose emotions are labile, strong, and easily aroused; they are moody, touchy, anxious, restless, and so forth. At the other extreme we have the people whose emotions are stable, less easily aroused, people who are calm, even-tempered, carefree, and reliable. Neurotics, needless to say, would be expected to have characteristics typical of the unstable type, normal persons typical of the stable type. (Eysenck & Rachman, 1965)

For extraversion, the same writers say this:

The typical extravert is sociable, likes parties, has many friends, needs to have people to talk to, and does not like reading or studying by himself. He craves excitement, takes chances, often sticks his neck out, acts on the spur of the moment, and is generally an impulsive individual. He is fond of practical jokes, always has a ready answer, and generally likes change; he is carefree, easygoing, optimistic, and 'likes to laugh and be merry'. He prefers to keep moving and doing things, tends to be aggressive and loses his temper quickly; altogether his feelings are not kept under tight control, and he is not always a reliable person.

The typical introvert is a quiet, retiring sort of person, introspective, fond of books rather than people; he is reserved and distant except to intimate friends. He tends to plan ahead, 'looks before he leaps', and mistrusts the impulse of the moment. He does not like excitement, takes matters of everyday life with proper seriousness, and likes a well-ordered mode of life. He keeps his feelings under close control, seldom behaves in an aggressive manner, and does not lose his temper easily. He is reliable, somewhat pessimistic, and places great value on ethical standards. (Eysenck & Rachman, 1965)

It seems reasonable to suppose that an understanding of the nature of these two important dimensions of personality would enable us to make some advance in understanding both what causes neurotic behaviour in general and what determines which kind of neurotic behaviour a particular individual displays. Can we make any progress in this direction?

As we have just seen, neuroticism is above all characterised by strong emotions which are easily aroused. Now we have been working in this book towards a view of the emotions as essentially dependent on the organisation of the CNS (an abbreviation which now has a deliberate ambiguity between 'central nervous system' and 'conceptual nervous system') for responding to various kinds of reinforcing events: reward, punishment, omission of reward, omission of punishment, and signals of these events. We have considered in particular fear and frustration, but we have touched at various points on guilt, hope, relief, anger, and the 'rage' of defensive attack. Thus it is a simple step to translate 'strong emotions, easily aroused' into 'a general sensitivity to all reinforcing events, whether rewarding or punishing'.

The nature of extraversion

What about extraversion? What is it that underlies this all-pervading dimension of personality? Let us begin our search for an answer to this question by considering a little more closely the symptoms we encounter in the dysthymic and psychopathic neuroses respectively.

We start with the dysthymic neuroses, and with the anxiety state in particular. This consists, subjectively, of a feeling of apprehension, ranging from mild uneasiness to panic; the patient is unable to give any reason for his feelings or to point to any stimulus as their object. Related equally obviously to the experience of fear are the phobias. These are fears, often very intense, of particular classes of stimuli which are not objectively dangerous or, at least, not as dangerous as they seem to the patient. Common examples are fears of heights, of closed spaces ('claustrophobia'), of open spaces or crowded places ('agoraphobia'), of social gatherings, of snakes, and of sharp objects. Less obviously related to fear is the obsessive-compulsive neurosis, in which the patient feels compelled to go through some idiosyncratic ritual or to think some repetitive train of thought over and over again on pain of something dreadful happening. The link between the obsession and fear can, however, be easily discovered: it suffices to prevent the patient from going through the ritual at the time he feels compelled to do so, whereupon he feels an overwhelming sense of fear. In neurotic depression, the link with fear is not always obvious; but in many cases it is so strong that it is difficult to decide between a diagnosis of depression and anxiety state. Also among the dysthymic syndromes is the psychogenic ulcer; we have seen enough of the causation of ulcers (Chapter 9) to concur in the clinical suspicion that this results from long-drawn-out conflict.

In psychopathy there are, in a sense, no 'symptoms', for these people do not themselves suffer, they make society suffer. The psychopath is given to stealing, lying, sexual delinquency, accidents, and traffic offences. He takes no account of the rights of others and appears to suffer from no guilt feeelings for his misdemeanours. If he keeps clear of actual crime, he may nevertheless fall constantly into debt, become a gambler or an alcoholic, or fail to hold a steady job. Mayer-Gross, Slater & Roth say of the psychopath in their textbook on psychiatry that 'he lives in the present only. His immediate wishes, affections or disgruntlements rule him completely, and he is indifferent about the future and never considers the past.' In short, there has been a failure of 'socialisation'; the psychopath has no 'conscience'.

The difference between dysthymic and psychopathic disorders also comes out in the investigation of children. Fig. 14.5 displays the results of a factor-analytic study of various 'personality' and 'conduct' problems shown by children referred to a child guidance clinic. The two factors of neuroticism and extraversion again appear. If they are neurotic, 'extraverted children swear, fight, are disobedient, destructive, play truant, steal, lie, are violent, rude, and egocentric, whereas introverted children are sensitive, absent-minded, depressed, seclusive, inefficient, have inferiority feelings, daydream, and are nervous' (Eysenck & Rachman, p. 23).

The extemely extraverted neurotic adult, then, has failed to develop a conscience, and the child of a similar personality make-up seems to have difficulty in developing one. What, then, is the conscience? We have already come across Eysenck's answer to this question (Chapter 9): conscience is a set of conditioned fear reactions. That is, in early life we learn to associate (by classical conditioning) fear of impending punishment with stimuli associated with the commission of socially disapproved acts. When in adult life we feel the impulse to commit such acts, the occurrence of a conditioned fear reaction prevents us: conscience. If we do succeed in carrying them out, we may nevertheless feel afraid of the consequences: guilt. It is difficult at the present stage of development in psychology to evaluate this suggestion with any precision; but there may well be much truth in it. Assuming for the moment that there is, our next question must be, what is the connection between the formation of conditioned fear reactions and the individual's degree of extraversion?

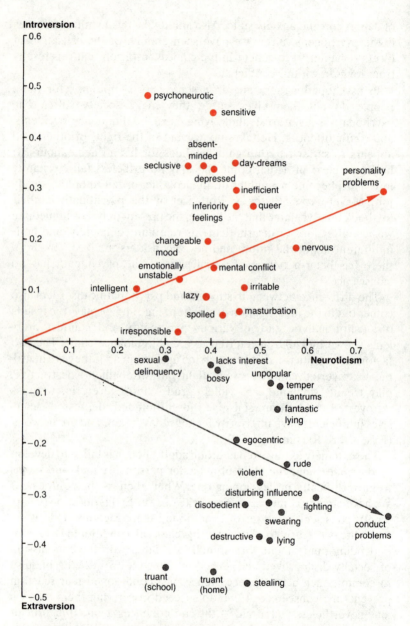

Fig. 14.5. Results of a factor analysis of various conduct and personality problems shown by children. The two factors of neuroticism and introversion–extraversion appear again.

Eysenck's own answer to this question (based on general theoretical arguments which are beyond the scope of this book) is that high extraversion is constitutionally accompanied by low conditionability: that is, the extravert has difficulty in forming *any* kind of conditioned reflex. The evidence relating to this hypothesis is conflicting. There are some conditions under which extraverts show the poorer conditionability postulated by Eysenck. However, it is now clear that these conditions are insufficiently wide for the conditionability postulate to bear the full weight that Eysenck's theory places upon it. Furthermore, the evidence demonstrating poorer conditionability in extraverts has been gathered only in experiments using aversive reinforcement (mainly, conditioning of the eyeblink reflex with a puff of air as the unconditioned stimulus). This evidence is therefore (as we shall now see) equally in agreement with an alternative hypothesis that I proposed in 1970.

Sensitivity to signals of punishment and reward

The extravert forms conditioned fear responses with difficulty. Eysenck's suggestion is that he is bad at *conditioning*; my alternative suggestion is that he is bad at *fear*. That is, the extravert is relatively insensitive to the threat of punishment. This hypothesis has a certain degree of 'face validity', it *looks* like a plausible account of the most obvious facts we have to explain.

It leads immediately to many of the obvious differences between introverted and extraverted neurotics which we have already noted: the presence of a subjective feeling of fear in the introverted neurotic, as in the anxiety state; the development of overt avoidance behaviour in the introverted phobic or obsessional; and, in the psychopath, the tendency to take a reward (by, say, stealing or sexual gratification) without thought for the consequences, the insensitivity to repeated punishment by society, and the lack of strong feelings of fear and guilt. It continues to predict the same failure of socialisation that Eysenck's hypothesis predicts. Like Eysenck's hypothesis (but for different reasons) it can account for the nature of the questionnaire and rating scale items which define the extraversion dimension. This dimension is in fact made up of two highly correlated sub-factors, one of 'impulsiveness' and one of 'sociability' or 'social extraversion'. We would propose that the extravert acts on the spur of the moment because his behaviour is relatively more determined by potential rewards in his environment, and he is relatively less likely to operate the 'stop' of passive avoidance in the face of potential punishment.

His greater liking for people can be understood if we recall that people are the most important dispensers of both rewards and punishments for other people; therefore, those who are less sensitive to punishment are more likely to seek them out. The optimism of the extravert, contrasted with the introvert's pessimism, is equally understandable as an example of insensitivity to potential punishment.

Indirect evidence for differences between introverts and extraverts in their modes of reaction to threatening stimuli comes from experimental work on personality differences on a 'repression–sensitisation' trait. It has been found that there are rather consistent individual differences in the perception, recognition, learning, and recall of threatening stimuli. Some individuals (called 'repressors') do worse in all these tasks when threatening stimuli are used than when neutral stimuli are involved; others ('sensitisers') do better with threatening stimuli than with neutral ones. The former have been shown to be extraverts and optimists; the latter, introverts and pessimists. Thus it seems that the introvert is more likely to notice threatening stimuli, recognise them subsequently, learn about them, and recall them.

More direct evidence favouring my hypothesis over Eysenck's comes from a number of experiments which have specifically examined the roles played by reward and punishment in determining learning and behaviour in introverts and extraverts. Both hypotheses predict an introvert superiority in learning when aversive reinforcement is used, Eysenck's by way of the putative superior conditionability of the introvert, mine by way of his putative greater fearfulness. But the two hypotheses part company if rewards are used: here, Eysenck continues to predict that introverts will condition better than extraverts, whereas I predict the reverse. However, to see that this is so, it is first necessary to consider my hypothesis in a little more detail.

This hypothesis is presented graphically in Fig. 14.6. This figure may be looked at from several different points of view, depending upon which dimension of personality one puts in the foreground.

(1) Increasing degrees of neuroticism represent increasing sensitivity to reinforcing events in general. (2) Increasing degrees of introversion represent increasing sensitivity to signals of punishment rather than signals of reward (Fig. 14.6). Now suppose we measure susceptibility to fear, anticipatory frustration, and anxiety directly. Where would we expect a factor describing this personality dimension to lie in the two-dimensional space defined by the neuroticism and extraversion axes? (3) Clearly, the most rapid increase in sensitivity to signals of punishment will be produced by moving from the stable extravert quadrant (low sensitivity to reinforcing events in general and to signals

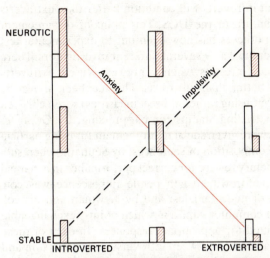

Fig. 14.6. Rotation of Eysenck's dimensions of neuroticism and introversion–extroversion proposed by Gray (1970b). The dimension of trait Anxiety represents the steepest rate of growth in susceptibility to signals of punishment (and other adequate inputs to the behavioural inhibition system); the dimension of Impulsivity represents the steepest rate of growth in susceptibility to signals of reward. Introversion–extroversion now becomes a derived dimension, reflecting the balance of susceptibility to signals of punishment and reward, respectively; and neuroticism similarly reflects the sum of these two types of susceptibility to signals of reinforcement. Open blocks, sensitivity to stimuli associated with reward; red blocks, sensitivity to stimuli associated with punishment.

of punishment in particular) towards the neurotic introvert quadrant (high sensitivity to reinforcing events in general and to signals of punishment in particular). The extensive work carried out by Janet Taylor Spence and Kenneth W. Spence on the measurement of just such a dimension of personality has produced just this result, with the exception that the dimension ('trait anxiety') is related slightly more closely to neuroticism (correlation = 0.7) than to introversion (0.4). (4) Corresponding to a dimension that directly reflects the most rapid increase in sensitivity to signals of punishment, there must also be (given the assumptions upon which Fig. 14.6 is based) a dimension reflecting the most rapid growth in sensitivity to signals of reward. This dimension ('impulsivity') has been drawn at right angles to trait anxiety, preserving the simplicity of two orthogonal factors defining a two-dimensional space.

The hypothesis depicted in Fig. 14.6 clearly predicts, as noted above, that extraverts will condition better with rewarding UCSs, whereas

Eysenck predicts that introverts will condition better than extraverts irrespective of the nature of the UCS. This point of disagreement between the two hypotheses has now been put to experimental test in a variety of different tasks by several different investigators. There is a rare unanimity in their results: whenever reward is used, extraverts condition and learn better than introverts. This has been shown by Nicholson & Gray (but using rather indirect measures) when children were rewarded with trading stamps for bar-pressing; by Gupta & Nagpal, when adults were given social reinforcement for using particular pronouns in the composition of sentences; by Seunath, when subjects were given monetary reward for correct performance in a manual tracking task; by Kantorowitz, when penile tumescence was conditioned to pictures of nude females; and by Newman and his colleagues in a series of computer games in which points (exchangeable for money) were gained for appropriate responses. In each of these experiments, the relationship between introversion–extraversion and performance was reversed when punishment was used in place of reward: now it was the introverted subjects who learned better. Perhaps most striking is a real-life experiment carried out in a California school by McCord & Wakefield. The children were given a pre-term arithmetic test and measures of personality. During term, the experimenters sat in the classroom and rated the different teachers of arithmetic for the degree to which they used rewards (e.g. praise) or punishments (e.g. blame) in the conduct of their class. Then, at the end of term, performance on an arithmetic test was measured again. It transpired – as predicted by the model shown in Fig. 14.6 – that in relatively introverted children improvement in arithmetic was greater the more the teacher tended to use punishments; whereas in relatively extraverted children improvement was greater the more the teacher tended to use rewards.

The relationships depicted in Fig. 14.6, then, have a large measure of empirical support. This support gives one license to ask which pair of axes contained in the figure (neuroticism and extraversion, or anxiety and impulsivity) reflects the activity of genuine, underlying, causally efficient influences on behaviour (assuming that either pair does). One may ask this question because the technique of factor analysis (upon which, as we have seen, Eysenck's dimensions are based) can determine how many independent dimensions are required to describe a matrix of correlations between different tests of individual differences, but not where they should be located in the resulting *n*-dimensional space. Any rotation of Eysenck's two dimensions that preserves orthogonality is as acceptable, mathematically speaking, as any other in the description of the data upon which they are based. To determine which rotation is the best, other criteria, unrelated to factor analysis, must be used.

A number of arguments suggests that the axes labelled 'anxiety' and 'impulsivity' in Fig. 14.6 have a better claim to reflecting biologically real lines of causation than any other rotation of these axes, including the original rotation used by Eysenck. To see that this is so, let us first ask where what I have called 'biologically real causation' could come from. The answer to this question must surely be: from the action of relatively separate sub-systems in the brain. Consider the analogous case of individual differences in intellectual abilities. Factor-analytic studies have regularly demonstrated that individuals differ from one another in verbal ability rather independently from the way in which they differ from one another in visuo–spatial ability, and these patterns of individual differences give rise to two major ability factors. Only much later did it become clear that verbal ability is critically dependent upon one, 'dominant' cerebral hemisphere (usually the left), while visuo–spatial ability depends upon the other. It now seems likely that the verbal and visuo–spatial ability factors reflect individual differences in the efficiency of functioning of the dominant and non-dominant hemispheres, respectively. Generalising from this inference, we may suppose that major personality factors similarly reflect individual differences in the functioning of specialised systems in the brain that deal with the kind of behaviour that determines personality.

One such specialised system has repeatedly come to our attention throughout this book: the one I have termed the 'behavioural inhibition system' (Fig. 12.4). This system responds to just those stimuli (conditioned fear and frustrative stimuli) reactivity to which gives rise to trait anxiety, as this is modelled in Fig. 14.6. This is hardly surprising, since the same set of concepts has given birth to both the behavioural inhibition system and this view of trait anxiety. But the independent evidence for the existence of a separable brain system that mediates the functions of the behavioural inhibition system is now rather compelling. We have, as it were, tugged at different parts of this brain system at different points in the book, and each time come out with most of the rest of it. Thus, Broadhurst selectively bred the Maudsley Reactive and Nonreactive rats to be high and low in fearfulness on the basis of Open Field defecation scores (Chapter 4). These strains subsequently turned out (Chapter 7): (1) to differ in their behaviour in ways that parallel the differences between undrugged rats and rats given an anti-anxiety drug; (2) to possess different densities of benzodiazepine receptors in the brain; (3) to show differences in the electrical activity of the septo–hippocampal system; and (4) to be different in the functioning of both noradrenergic and serotonergic systems in the brain. Similarly, in the rat, the two sexes (Chapters 6 and 7) differ from each other in Open Field defecation, in behaviour that is affected by the anti-anxiety drugs, in the functioning of the septo–hippocampal system, and in the densities of brain benzodiazepine receptors. A final example lies in the technique of early handling (Chapter 8), which appears to lower the level of fearfulness for the rest of the animal's life. This technique, too, turns out to affect the hippocampus or more specifically the density of corticosterone receptors in this organ.

These recurring patterns in the data we have reviewed in this book make it almost certain that the brain contains a relatively separate and relatively unified sub-system that is specialised to mediate the kind of behaviour which, in human beings, is experienced as anxiety (since it is counteracted by the anti-anxiety drugs). Individual differences in the functioning of this system would be expected, therefore, to give rise to a strong factor in studies of personality traits related to anxiety, just as individual differences in the functioning of the non-dominant hemisphere give rise to a strong visuo–spatial ability factor. There is evidence from human studies that this is indeed the case.

This evidence comes from the clinical investigations, described above, in which cingulectomy and frontal leucotomy were shown to alleviate anxiety and depression. As already noted, one of these operations interrupts the Papez circuit (which, in Chapter 13, was allotted the function of generator of predictions in the circuits presumed to mediate the cognitive features of anxiety), while the other interrupts the connections between the prefrontal cortex, on the one hand, and the Papez circuit and septo–hippocampal system, on the other. In several of the studies concerned, measures of personality were taken before and after surgery. These measurements showed that the operation reduced both neuroticism and introversion, but that it reduced neuroticism more; at the same time, of course, anxiety was also reduced. Now, a glance at Fig. 14.6 will show that these results are exactly as this model predicts, once one adds the further assumption that individual differences in trait anxiety are due to individual differences in the functioning of the neural system delineated in the previous chapter as mediating anxiety. If, on the other hand, one tries to apply Eysenck's theory to these observations, one must say that cingulectomy and frontal leucotomy each happen to reduce *both* neuroticism *and* introversion; and that these two effects are independent (since Eysenck holds neuroticism and introversion–extraversion to be independent dimensions of personality, causally unrelated to each other). This position is at once weak and lacking in parsimony.

These arguments and the data upon which they are based make it reasonable, then, to adopt the hypothesis that individual differences in susceptibility to anxiety give rise directly to the dimension of personality shown along one diagonal in Fig. 14.6; this dimension is not secondary to, nor an amalgam of, neuroticism and introversion–extraversion. Furthermore, we may now state this hypothesis in a different manner: individual differences in susceptibility to anxiety are due to differences in the functioning of a unified neuropsychological system, whose inputs and outputs are as defined by the behavioural inhibition sytem (Fig. 12.4), and whose neural substrates and modes of information processing are as described in Chapter 13 and amplified in the early part of this chapter. This formulation of the hypothesis has already received direct support from data, considered earlier in the chapter, which show that patients who suffer from spontaneous panic attacks have abnormal functioning in brain regions closely related to the septo–hippocampal system; and that this same group of patients has a highly reactive locus coeruleus system.

Other evidence that is consistent with Fig. 14.6 has come from studies of psychopathy. It is now clear that there are two kinds of psychopath, both with high extraversion scores, but one with low and one with high neuroticism scores, termed 'primary' and 'secondary' psychopaths, respectively. A similar distinction has emerged in questionnaire studies of self-reported impulsive behaviour: some individuals (high on neuroticism) merely tend to act on the spur of the moment; but others (low on neuroticism) are prepared to take real risks (e.g. by parachute-jumping, scuba-diving, or taking drugs). The latter, in short, are lacking in fear, in agreement with the prediction (Fig. 14.6) that the lowest sensitivity to threats of punishment will be found among individuals high in extraversion and low in neuroticism. Direct investigations of passive avoidance learning among primary psychopaths confirms these questionnaire studies. Most kinds of learning are perfectly normal in these individuals, but they show a specific deficit in refraining from making responses that are punished, e.g. by electric shock. In addition, primary psychopaths are deficient in the formation of conditioned skin conductance responses, leading Don Fowles to propose that there may be a particularly intimate relationship between this response and the behavioural inhibition system.

On this view, then, we have one dimension in Eysenck's two-dimensional space which runs from high susceptibility to anxiety at one pole to primary psychopathy at the other. (Note the concordance between this conclusion and the evidence, considered above, for lowered serotonergic transmission in violent criminals.) But a two-dimensional space requires a second dimension placed orthogonally to the first. This is the axis labelled impulsivity in Fig. 14.6. According to the model I proposed in 1970, this axis reflects increasing sensitivity to signals of reward, and therefore the activity of the brain systems that mediate rewarded behaviour. This hypothesis fits well with the behaviour of secondary psychopaths: one needs simply to suppose that, while these individuals are not completely lacking in fear, their behaviour is relatively more determined by the rewards than by the punishments available in their environment.

It should be noted, however, that other work has identified rather different biological correlates of the combination of high neuroticism and high extraversion. In particular, William Revelle has shown that individuals with this type of personality are relatively under-aroused in the early morning and relatively over-aroused late in the evening, whereas individuals low in neuroticism and high in introversion have the reverse diurnal rhythm. It is not known how these diurnal rhythms in arousal relate to sensitivity to reward or punishment. Nonetheless, Revelle's observations (and others showing a variety of biological correlates of high impulsivity) strengthen the case that the rotation of Eysenck's axes proposed in Fig. 14.6 gives the best fit to underlying biologically causal mechanisms.

Anxiety and depression

Another issue which the study of personality helps elucidate is the relationship between anxiety and depression. I said above that the best available definition of neurotic behaviour is probably 'behaviour to which neurotic people are especially prone'. In effect, we have slightly modified this definition by replacing the term 'neurotic people' with 'people high on trait anxiety', that is, Eysenckian neurotic introverts. But these individuals are especially prone to certain kinds of depressive behaviour, as well as to anxiety. Does this mean that depression and anxiety are one and the same thing?

Before attempting to answer this question we need to tackle a closely related issue: the heterogeneity of the conditions termed 'depression'. We touch here on an argument which has exercised psychiatrists and psychologists alike for decades. Roughly speaking, three positions have been adopted. According to the first, there is one kind of depression and the differences seen between different kinds of patient are only superficial. According to the second, there are quite distinct categories of depression (though there is far from universal agreement as to the categories). According to the third, there is a smooth continuum along which depressions and depressive individuals vary. Those who hold this last position give different names to the poles which define the extremes of the continuum, but there is a good deal of agreement about the characteristics to be found at each pole.

This problem has been given sustained experimental attention by Roth's group in Newcastle. Using factor analysis and other, similar mathematical techniques, they have been able to reject conclusively the position that depression is a completely undifferentiated syndrome. It is harder to choose between the other two alternatives, since the best description of the available data is that they form a kind of lumpy continuum. Depressive patients or syndromes can be ordered along a continuum which has no definite breaks. The poles of this continuum are variously labelled 'neurotic' vs. 'psychotic' or 'reactive' vs. 'endogenous'. At the neurotic pole one finds such features as autonomic signs of anxiety, obsessional symptoms, irritability, restlessness, and difficulty in falling asleep; at the psychotic pole, marked slowing of responses ('retardation'), early morning wakening, depressed mood particularly pronounced in the morning, weight loss, feelings of hopelessness, and tendencies to be male and older. These differences are quantitative, not qualitative, as befits the notion of a continuum. However, there is some tendency for patients and syndromes to hug the extremes of the distributions of these symptoms; it is in this sense that the 'neurotic–psychotic depression continuum' is lumpy.

Now, the items that define the neurotic pole of this continuum clearly include symptoms found also in states of anxiety. Is neurotic depression, then, simply anxiety under another name? Not quite. For Roth's group has performed a similar exercise to distinguish between anxiety and depression. Again, there is a lumpy continuum, with no empty spaces, but a tendency to hug the extremes. When the symptoms which distinguish the two poles of this 'anxiety–depression continuum' are examined, however, they resemble very closely those that discriminate between neurotic and psychotic depression (Fig. 14.7). Thus it seems that our lumpy continuum stretches from anxiety through neurotic to psychotic depression.

This continuum may be placed within the broader framework provided by the description of normal personality. As shown in Fig. 14.7,

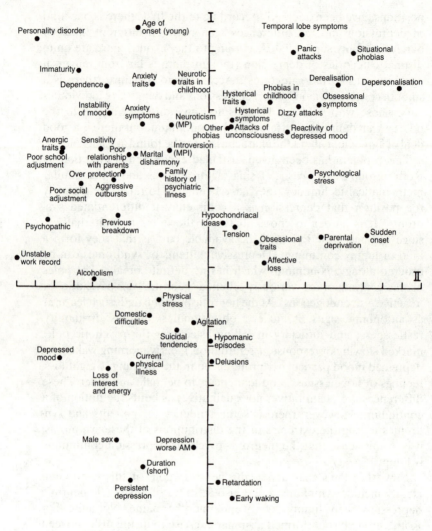

Fig. 14.7. Plot of 58 items on Roth *et al.'s* (1976) first two principal components derived from an analysis of the scores of 145 patients suffering from anxiety, depression, or both. The first principal component (I) extends from anxiety (up) to depression (down).

patients displaying symptoms of anxiety or neurotic depression tend to obtain high neuroticism and introversion scores, while those closer to the psychotic depression pole have the reverse personality characteristics. In addition, the anxiety–depression continuum is related to the third major dimension described by Hans and Sybil Eysenck in their work on normal personality. This dimension has been termed 'psychoticism' or more neutrally 'P' (Fig. 14.2); appropriately enough, patients with features of psychotic depression tend to obtain high P scores and patients with features of neurotic depression or anxiety obtain low P scores.

It would appear, then, that in its neurotic form depression is similar to anxiety and occurs in people with the same kind of personality as those who develop anxiety syndromes. These similarities suggest that similar psychological and neurological mechanisms mediate this form of depression and anxiety.

Psychologically speaking, the similarity we seek emerges naturally from the general framework of ideas built up in this book. Neurotic depression is often triggered by loss. Thus, in their well-known textbook of psychiatry, Mayer-Gross, Slater & Roth list as frequent causes of neurotic depression, the death of a wife, the loss of a job, or the breaking of a love affair. In keeping with our general strategy of seeking the causes of emotion in reinforcing events, we may therefore describe neurotic depression as a state induced by the loss of important sources of reward, or by stimuli associated with such loss (clinincal data do not allow one to distinguish between these two theoretically different descriptions). But, if we describe neurotic depression in the latter way, we see at once that it is precisely those individuals who have a highly reactive behavioural inhibition system (that is, Eysenckian neurotic introverts) who would be expected readily to enter states of this kind, as in fact they do.

There are also data that indicate a similar neurological basis to anxiety and neurotic depression. As noted above, the two operations used in psychosurgery, cingulectomy and prefrontal leucotomy, are effective in both these conditions. Patients treated in this way show not only alleviation of symptoms of both anxiety and depression, but also a reduction in introversion and neuroticism scores on personality inventories. Similar findings have emerged in studies of drug treatments. Thus such typical anti-anxiety drugs as Librium have been found to reduce self-reported depression as much as they reduce anxiety; and among out-patients (who would typically include largely neurotic rather than psychotic depressives) they produce much the same degree of improvement in patients whose major complaint is depression as in those whose major complaint is anxiety. But these drugs are of little value in cases of psychotic depression. Conversely, one class of anti-depressant drugs, the monoamine oxidase inhibitors, is effective in cases of depression with a prominent component of anxiety, but not in psychotic depression; and these drugs are also effective in the treatment of such classic symptoms of anxiety as phobias. The treatments that are most effective in psychotic depression (electroconvulsive therapy and the tricyclic anti-depressant drugs) complement this picture, for they are of little or no use in cases of anxiety.

This pattern of effects strongly suggests that, at the neurotic pole, depression and anxiety depend upon similar mechanisms, but that as one approaches the psychotic pole, depression comes to depend upon a separate or additional mechanism. Recent studies of the neural basis of helplessness (see Chapter 9), the animal model of human depression with perhaps the best claim to validity, have yielded data consistent with this conclusion.

As pointed out by Weiss, the effects of uncontrollable stress in animals map rather well onto the symptoms of human depression, including as they do reduced activity, loss of appetite, weight loss, reduced aggression and competitive behaviour, and sleep disturbance (Table 14.1). Furthermore, Sherman & Petty have shown that helplessness in animals responds to drugs in a manner that is similar to human depression. Depression, for example, responds to chronic but not acute treatment with monoamine oxidase inhibitors or tricyclic drugs, the therapeutic effects taking several weeks to become apparent. Helplessness, similarly, can be reversed by these substances, but only if the drug is administered for about a week. This effect is relatively specific to anti-depressants, since a range of other drugs (stimulants, sedatives, etc.) were ineffective. Anxiolytic drugs are capable of blocking helplessness if they are administered at the time of exposure to uncontrollable stress, but they do not reverse helplessness (as do the anti-depressants) once it has been induced. Electroconvulsive therapy also needs to be administered chronically (six or seven times over a period of days) to alleviate depression; and Dorworth & Overmier have shown a similar time course for the reversal of helplessness.

These findings suggest, then, that helplessness in animals is a model of at least some aspects of human depression. What is known about its neural basis? As we would expect from the arguments advanced above, recent evidence indicates that, in part, helplessness depends upon the same neural systems that mediate anxiety, but that in addition helplessness involves further neural changes that are not seen in states of anxiety.

Several lines of evidence implicate the septo–hippocampal system. Elmes & Jarrard showed that hippocampal lesions prevented helplessness while Kelsey & Baker have reported that medial septal lesions

Table 14.1. *Comparison of the effects of uncontrollable shock with symptomatic indications of depression from the third edition of the* Diagnostic and Statistical Manual for Mental Disorders *(DSM-III)*

Uncontrollable shock as a model of depression		DSM-III criteria for depression: four of the following
Uncontrollable shock produces the following symptomatology		
(1) Decreased food and/or water consumption	Brady, Thornton & Fisher, 1962 [ref. 17] Pare, 1964 [18] Pare, 1965 [19] *Weiss, 1968 [12] Ritter, Pelzer & Ritter, 1978 [13]	(1) poor appetite and significant weight loss
(2) Weight loss	Brady, Thornton & Fisher, 1962 [17] Pare, 1965, [19] *Weiss, 1968 [12]	(2) psychomotor alterations
(3) Poor performance in tasks requiring active motor behaviour (Shuttle advoidance-escape, lever-press escape, water-escape, Open Field activity, etc.)	Overmier & Seligman, 1967 [1] *Seligman & Maier, 1967 [2] Overmier, 1968 [20] Weiss & Glazer, 1975 [11] Weiss, Glazer, Pohorecky, Brick & Miller, 1975 [21] Weiss, Bailey, Korzeniowski, & Grillone, 1980 [22] *Weiss, Goodman, Losito, Corrigan, Harry & Bailey, 1981 [15] * Sutton, Coover & Lints, 1981 [23]	(3) loss of energy or fatigue (4) loss of interest in usual activities (5) sleep changes
(4) Loss of normal aggressiveness or competitiveness	Peters & Finch, 1961 [24] * Maier, Anderson & Leiberman, 1972 [25] * Corum & Thurmond, 1977 [26]	(6) indecisiveness, evidence of decreased ability to think
(5) Loss of normal grooming or play activity	* Redmond, Maas, Dekirmanjian & Schlemmer, 1973 [27] Stone, 1980 [14] *Weiss, Goodman, Losito, Corrigan, Harry & Bailey, 1981 [15]	(7) *feelings* of worthlessness
(6) Decreased sleep	*Weiss, Goodman, Losito, Corrigan, Harry & Bailey, 1981 – unpublished observation [16]	(8) recurrent *thoughts* of death and suicide

* Denotes study demonstrating that the effect depends on uncontrollability of shock
From Weiss *et al.* (1982), which should be consulted for the references given in the table.

(destroying the cholinergic projection that controls the hippocampal theta rhythm) reduce both helplessness and the stomach ulcers normally seen after exposure to inescapable shock. Thus an intact septo–hippocampal system appears to be a prerequisite for helplessness. To this evidence Petty & Sherman have added data from a variety of experiments suggesting that helplessness is mediated by increased activity in the hippocampus secondary to reduced GABAergic inhibition in this structure. Finally, Weiss has shown that helplessness is well correlated with a fall in the levels of noradrenaline in the region of the locus coeruleus. He suggests that this gives rise to lessened stimulation of noradrenergic auto-receptors in the locus coeruleus, with a consequent reduction in inhibitory feedback and so increased firing rates in noradrenergic fibres projecting to the forebrain, including the septo–hippocampal system. These various lines of evidence, then, implicate exactly the same structures as we saw in Chapter 13 to be important for anxiety.

So much for the similarities between the neural bases of anxiety and helplessness. What about the differences? These appear to consist in two additional phenomena that are related to helplessness but not to anxiety.

First, there are several reports that uncontrollable stress gives rise to a fall in the levels of noradrenaline that is particularly marked in the hypothalamus (whereas, in anxiety, there is evidence only for an increased release of noradrenaline, especially in the forebrain). Now, only about 30% of the noradrenergic terminals in the hypothalamus come from cell bodies in the locus coeruleus. The remainder derive from cells in a number of other brainstem nuclei whose axons travel to the hypothalamus in the ventral noradrenergic bundle (Fig. 13.3). A fall in the level of hypothalamic noradrenaline suggests a reduced level of noradrenergic transmission in this region. An effect of this kind might account for the impairments in eating and aggression that accompany helplessness, since both these forms of behaviour are in part controlled by mechanisms located in the hypothalamus. Ritter, for example, has shown that when the fall in hypothalamic noradrenaline induced by inescapable shock reaches 25%, there is a loss of the eating response to glucose deprivation, a response mediated by hypothalamic noradrenergic receptors. It is possible, therefore, that the deficits in eating and aggressive behaviour characteristic of human depression arise when the release of noradrenaline from hypothalamic terminals of the ventral bundle exceeds the capacity of these terminals to replace the lost transmitter.

Second, uncontrollable stress also gives rise to a fall in the levels of dopamine in both the dorsal and ventral striatum. This transmitter plays a vital role in the initiation of movement; indeed, it is degeneration in the dopaminergic projection from the substantia nigra to the dorsal striatum that gives rise to the distressing loss of voluntary movement seen in Parkinson's disease. The role played by dopamine in the initiation of movement is probably closely related to its role in the transmission of incentive motivation to motor systems; though this role may depend more upon the dopaminergic projection to the ventral striatum (see Chapter 13). Thus it is likely that the fall in striatal dopamine levels observed after uncontrollable stress in animals contributes to the impairments in activity and in the capacity for new instrumental learning that characterise helplessness (Chapter 9) and, by inference, also to the corresponding impairments seen in human depression (Table 14.1).

I began this section by asking about the relationship between anxiety and depression. We may now see in outline an answer to this question. We first identify anxiety with activity in the ramified system described in the last chapter and centering upon the hippocampus, its inputs and outputs. We next identify depression with impaired noradrenergic transmission in the hypothalamus and impaired dopaminergic transmission in the dorsal and ventral striatum. Note that though anxiety and depression are therefore seen as involving opposite changes in noradrenergic function (increased in anxiety, decreased in depression), these changes take place in different regions of the brain, thus allowing symptoms of anxiety and depression to co-exist (as of course they do, especially in neurotic depression).

We may now consider two ways in which anxiety and depression might shade into each other, so creating the anxiety–depression continuum considered earlier in the chapter.

Consider first a temporal continuum, as stress is continuously applied. At first there will be an overall increase in ascending noradrenergic (and serotonergic) function, especially in the forebrain, activating the septo–hippocampal system and related structures and appearing as a state of pure anxiety. As stress continues to be applied, functional exhaustion of noradrenaline stores occurs in the hypothalamus and of dopamine stores in the striatum, giving rise to neurotic depression (which is now seen to be a mixture of anxiety and depression, as these states have been defined here). Finally with further continued stress, there may occur additional functional exhaustion in the noradrenergic terminals in the septo–hippocampal system (a proposition that awaits experimental evidence), giving rise to a state of pure depression.

I have so far described this progression along a temporal continuum in the same individuals; and indeed such a progression is consistent with clinical impression. But one might also see a similar progression along a continuum of individual differences. Thus, under identical conditions of stress, we may suppose that some individuals have a relatively great functional capacity in noradrenergic and dopaminergic neurons and respond only with anxiety; others have a moderate functional capacity and respond with mixed anxiety and depression; and still others have a weak functional capacity and respond only with depression. In this way, one would arrive at Roth's lumpy continuum from anxiety through neurotic to psychotic depression.

The nature of phobic symptoms

We have so far considered some of the major neurotic syndromes primarily in an attempt to understand the nature of their relation to personality. Let us now look at them from another point of view: that of their functional significance.

The symptoms whose function is most obvious are the simple phobias. These look like straightforward passive avoidance behaviour: a person is afraid of something and therefore avoids going anywhere near it. In fact, monosymptomatic phobias (i.e. phobias directed towards only one class of objects, with no further psychiatric symptoms) are extremely common among the normal population and are not in themselves indicative of strong neurotic tendencies. Indeed, given the strong evidence for continuity in the objects feared by ourselves and the other mammals considered in Chapter 2, they do not even necessarily indicate special life experiences or the symbolic associations favoured by Freud. The simplest explanation of, say, the human fear of snakes is that it is an innate fear of a stimulus configuration which has been dangerous throughout the evolutionary history of our own species and its ancestors. Fears of enclosed spaces, of very wide-open spaces, and of heights may well have similar origins. Social phobia, or fear of public gatherings, is also no surprise, given the role played by social stimuli in the evocation of fear among many other species besides Man (Chapters 2 and 6). In mild degrees, as in the fear of public speaking, it is widespread among the normal population. It is interesting to note that the typical age of onset of the social phobias is in the late teens or early twenties, at a time when social interaction has ceased to be play and has started to be a genuine battle for rank in the mammalian status hierarchy. In contrast, fears of particular kinds of animals, also quite common among the general population, typically start during childhood; they seem, in fact, to be childhood fears which, for one reason or another, have not been outgrown.

The fears of the neurotic are distinctive either in the severity with which they are experienced or in the unusual nature of the objects which are feared. Social phobia, for example, may attain such severe proportions as to render normal social life impossible; and, indeed, this is one of the most crippling and most common neurotic complaints seen in the psychiatric clinic. In cases such as these, we appear to be dealing with a normal type of fear which has been magnified by the particular characteristics of the neurotic personality. Where unusual objects are feared, there are likely to be, as well as predisposing features of the personality, special incidents in the patient's life which have led to this particular class of objects becoming *conditioned* fear stimuli. Freeman & Kendrick, for example, describe a woman who was so afraid of cats that she had completely given up going out in the streets at night in case she encountered one. They were able to trace this fear to the distress suffered by the patient when, at the age of four, she had seen her father drown a kitten in a bucket. Once such conditioning has occurred, it is very likely to spread to other stimuli resembling the original CS, according to the principles of stimulus generalisation which we have encountered several times in this book. Thus, in the case just quoted, the patient's fear of cats had spread to include any cat-like fur, fur gloves, fur coats, pictures of cats, and her daughter's toy koala bear. There is no reason, of course, why stimulus generalisation should not take all the complex routes provided by human verbal associations. There is laboratory evidence for such associational and semantic generalisation of simple conditioned responses. So, when such associations are found, we still do not need to call in any further principles than those of stimulus generalisation.

The condition of agoraphobia calls for special comment. This is the most destructive of the phobias. The patient experiences such severe fears of public places (e.g. supermarkets, city streets, buses, or trains) that he, or more commonly she, may remain house-bound for years. Frequently such patients also experience sporadic panic attacks. These occur without apparent good reason, and involve the whole gamut of autonomic changes outlined in Chapter 5 (pounding heart, hyperventilation, cold sweats, etc.). It is possible that in some cases of agoraphobia these panic attacks act as UCSs for the conditioning of anxiety to other stimuli. Suppose, for example, that the first panic attack occurs in a department store (as reported by some patients); then the multifarious cues that go to make up the complex stimulus 'department store' will come to be associated with the panic attack and the patient will in future avoid them. Agoraphobia is not, however, invariably associated with panic attacks, or panic attacks with goraphobia (nor, on the above account, would the association betwen the two be expected to be invariable). Panic attacks without agoraphobia constitute a major symptom in the so-called 'anxiety state'. In this condition physiological measures reveal, in addition to acute episodes of panic, chronically raised levels of activity in the autonomic nervous system.

The distinction between the panic attack as an unconditioned stimulus and the anticipatory anxiety that comes to be conditioned to the environment in which the panic attack occurs makes sense out of some otherwise puzzling pharmacological data reported by Klein. He finds that, whereas anti-anxiety drugs reduce anticipatory anxiety in agoraphobics (as, of course, we would expect), these drugs do not affect panic attacks, which are in contrast reduced by certain anti-depressants. Now, Graeff points out that the autonomic changes seen in panic attacks resemble those that are produced in animals by electrical stimulation of the central grey component of the fight/flight system (Chapter 13); and also that the drugs which are effective in controlling panic attacks include anti-depressants whose primary mode of action is to inhibit the re-uptake of serotonin, so facilitating serotonergic transmission (see above). Further, as noted in Chapter 13, the behavioural inhibition system apparently inhibits activity in the central grey by way of a serotonergic projection. Graeff proposes therefore that anti-depressants reduce panic attacks by facilitating the action of this inhibitory pathway. This proposal is consistent with the suggestion, made above, that panic attacks act as UCSs for the conditioning of anxiety, since it is precisely the fight/flight system that has been charged with mediating the effects of unconditioned aversive stimuli (Chapter 13). The cause of the panic attacks in predisposed patients, however, remains unknown.

The obsessive–compulsive syndrome

The obsessive-compulsive syndrome calls for rather different explanatory principles than those which have served in the case of the phobias. This syndrome involves both cognitive and behavioural phenomena: the patient feels impelled to think certain thoughts, to feel certain impulses, or to conjure up certain images, and also to carry out certain behavioural rituals.

Not surprisingly, behavioural rituals have proved more amenable to behavioural analysis than the cognitive symptoms of the obsessive patient. Typically, such a patient will, say, wash his hands compulsively and repetitively whenever he is faced with certain stimuli. If he is prevented from carrying out this ritual, he experiences intense fear. He reports that carrying out the ritual alleviates the fear caused by the stimuli. Should we look for any more recondite explanation of this behaviour than that the patient has acquired, through the vagaries of conditioning and learning, an active avoidance response not different in principle from the kind that carries a rat over a hurdle to avoid a potential electric shock? The fact that the ritual is not *instrumental* in avoiding a danger, even assuming there is an objective danger to avoid, is neither here nor there. There are many examples of such superstitious learning in the experimental literature. The things we learn are useful to us because evolution has guided our learning capacities into channels which normally *do* lead to useful results; but there is no guarantee that they will always lead to useful results no matter how our environment is structured.

The frequency with which hand-washing is chosen as the obsessional ritual gives us some clue as to how such superstitious avoidance learning might take place. It is extremely common in our culture for parents to punish their children for getting dirty or coming into contact with possible sources of infection. Washing the hands is a sure way of avoiding or mollifying the parental ire under these conditions and, in the right personality (a neurotic introvert), could become a well-established way of relieving fear. Once established, it might well transfer to other situations which also elicit fear or which are associated with dirt. The extremely formalised nature of the obsessional ritual, the need the patient feels to perform it in a definite, prescribed manner, is also not unexpected if this is a form of avoidance behaviour. It is a common observation in animal learning experiments that after an initial period of trial and error the subject settles down to performing a successful active avoidance response in an identical manner from trial to trial, even though the experimental arrangements actually leave considerable latitude for variation.

This acount of compulsive rituals as active avoidance behaviour has much to recommend it, and direct tests have provided experimental support for certain of its major assumptions. Thus, as Jack Rachman has shown, it is indeed possible in many cases to identify environmental stimuli which provoke the ritual; exposure of these stimuli without performance of the ritual increases subjective anxiety; and performance of the ritual is followed by a decrease in anxiety. But extension of the active-avoidance model to obsessional cognitive phenomena is more difficult. Rachman & Hodgson have attempted to treat these phenomena – e.g. the *thought* of homosexual acts, the *impulse* to jump out of the window, or the *image* of dead people in an open coffin – as noxious internal stimuli which, like phobic objects, give rise to anxiety and attempts at avoidance behaviour. Such behaviour may be overt. So, one who is troubled by an impulse to jump out of windows may (passively) avoid windows. Or a compulsive ritual may be used as an active avoidance response. Rachman, for example, describes a patient who was troubled by intrusive aggressive and sexual thoughts and was able to gain relief by going through a hand-washing ritual. Alternatively, the avoidance behaviour, like the obsessions, may remain internal. Thus another patient described by Rachman, troubled by an image of a child in danger, was able to 'put things right' by forming a counter-image of the child safe and sound.

The difficulty with this account is that it fails to provide an explanation for the intrusive nature of obsessional cognitions. It is reasonable to describe these phenomena as 'noxious', and there is experimental evidence that the physiological reactions that accompany them are similar to those elicited by phobic stimuli. But in that case why do they occur at all? If we treat internal events (though we should not) as stimuli, would we not expect noxious thoughts to be avoided passively, that is, not occur? Contrary to this expectation, of course, obsessions occur not only once but repeatedly and intrusively. And, since they are precisely *not* external stimuli, this can happen only because the patient's brain is actively producing them; but the active-avoidance hypothesis does not tell us why. A similar but less acute problem arises in the application of the active-avoidance model to compulsive behavioural rituals: why do these too occur not once but repeatedly? It is commonly found that rituals like hand-washing, checking that the door is locked, or that there is no glass splinter in the food, must be performed tens or even hundreds of times before the patient feels relief. This similarity between the repetitiveness of compulsive rituals and obsessional cognitions begs for a common explanation; but the active-avoidance hypothesis does not offer one.

Theories of anxiety that are based purely on avoidance behaviour, then, experience increasing strain as they move from phobias to compulsive behavioural rituals and finally to obsessional cognitions. In contrast, the account of anxiety developed in the previous chapter in terms of brain function is in some ways applied most naturally to the obsessive–compulsive syndrome. As we have seen, the chief function of the behavioural inhibition system is to monitor ongoing behaviour, checking continuously that outcomes coincide with expectations. In this role, it scans incoming sensory information for threatening or unexpected events and, if they occur, brings all other behaviour to a halt so as to evaluate the nature of the threat. Now, if such a system becomes hyperactive, if it treats too many stimuli as potentially threatening, if it searches for them too persistently, is it not obsessive–compulsive behaviour that it will produce?

Had it not been derived from quite independent sources of data, this description might seem too close to the observed phenomena to count as an explanation at all. It treats obsessive–compulsive behaviour much as does common sense. The patient scans his environment to an excessive degree for potential threats: dirt, bacteria, sharp objects, and the like. Much of the scan is carried out overtly, in the form of checking rituals. The exact form of the ritual depends, naturally, on the threat the patient is attempting to exclude. If this is a scratch, he searches for sharp objects; if it is the more abstract threat of 'disorder', he searches for objects out of place. Some rituals may simultaneously serve this kind of checking function and act as avoidance responses. Thus that most common of rituals, hand-washing, is at once an effective means of searching for dirt and a way to remove it. But, if it were only an active avoidance response, it would not be expected to occur over and over again; its repetitiveness derives from its checking function.

The cognitive symptoms of the obsessive–compulsive patient also find a natural explanation along these lines. There are two ways in which such symptoms can arise. First, the scan for potential environmental threats can extend to internal respositories of information concerning such threats. These are likely to be verbally coded and stored in the language areas of the temporal lobe. The subicular area has easy access to this region of the neocortex by way of its projection to the entorhinal area (Fig. 13.15). This is therefore perhaps the route used when one who is anxious, say, about cutting himself checks his memory to verify where he disposed of a razor blade. It is easy to see how an internal scan of this kind could also involve imagining the relevant scene. In this way, a mother whose greatest fear is of harming her own child might imagine scenes in which she has done just that. Second, there are some threats which are themselves of purely internal origin. Take someone who is afraid that he may experience a sudden impulse to jump out of a window. How is the behavioural inhibition system to check on threats of this nature? Evidently, this can be achieved only by some sort of internal scan (controlled perhaps by the descending projection from the prefrontal cortex discussed above) of the systems that produce the behaviour of which the patient is afraid. But at this point the principle of ideo-motor action described by William James in 1890 comes into play: that is, the thought of a particular action itself primes the systems that produce it. Thus checking whether one has an impulse to jump out of a window will increase the probability of actually feeling that impulse. The intrusive and repetitive nature of obsessional impulses could in this way arise from the very checking process which attempts to ensure that they are absent.

Behaviour therapy

At several points in this book (Chapters 9, 10, 12, and 13) we have discussed ways in which an initially aversive stimulus, conditioned or unconditioned, can be stripped of some of its aversive properties. A number of separate mechanisms appears to be able to produce this kind of effect: a non-associative process ('toughening up') dependent simply upon repeated exposure to the aversive stimulus; Pavlovian counter-conditioning, in which the aversive stimulus is used as a CS for an appetitive UCS; Capaldi's process of instrumental counter-conditioning, in which the aversive stimulus comes to act as a discriminative stimulus for the performance of a rewarded instrumental response; Amsel's process, in which cues from the internal state of anxiety similarly come to act as discriminative stimuli for rewarded behaviour; and there may be others so far unidentified. The existence of these diverse mechanisms suggests that one ought to be able to use behavioural methods to weaken the anxiogenic impact of the stimuli that elicit neurotic behaviour in Man. Much clinical research, conducted over the last three or four decades, demonstrates that this is indeed the case.

There has been wide variation both in the particular procedures developed for carrying out 'behaviour therapy' of this kind, and in the theoretical justifications offered for them. This diversity of theoretical starting points has been seen as a weakness in the scientific credentials of the behaviour therapy movement. We might now see it, however, as a strength: since more than one mechanism for increasing behavioural tolerance for stress has been identified in the laboratory, so should it be in the clinic. Unfortunately for this complacent conclusion, there has been little concordance between the detailed theoretical predictions made concerning the optimal parameters for a particular procedure and actual therapeutic effectiveness. Indeed, to a large extent these details – e.g. the ordering of the sequence of presentation of phobic items, the occurrence or otherwise of relaxation after presentation of a particular item (see below) – turn out to play an insignificant role in therapeutic effectiveness. All that seems to matter (to a first approximation) is the total amount of time for which the patient is exposed to the anxiogenic stimulus. This suggests that the major therapeutic ingredient in most existing procedures consists in engagement of the non-associative process distinguished above, rather than any of the associative processes; for the latter ought to show a greater dependence upon specific procedures than has been observed.

Even though research has shown that the particular details of the procedures used are relatively unimportant, a brief description of two of the more widely used methods will perhaps help the reader to form a clearer idea of what is involved in behaviour therapy.

In systematic desensitisation, stimuli are first graded into a hierarchy according to their capacity to elicit fear. They are then presented to the subject, sometimes in reality but often by asking him to imagine them, in a sequence which corresponds to gradually increasing anxiogenic power. Each stimulus presentation is typically short, since it is terminated as soon as the patient signals that he is beginning to feel anxious. As soon as the stimulus is terminated, the patient is instructed to relax deeply, using techniques in which he has been trained before therapy commences. This, essentially, is the method introduced by Joseph Wolpe. It is based on the concept of 'reciprocal inhibition' of anxiety by relaxation, a notion that is closely related to Amsel's concept of counter-conditioning of anticipatory frustration by food reward (Chapter 10) and to the Pavlovian counter-conditioning from which both concepts originally spring. The graded hierarchy of anxiogenic stimuli and the short stimulus presentations are intended to keep the level of anxiety down so that counter-conditioning is facilitated.

In flooding or implosive therapy, the patient is again presented with real or imagined aversive stimuli. But now the therapist attempts to maximise emotional arousal by concentrating from the start on strongly anxiogenic stimuli and by opposing any attempt on the part of the patient to elude them. This is tantamount to throwing a child who is afraid of water into the deep end of a swimming pool; in this respect flooding is diametrically opposed to the step-by-step gradualism of systematic desensitisation. The idea behind the technique is to allow the extinction of conditioned fear by eliciting it without reinforcement. From this point of view, the brief and mild elicitation of fear that takes place during systematic desensitisation trials would be a slow and inefficient way to produce extinction.

Given these very different procedures and the different theoretical analyses on which they are based, it would have been reasonable to suppose that if one worked well the other would not. However, as already noted, attempts to discriminate between their therapeutic effectiveness have by and large failed. Fortunately, this is not because neither works. On the contrary, using these and other methods more or less like them, behaviour therapists are now able to eliminate even severe phobias (including agoraphobia, the most crippling of them all) in over 90% of cases, as shown in well-controlled clinical trials. These successes are obtained after only a few weeks of therapy for each patient; moveover, during much of this time the patient treats himself, following a few simple principles in which he is initially instructed by the therapist. Remember that many of the patients treated in this way have been severely phobic for years, and would be likely to continue so indefinitely without the aid of behaviour therapy; remember also that no other treatment for these conditions existed before the advent of behaviour therapy (or since): you will then appreciate what a considerable achievement a 90% success rate represents. Furthermore, similar rates are now being achieved with obsessive–compulsive patients who are exposed to the stimuli which trigger their compulsive ritual while being prevented from carrying the ritual out.

Drug therapy

The other major method of treating anxiety is with drugs, especially the benzodiazepines. It is consistent with the evidence concerning the effects of these drugs in animals (Chapter 12) that they act only temporarily to suppress anxiety; they do not eliminate the patient's reaction to anxiogenic stimuli as, in the best of cases, does behaviour therapy. Indeed, it is even possible that under some conditions the anti-anxiety drugs antagonise the long-term beneficial effects of behaviour therapy.

As we saw in Chapter 10, it is possible to increase resistance to frustrative non-reward (extinction) by training animals on a partial reinforcement schedule; and it is similarly possible to increase resistance to punishment by training them on a partial punishment schedule. These two phenomena bear at least a general similarity to the methods used in behaviour therapy to eliminate human fears. But we also saw in Chapter 10 that both the partial reinforcement extinction effect and the partial punishment effect may sometimes be blocked if training on the relevant schedules is conducted while the animal is drugged with Librium. This blockade is complete if the interval between trials is of the order of 24 hours, but state dependent (that is, the animal shows increased resistance if both trained and tested under the drug, but not if trained under the drug and tested without it) if the inter-trial interval is a few minutes. Now, either type of blockade would be undesirable if similar interactions occur when behaviour therapy and drug therapy are combined with human patients. Complete blockade would entirely prevent the patient from benefitting from behaviour therapy. And, while state-dependent blockade would allow the patient to benefit from behaviour therapy, the benefit would only be manifest while he continued to take drugs; in this way, behaviour therapy might inadvertently contribute to the development of drug dependence.

Existing clinical data are insufficient to decide whether effects of this kind do indeed take place when behavioural and pharmacological treatments for human anxiety are combined; and we are therefore conducting a major trial to investigate the issue further. But note that behaviour therapy may be regarded as a formal means of arranging a contingency, namely, repeated exposure to anxiogenic events, which the normal everyday environment provides anyway, though in a more haphazard manner. Such environmental contingencies may be expected to lead to toughening up in many people exposed to them; and this is perhaps the reason why the spontaneous rate of recovery from anxiety symptoms is so high (nearly 70% in 2 years). Now, the same arguments that led to the prediciton that anxiolytic drugs may interfere with the effectiveness of behaviour therapy lead also to the fear that they may hamper the spontaneous process of toughening up. Since in any case, as noted above, anxiolytic drugs only temporarily alleviate anxiety, long-term maintenance on them is usually contra-indicated as a therapeutic strategy. The best use of these drugs is probably as a crutch to aid coping with a particularly threatening but short-lived situation.

Conclusion

The arguments deployed in this book have focussed on a relatively delimited topic: the psychology of fear and stress. Nonetheless, we have needed to range over a wide range of data, disciplines, and concepts. The material we have considered in this, our final chapter, will, I hope, have convinced you that our journey, taking us as it has through an experimental literature largely concerned with the behaviour of animals and the brain mechanisms that control their behaviour, has sent us back home able to view that prime object of our concern, Man, from a new and valuable perspective. It is a perspective that is increasingly paying off in the clinic. For the first time in psychiatric history we are witnessing the creation of a range of therapeutic techniques which are based on the fundamental knowledge of behaviour acquired in the laboratory. The most relevant sections of this knowledge are described by experimental psychologists in terms of the concepts I have tried to piece together in this book. Some familiarity with these concepts and, even more importantly, with the experimental evidence on which they are based, is likely to be of growing value for anyone who wishes to understand and evaluate what psychiatrists and clinical psychologists are trying to do for the neurotic patients who come into their care. For there can be little doubt that the experimental study of behaviour is going to be as important for the psychiatry of the future as the study of physiology has been for the physical medicine of the past.

Sources and References

1 Introduction

On the emotions treated from a philosophical point of view, see Kenny (1963), some of the papers in Mischel (1969), and Lyons (1980). On the treatment of the emotions within learning theory, see Mowrer (1960), Millenson (1967), and Gray (1972a, 1975).

2 Fears, innate and acquired

Hebb (1946a, 1946b, 1949), Valentine (1930), Tinbergen (1948, 1951), Seligman (1971), and Mineka (1987) are all very good value on this subject. On the experimental fate of the hawk–goose shape, see McNiven (1960), Martin & Melvin (1964), Melzack *et al.* (1959), and Schleidt (1961). The sex differences in the fears reported by human beings are given by Geer (1965). On the fear of strangers in human infants, see Ricciuti (1974). On pain, see Bonica (1980), Melzack & Wall (1983), and Bolles & Fanselow (1980). The observations of the emotional expressions elicited by pain in human infants are reported by Izard *et al.* (1983). The experiments on the effects of a fear CS and fear odours on responses to pain were reported by Fanselow & Baackes (1982) and Fanselow (1985). The experiment on the cultural transmission of mobbing in blackbirds is due to Vieth *et al.* (1980).

3 The expression of fear

I have followed Andrew's three papers closely in this chapter. Chapter 5 in Marler & Hamilton (1966) is a good review of the ethological literature.

4 The inheritance of fear

I have followed Broadhurst (1960a) closely throughout much of this chapter. Robinson (1965) gives a very thorough review of both Hall's and Broadhurst's work. More recent studies of the Maudsley strains are summaried by Broadhurst (1975) and Blizard (1981). The questionnaire studies of behaviour in the Second World War are by Stoufer *et al.* (see Broadhurst, 1960a). Parker's work is described by Hall (1941). For a mathematical treatment of factor analysis, see Harman (1960). Data on heritability estimates can be found in Eysenck (1967) and Floderus-Myrhed *et al.* (1980). On patterns of emotional behaviour in mice, see Van Abeelen (1966). Evidence discounting the possibility that the difference in defecation between the Maudsley strains reflects merely metabolic factors is presented by Blizard (1970). The experiment on rodent analogues of phobic and psychopathic behaviour is described in Eysenck (1967, p. 212).

5 The physiology of the emotions: fear and stress

For a general introduction to endocrinology, see Turner & Bagnara (1976); for an introduction to behavioural aspects of endocrinology, see Leshner (1978). Recent reviews of catecholamine, glucocorticoid, and adreno-pituitary peptide functions relevant to the material in this chapter can be found in Axelrod & Reisine (1984), Munck *et al.* (1984), and De Wied & Jolles (1982), respectively. Munck *et al.* advance the

375

interesting hypothesis that stress-induced increases in glucocorticoid levels protect 'not against the source of stress itself but rather against the body's normal reactions to stress'; their paper provides a useful review of the data from this point of view. Research on the effects of facial expression on felt emotion is discussed by Laird (1984). King's experiments on the behavioural effects of destruction of the sympathetic nervous system are described in Lord *et al.* (1976) and Oei & King (1978). The experiments on adrenal demedullation are by Caldwell (1962) and Levine & Soliday (1962); those on injections of ACTH are by De Wied (1964, 1966), Miller & Ogawa (1962), and Applezweig & Baudry (1955) (and see De Wied & Jolles, 1982). ACTH secretion under stress was measured in monkeys by Sidman *et al.* (1962), and the corresponding data on humans are reviewed by Mason (1959). The experiments on the effects of frustration and feeding on corticosterone levels were reported by Coover & Levine (1976) and Goldman, Coover & Levine (1973).

6   An excursion into social biology: fear and sex

Richter's work is reviewed by Robinson (1965, Chapter 5); see also Richter (1959). In the section on the general adaptation syndrome I have followed Christian (1959, 1963). Two later papers by Christian (1970 and 1975) should also be consulted. The whole area of social biology has become much more active since the first edition of this book, especially after the publication of E.O. Wilson's *Sociobiology* (1975). The general problem posed for genetic theory by apparently altruistic behaviour (of which the intrinsic controls over population density discussed in this chapter constitute one example) is discussed in Boorman & Levitt (1980). On wartime sex ratios, see Colombo (1957). On the factors that determine social rank and the consequences of rank for dispersal, see Bernstein & Gordon (1980) and Gauthreaux (1978).

7   The route from gene to behaviour: sex differences and fear

For major sources for Broadhurst's and Christian's work, see above (notes to Chapters 4 and 6). The view of sex differences in the emotional behaviour of rodents presented here has been documented more fully by Gray (1971) and Gray & Buffery (1971), and criticisms of this view (e.g., Archer, 1975) have been answered by Gray (1979). On the topic of sexual differentiation, see Harris (1964), Levine (1966), Beach (1971), Knobil (1974), and the issue of *Science* for 20 March 1981 (Vol. 211, no. 4488). On the sex difference in passive avoidance, see (besides the references given by Gray, 1979) Drago *et al.* (1980), Joseph & Gallagher (1980), and van Oyen *et al.* (1981*a*, *b*). For sex differences in ulcer formation in the rat, see Ader (1962) and Sawrey & Long (1962). Note, however, that when enforced immobilisation rather than psychological conflict has been used as the means of inducing ulcers, the sex difference is reversed (Ader, Beels & Tatum, 1960*a*; Sines, 1961). On sex differences in endocrine organs other than sex glands, see Kitay (1963) for the adrenocortical system, Bottari (1960) and Brown-Grant (1965) for the thyroid, and Green & Miller (1966) for adrenaline and noradrenaline secretion. The work from my own laboratory relating to sex and strain differences in the electrophysiology of the septo–hippocampal system is reported by Drewett *et al.* (1977). For Blizard's work on noradrenergic function in the two Maudsley strains, see Liang & Blizard (1978), Blizard & Liang (1979), Blizard *et al.* (1980), and Liang *et al.* (1982). For a recent report of the effects of neonatal testosterone on adult open-field behaviour, see Stevens & Goldstein (1981). The human sex differences in the incidence of ulcers are from Halliday (1945), and of phobias, from Marks (1969). For a review of sex differences in children's fears, see Graziano *et al.* (1979); and for an empirical study of possible artefacts affecting the sex difference in depression, see Amenson & Lewinsohn (1981). On sex differences in disturbances of sexual behaviour, see Kinsey *et al*. (1953). On sex differences in aggression, see Gray & Buffery (1971) and Buffery & Gray (1972). Induction of aggressive behaviour by *prolonged* administration of testosterone to adult female mice has been reported by Simon *et al*. (1984). For a recent report on sex differences in emotional behaviour in the chimpanzee, confirming the generalisation (Buffery & Gray, 1972) that females are more timid and less aggressive than males among primates, see Buirski *et al*. (1978).

8   The early environment and fearfulness

Major sources for this chapter have been Levine (1962*a*, *b*), Levine & Mullins (1966), and Denenberg (1964). There is a good review of the early handling literature by Russell (1971). Ader (1975) also provides an excellent review of the effects of early experience and the possible involvement in these of hormonal pathways. The relevant Ader studies discussed here are: on social isolation, Ader, Kreutner & Jacobs (1963) and Ader & Friedman (1964); on early weaning, Ader (1962); and on removal of the mother, Ader, Beels & Tatum (1960*b*). On the effects of social isolation in ferrets, see Chivers & Einon (1982). The experiments showing that rat mothers treat handled and non-handled pups differently are reported by Brown, Smotherman & Levine (1977); those showing that the effects of early handling may be modified by the strain of foster-mother, by Hennessy, Vogt & Levine (1982). An effect of the pre-conceptual environment like Joffe's had previously been reported by Denenberg & Whimbey (1963). The 'sibling effect' illustrated in Fig. 8.5 is reported in detail by Gray *et al*. (1969). The electroencephalographic changes caused in dogs by neonatal treatment were reported by Fox & Stelzner (1966).

9   Punishment and conflict

Major sources for this chapter include N.E. Miller (1951, 1959, 1976), Solomon (1964), Azrin & Holz (1966), Hinde (1966), Gray (1975), and Mackintosh (1983). The literature on experimental neurosis is reviewed by Broadhurst (1960*b*) and Mineka & Kihlstrom (1978); on ulcer formation, by Weiss (1977); on the effects of predictable versus unpredictable shock, by Abbott, Schoen & Badia (1984); on displacement activities, by Zeigler (1964) and McFarland (1966); on the role of approach–avoidance conflict in human social encounters, by Patterson (1973); and on counter-conditioning, by Dickinson & Pearce (1977). Seligman (1975) and Weiss *et al*. (1976*a*, 1982) offer very different views of the nature of 'learned helplessness' (see Chapter 12 in Gray, 1982); and the Weiss *et al*. (1976*a*, 1982) papers, as well as Miller (1976) and Sklar & Anisman (1981), describe the phenomenon of 'toughening up'. The experiment showing that the incidence of ulcers after inescapable shock can be reduced if a rat is allowed to attack another rat is reported by Weiss *et al*. (1976*b*); and the demonstration that this manipulation also attenuates the adrenocortical response to shock is due to Weinberg *et al*. (1980). The experiment from Maier's laboratory which demonstrates blocking ('immunization') of shock-induced and drug (morphine)-induced analgesia by prior exposure to escapable shock was reported by Moye *et al*. (1983); and the experiment in which this group described the effects of hypophysectomy and the administration of dexamethasone on shock-induced analgesia and escape deficits was reported by Mac-Lennan *et al*. (1982). Maier *et al*. (1979) showed that shock-induced opioid analgesia follows a similar time-course to shock-induced escape and activity deficits; while Mah *et al*. (1980) demonstrated that different experimental parameters affect a non-opioid form of shock-induced analgesia and the escape deficit, respectively. The experiment in which inescapable shock was shown to impair choice behaviour in a subsequent Y-maze shock-escape task is due to Jackson, Alexander & Maier (1980).

10   *Fear and frustration*

Major sources for this chapter are Miller (1959, 1964), Wagner (1966), Amsel (1962), Mackintosh (1974), and Gray (1967, 1975, 1977). Evidence that non-reward gives rise to an elevation in blood levels of corticosterone is presented by Goldman, Coover & Levine (1973). The drive-summation experiments summarised in connection with Fig. 10.1 include Amsel & Cole (1953) and Siegel & Brantley (1951). The Meryman experiment (Fig. 10.4) is described in Brown (1961). On the double-runway frustration effect, see Wagner (1959). Analogous apparent effects of frustration have been reported for pigeons by Staddon & Innis (1966), for monkeys by Davenport & Thompson (1965), and for children by Ryan & Watson (1968) and Nicholson & Gray (1972), although the interpretation of some of these effects is in doubt. For the ingestion of alcohol by animals under stress, see Masserman & Yum (1946); and for amytal, Davis & Miller (1963). Feldon's experiments on the influence of anti-anxiety drugs on the partial reinforcement extinction effect (Fig. 10.8) are reported in Feldon *et al.* (1979) and Feldon & Gray (1981). On the clinical implications of this type of finding, see Gray (1987). On the failure of amytal and other anti-anxiety drugs to alter the double-runway frustration effect, see Ison, Daly & Glass (1967), Gray (1969), Gray & Dudderidge (1971), Freedman & Rosen (1969), and further references given by Gray (1977). Increased response force during extinction has been reported by Notterman (1959) and increased response duration, by Margulies (1961).

11   *The learning of active avoidance*

Major sources for Chapter 11 include Miller (1951, 1959), Mowrer (1960), Rescorla & Solomon (1967), Herrnstein (1969), Gray (1975), Mineka (1979), and Mackintosh (1983). The persistence of avoidance behaviour (with which much of the chapter is concerned) offered no problems for Miller's version of the Miller–Mowrer theory, but this version has been shown to be wrong (Hilgard & Marquis, 1961). I have therefore considered only Mowrer's version, which treats classical and instrumental conditioning as two distinct processes (see Gray, 1975). On the aversiveness of unsignalled relative to signalled shock, see Lockard (1963), Perkins *et al.* (1966), R. Brown (1965), and the review by Badia, Hursh & Abbott (1979). The experiments on current flow and grid contact time as a function of signalled versus unsignalled shock are by Imada *et al.* (1981) and Badia & Abbott (1980), respectively. The influential Rescorla–Wagner theory (and later modifications upon its basic theme) is described in many sources, e.g., Wagner & Rescorla (1972), Dickinson (1980), Mackintosh (1983).

12   *A conceptual nervous system for avoidance behaviour*

The central ideas in this chapter originated in a paper by Gray & Smith (1969), who give them a mathematical treatment, and in the first edition of this book; and they were developed further by Gray (1972*a*, 1975, 1976, 1977, 1982). Many of them have their precursors in Mowrer's (1960) stimulating text on learning theory. On the fight/ flight system, see Adams (1979), Panksepp (1982), and Graeff (1987). The early work on reinforcing electrical stimulation of the brain was ably reviewed by Olds & Olds (1965); for a more recent review, see Olds & Fobes (1981). On the motivational and reinforcing properties of central stimulation which produces predatory attack, see Roberts & Kiess (1964) and Hutchinson & Renfrew (1966). Work on the behavioural effects of anti-anxiety drugs is reviewed by Gray (1977, 1982), Dantzer (1977), and Graeff (1981). On the differences in emotionality between the Roman high and low avoidance strains, see Imada (1972), Chamove & Sanders (1980), and Drewek & Broadhurst (1979). Guillamon's observation that amytal selectively increases running speed on the non-rewarded trials of a single alternation schedule is reported in Feldon *et al.* (1979). For a recent report that chlordiazepoxide, like sodium amytal, reduces the response to stimulus change in Ison's T-maze task, see Quintero *et al.* (1985*a*). For the results of recent 'retardation' experiments along the lines of Konorski and Szwejkowska's, see Scavio & Gormezano (1980) and De Vito & Fowler (1982), who studied appetitive to aversive transfer, and Bromage & Scavio (1978), who studied the reverse direction of transfer.

13   Fear and the central nervous system

Some of the earlier material in this chapter was culled from the textbooks of physiological psychology written by Grossman (1967) and Deutsch & Deutsch (1966). However, knowledge of the brain has progressed enormously since the first edition of this book. In consequence, Chapter 13 has required extensive revision and expansion. Major sources for this edition include Gray (1982), Gray *et al.* (1982), Rawlins (1985), Gray & Rawlins (1986), Adams (1979), Panksepp (1982), Graeff (1987), and Melzack & Wall (1983). For a general introduction to the limbic system, see Isaacson (1982). A view of the functions of portions of the limbic system that bears much similarity to my own was independently proposed by Numan (1978). For a review of research on the general pharmacology of the benzodiazepines, see Haefely *et al.* (1981); for a summary of recent work on the GABA/benzodiazepine receptor complex, see Haefely *et al.* (1985); and for interactions between the barbiturates and this receptor complex, see Olsen (1981). On the locus coeruleus and the dorsal ascending noradrenergic bundle, see McNaughton & Mason (1980) and Redmond (1979). Some of the evidence that this pathway has the function of increasing the signal-to-noise ratio in the brain areas that it innervates comes from experiments by Segal (1977*a*, *b*). On the general psychological notion of 'arousal', see Gray (1964) and Eysenck (1967). The work of Kandel's group on classical fear conditioning in *Aplysia* is reported by Carew *et al.* (1981) and Walters *et al.* (1981). For some (widely differing) contemporary views of the functions of the hippocampus, see (besides Gray, 1982, and among others) O'Keefe & Nadel (1978), Olton *et al.* (1979), and Rawlins (1985). The experiments from my own laboratory in which we have tested the GABAergic hypothesis of anti-anxiety drug action are reported by Quintero *et al.* (1985*a*, *b*, *c*) and Buckland *et al.* (1986). The experiments conducted by Fuxe's group on stress-induced increases in noradrenaline turnover are reported by Lidbrink *et al.* (1973). The experiments on the influence of anti-anxiety drugs on the theta-driving curve (Fig. 13.17), and on the role of noradrenergic neurons in controlling the shape of this curve (Fig. 13.18), are reported by Gray & Ball (1970), Gray *et al.* (1975), and McNaughton *et al.* (1977). The projection from the subiculum to the nucleus accumbens and other regions of the ventral striatum has been described by Kelley & Domesick (1982). For an integrative summary of the behavioural functions of the dopaminergic projections to these same brain regions, see Willner (1985). The experiments describing the effects of hippocampal and fornix-fimbria lesions on the PREE (see Fig. 13.21) have been reported by Rawlins *et al.* (1980*a*, 1985) and Feldon *et al.* (1985). For reviews of the anatomy, physiology, and functions of serotonergic systems, see Azmitia (1978), Gray (1982, Chapter 11), and Soubrié (1986); Soubrié's view of the behavioural functions of these systems is very close to my own.

14   Man: neurosis, neuroticism, therapy

Much of this chapter is based on Gray (1970*a*, 1972*a*, 1981, 1982, 1984). On the general approach to personality and abnormal behaviour it exemplifies, see Eysenck (1967), Eysenck & Eysenck (1969), and Gray (1972*b*). The important work of Cattell's (e.g., 1965) group on the factor-analytic description of personality has been marked by a number of differences of approach compared to Eysenck's work, but Eysenck's and Cattell's findings agree about those major results which are dealt with here. For a review of the involvement of serotonergic systems in human psychiatric disorder, see Soubrié (1986) and Graeff (1987); and on the role of the serotonergic projection to the substantia nigra in punishment-induced response suppression, see Thiébot, Hamon & Soubrié (1983). For references to studies of the role in anxiety of the dopaminergic projections to the frontal and cingulate cortices, see Gray (1982, p. 422). For reviews of the effects of psychosurgery, see Willett (1960), Powell (1979), and Gray (1982, Chapter 13). Research on the functions of the frontal cortex in animals is reviewed by Gray (1982, Chapter 13). For a good general textbook of psychiatry, see Mayer-Gross, Slater & Roth (1979). On the question of psychiatric classification, see Eysenck (1960a) and Eysenck & Rachman (1965). On phobias, see Marks (1969); on criminality and psychopathy, see Eysenck (1977), Passingham (1972), and Hare & Schalling (1978). On the repression–sensitisation scale, see Eriksen (1966) and Byrne (1964). Gorenstein & Newman (1980) have advanced very similar views to my own concerning the relationships between personality, sensitivity to reward and punishment, and psychopathology. An additional, recent experiment that supports these views has been reported by Boddy *et al.* (1986). McFie (1972) has reviewed the evidence relating factor-analytic studies of individual differences in intellectual abilities to the organisation of the brain. The report that the Maudsley strains differ in the densities of benzodiazepine receptors in the brain is due to Robertson *et al.* (1978). The relationship between anxiety and depression, and the possible neural bases for this relationship, are discussed in Gray (1982, Chapter 12). Willner (1985) provides an excellent overview of the general literature on the neurobiology of depression. On the success of behaviour therapy, see e.g., Rachman (1967), Mathews (1978), and Mathews *et al.* (1981). On the spontaneous recovery rate for neurosis, see Eysenck (1960b). For a detailed discussion of the possible interactions between pharmacological and behavioural treatments for phobias, see Gray (1987).

References

Abbott, B.B., Schoen, L.S. & Badia, P. (1984). Predictable and unpredictable shock: behavioural measures of aversion and physiological measures of stress. *Psychological Bulletin*, **96**, 45–71.

Adams, D. & Flynn, J.P. (1966). Transfer of an escape response from tail shock to brain-stimulated attack behaviour. *Journal of the Experimental Analysis of Behaviour*, **9**, 401–8.

Adams, D.B. (1979). Brain mechanisms for offence, defense, and submission. *Behavioural and Brain Sciences*, **2**, 201–41.

Adelman, H.M. & Maatsch, J.L. (1956). Learning and extinction based upon frustration, food reward and exploration tendency. *Journal of Experimental Psychology*, **52**, 311–15.

Ader, R. (1962). Social factors affecting emotionality and resistance to disease in animals. III. Early weaning and susceptibility to gastric ulcers in the rat: a control for nutritional factors. *Journal of Comparative and Physiological Psychology*, **55**, 600–2.

Ader, R. (1975). Early experience and hormones: emotional behaviour and adrenocortical function. In *Hormonal Correlates of Behaviour*, ed. B.E. Eleftheriou & R.L. Sprott, pp. 7–33. New York: Plenum.

Ader, R., Beels, C.C. & Tatum, R. (1960*a*). Blood pepsinogen and gastric erosions in the rat. *Psychosomatic Medicine,* **22**, 1–12.

Ader, R., Beels, C.C. & Tatum, R. (1960*b*). Social factors affecting emotionality and resistance to disease in animals. II. Susceptibility to gastric ulceration as a function of interruptions in social interactions and the time at which they occur. *Journal of Comparative and Physiological Psychology,* **53**, 455–8.

Ader, R. & Friedman, S.B. (1964). Social factors affecting emotionality and resistance to disease in animals. IV. Differential housing, emotionality, and Walker 256 carcinosarcoma in the rat. *Psychological Reports,* **15**, 535–41.

Ader, R., Kreutner, A., Jr & Jacobs, H.L. (1963). Social environment, emotionality and alloxan diabetes in the rat. *Psychosomatic Medicine,* **25**, 60–8.

Albert, D.J. & Chew, G.L. (1980). The septal forebrain and the inhibitory modulation of attack and defence in the rat: a review. *Behavioural and Neural Biology,* **30**, 357–88.

Amenson, C.S. & Lewinsohn, P.M. (1981). An investigation into the observed sex differences in prevalence of unipolar depression. *Journal of Abnormal Psychology,* **90**, 1–13.

Amsel, A. (1962). Frustrative nonreward in partial reinforcement and discrimination learning: some recent history and a theoretical extension. *Psychological Review,* **69**, 306–28.

Amsel, A. & Cole, K.F. (1953). Generalization of fear-motivated interference with water intake. *Journal of Experimental Psychology,* **46**, 243–7.

Amsel, A. & Roussel, J. (1952). Motivational properties of frustration. I. Effect on a running response of the addition of frustration to the motivational complex. *Journal of Experimental Psychology,* **43**, 363–8.

Amsel, A. & Surridge, C.T. (1964). The influence of magnitude of reward on the aversive properties of anticipatory frustration. *Canadian Journal of Psychology,* **18**, 321–7.

Anderson, E.E. (1938*a*). The interrelationship of drives in the male albino rat. II. Intercorrelations between 47 measures of drive and learning. *Comparative Psychology Monographs,* **14** (6), 1–119.

Anderson, E.E. (1938*b*). The interrelationship of drives in the male albino rat. III. Interrelations among measures of emotional, sexual, and exploratory behaviour. *Journal of Genetic Psychology,* **53**, 335–52.

Anderson, E.E. (1940*a*). The sex hormones and emotional behaviour. I. The effect of sexual receptivity upon timidity in the female rat. *Journal of Genetic Psychology,* **56**, 149–58.

Anderson, E.E. (1940*b*). The sex hormones and emotional behaviour. III. The effect of castration upon timidity in male and female rats. *Journal of Genetic Psychology,* **56**, 169–74.

Anderson, E.E. & Anderson, S.F. (1940). The sex hormones and emotional behaviour. II. The influence of the female sex hormone upon timidity in normal and castrated female rats. *Journal of Genetic Psychology,* **56**, 159–68.

Andrew, R.J. (1963). Evolution of facial expression. *Science,* **143**, 1034–41.

Andrew, R.J. (1964). The displays of the primates. In *Evolutionary and Genetic Biology of Primates,* vol. 2, ed. J. Buettner-Janusch, pp. 227–309. New York: Academic Press.

Andrew, R.J. (1965). The origins of facial expression. *Scientific American,* **213**, 88–94.

Anisman, H., De Catanzaro, D. & Remington, G. (1978). Escape performance following exposure to inescapable shock: deficits in motor response maintenance. *Journal of Experimental Psychology: Animal Behaviour Processes,* **4**, 197–218.

Anisman, H., Remington, G. & Sklar, L.S. (1979). Effect of inescapable shock on subsequent escape performance: catecholaminergic and cholinergic mediation of response initiation and maintenance. *Psychopharmacology,* **61**, 107–24.

Applezweig, M.H. & Baudry, F.D. (1955). The pituitary-adrenocortical system in avoidance learning. *Psychological Reports,* **1**, 417–20.

Archer, J. (1975). Rodent sex differences in emotional and related behaviour. *Behavioural Biology,* **14**, 451–79.

Argyle, M. & Dean, J. (1965). Eye-contact, distance and affiliation. *Sociometry,* **28**, 289–304.

Ax, A.F. (1953). The physiological differentiation between fear and anger in humans. *Psychosomatic Medicine,* **15**, 433–42.

Axelrod J. & Reisine, T.D. (1984). Stress hormones: their interaction and regulation. *Science,* **224**, 452–9.

Azmitia, E.C. (1978). The serotonin producing neurons of the midbrain median and dorsal raphe nuclei. In *Handbook of Psychopharmacology,* vol. 9, ed. L.L. Iversen, S.D. Iversen & S.H. Snyder, pp. 233–314. New York: Plenum Press.

Azrin, N.H. (1967). Pain and aggression. *Psychology Today,* **1**, (1), 26–33.

Azrin, N.H. & Holz, W.C. (1966). Punishment. In *Operant Behaviour: Areas of Research and Application,* ed. W.K. Honig, pp. 380–447. New York: Appleton-Century-Crofts.

Badia, P. & Abbott, B. (1980). Does shock modifiability contribute to preference for signaled shock? *Animal Learning and Behaviour,* **8**, 110–15.

Badia, P., Abbott, B. & Schoen, L. (1984). Choosing between predictable shock schedules: long- versus short-duration signals. *Journal of the Experimental Analysis of Behaviour,* **41**, 319–27.

Badia, P., Hursh, J. & Abbott, B. (1979). Choosing between predictable and unpredictable shock conditions: data and theory. *Psychological Bulletin,* **86**, 1107–31.

Baker, A.G. (1976). Learned irrelevance and learned helplessness: rats learn that stimuli, reinforcers and responses are uncorrelated. *Journal of Experimental Psychology: Animal Behaviour Processes,* **2**, 130–41.

Ball, G.G. & Gray, J.A. (1971). Septal self-stimulation and hippocampal activity. *Physiology and Behaviour,* **6**, 547–9.

Banks, R.K. (1966). Persistence to continuous punishment following intermittent punishment training. *Journal of Experimental Psychology,* **71**, 373–7.

Barlow, D.H., Sakheim, D.K. & Beck, J.G. (1983). Anxiety increases sexual arousal. *Journal of Abnormal Psychology,* **92**, 49–54.

Barnett, S.A. (1955). Competition among wild rats. *Nature,* **175**, 126–7.

Basbaum, A.I. & Fields, H.L. (1978). Endogenous pain control mechanisms: review and hypothesis. *Annals of Neurology,* **4**, 451–62.

Beach, F.A. (1947). A review of physiological and psychological studies of sexual behaviour in mammals. *Physiological Reviews,* **27**, 240–307.

Beach, F.A. (1971). Hormonal factors controlling the differentiation, development and display of copulatory behaviour in the ramstergig and related species. In *The Biopsychology of Development,* ed. E. Tobach, L.R. Aronson & E. Shaw, pp. 249–96. New York: Academic Press.

Beach, F.A. & Fowler, H. (1959). Effects of 'situational anxiety' on sexual behaviour in male rats. *Journal of Comparative and Physiological Psychology,* **52**, 245–8.

Berger, D.F., Starzec, J.J. & Mason, E.B. (1981). The relationship between plasma corticosterone levels and leverpress avoidance vs escape behaviour in rats. *Physiological Psychology,* **9**, 81–6.

Bernstein, J.S. & Gordon, T.P. (1980). The social component of dominance relationships in rhesus monkeys (*Macaca mulatta*). *Animal Behaviour, 28*, 1033–9.

Bertenthal, B.I., Campos, J.J. & Barrett, K.C. (1984). Self-produced locomotion: an organizer of emotional, cognitive and social development in infancy. In *Continuities and Discontinuities in Development*, ed. R.N. Erride & R.J. Harmon, pp. 175–210. New York: Plenum.

Bignami, G. (1965). Selection for high rates and low rates of avoidance conditioning in the rat. *Animal Behaviour, 13*, 221–7.

Black, A.H. (1958). The extinction of avoidance responses under curare. *Journal of Comparative and Physiological Psychology, 51*, 519–24.

Blizard, D.A. (1970). The Maudsley strains: the evaluation of a possible artifact. *Psychonomic Science, 19*, 145–6.

Blizard, D.A. (1981). The Maudsley Reactive and Nonreactive strains: a North American perspective. *Behaviour Genetics, 11*, 469–89.

Blizard, D.A. & Chai, C.K. (1972). Behavioural studies in mice selectively bred for differences in thyroid function. *Behaviour Genetics, 2*, 301–9.

Blizard, D.A. & Liang, B. (1979). Plasma catecholamines under basal and stressful conditions in rat strains selectively bred for differences in response to stress. In *Catecholamines: Basic and Clinical Frontiers*, vol. 2, ed. E. Usdin, I.J. Kopin & J. Bachas, pp. 1795–7. New York: Pergamon.

Blizard, D.A., Liang, B. & Emmel, D.K. (1980). Blood pressure, heart rate, and plasma catecholamines under resting conditions in rat strains selectively bred for differences in response to stress. *Behavioural and Neural Biology, 29*, 481–92.

Boddy, J., Carver, A. & Rowley, K. (1986). Effect of positive and negative verbal reinforcement on performance as a function of extraversion-introversion: some tests of Gray's theory. *Personality and Individual Differences, 7*, 81–8.

Bolles, R.C. (1971). Species-specific defence reactions. In *Aversive Conditioning and Learning*, ed. F.R. Brush, pp. 183–233. New York and London: Academic Press.

Bolles, R.C. & Fanselow, M.S. (1980). A perceptual defensive-recuperative model of fear and pain. *Behavioural and Brain Sciences, 3*, 291–323.

Bolles, R.C., Holtz, R., Dunn, T. & Hill, W. (1980). Comparison of stimulus learning and response learning in a punishment situation. *Learning and Motivation, 11*, 78–96.

Bolles, R.C., Stokes, L.W. & Younger, M.S. (1966). Does CS termination reinforce avoidance behaviour? *Journal of Comparative and Physiological Psychology, 62*, 201–7.

Bonica, J.J. (ed.) (1980). *Pain*. New York: Raven Press.

Boorman, S.A. & Levitt, P.R. (1980). *The Genetics of Altruism*. New York: Academic Press.

Bottari, P.M. (1960). The relationship between pituitary and thyroid gland in health and disease. In *Progress in Endocrinology*, part I, ed. K. Fotherby, J.A. Lovaire, J.A. Strong & P. Eckstein, pp. 74–88. Cambridge University Press.

Bottjer, S.W. (1982). Conditioned approach and withdrawal behaviour in pigeons: effects of a novel extraneous stimulus during acquisition and extinction. *Learning and Motivation, 13*, 44–67.

Brady, J.V. (1958). Ulcers in 'executive' monkeys. *Scientific American, 199* (4), 95–100.

Broadhurst, P.L. (1960a). Application of biometrical genetics to the inheritance of behaviour. In *Experiments in Personality*, vol. 1, ed. H.J. Eysenck, pp. 3–102. London: Routledge & Kegan Paul.

Broadhurst, P.L. (1960b). Abnormal animal behaviour. In *Handbook of Abnormal Psychology*, ed. H.J. Eysenck, pp. 726–63. London: Pitman.

Broadhurst, P.L. (1975). The Maudsley Reactive and Nonreactive strains of rats: a survey. *Behaviour Genetics*, **5**, 299–319.

Bromage, B.K. & Scavio, M.J., Jr (1978). Effects of an aversive CS+ and CS− under deprivation upon successive classical appetitive and aversive conditioning. *Animal Learning and Behaviour*, **6**, 57–65.

Bronson, F.H. & Eleftheriou, B.E. (1963). Adrenal responses to crowding in *Peromyscus* and C57 BL/10J mice. *Physiological Zoology*, **36**, 161–6.

Brookes, S., Rawlins, J.N.P. & Gray, J.A. (1983). Hippocampal lesions do not alter the partial punishment effect. *Experimental Brain Research*, **52**, 34–40.

Brown, C.P., Smotherman, W.P. & Levine, S. (1977). Interaction-induced reduction in differential maternal responsiveness: an effect of cue-reduction or behaviour? *Developmental Psychobiology*, **10**, 273–80.

Brown, J.H.U. & Barker, S. (1966). *Basic Endocrinology*. Oxford: Blackwell.

Brown, J.S. (1961). *The Motivation of Behaviour*. New York: McGraw-Hill.

Brown, J.S., Kalish, H.I. & Farber, I.E. (1951). Conditioned fear as revealed by magnitude of startle response to an auditory stimulus. *Journal of Experimental Psychology*, **41**, 317–28.

Brown, R. (1965). Discrimination of avoidable and unavoidable shock. *British Journal of Psychology*, **56**, 275–83.

Brown, R.T. & Wagner, A.R. (1964). Resistance to punishment and extinction following training with shock or nonreinforcement. *Journal of Experimental Psychology*, **68**, 503–7.

Brown-Grant, K. (1965). The effect of testosterone during the neonatal period on the thyroid gland of male and female rats. *Journal of Physiology*, **176**, 91–104.

Bruell, J.H. (1969). Genetics and adaptive significance of emotional defecation in mice. *Annals of the New York Academy of Science*, 159, 825–30.

Brush, F.R., Baron, S., Froelich, J.C., Ison, J.R., Pellegrino, L.J., Phillips, D.S., Sakellanis, P.C. & Williams, V.W. (1985). Genetic differences in avoidance learning by *Rattus norvegicus*: escape/avoidance responding, sensitivity to electric shock, discrimination learning, and open-field behaviour. *Journal of Comparative Psychology*, **99**, 60–73.

Buckland, C., Mellanby, J. & Gray, J.A. (1986). The effects of compounds related to γ-aminobutyrate and benzodiazepine receptors on behavioural responses to anxiogenic stimuli in the rat: extinction and successive discrimination. *Psychopharmacology*, **88**, 285–95.

Buffery, A.W.H. & Gray, J.A. (1972). Sex differences in the development of spatial and linguistic skills. In *Gender Differences, Their Ontogeny and Significance*, ed. C. Ounsted & D.C. Taylor, pp. 123–57. London: Churchill.

Buirski, P., Plutchik, R. & Kellerman, H. (1978). Sex differences, dominance, and personality in the chimpanzee. *Animal Behaviour*, **26**, 123–9.

Burke, A.W. & Broadhurst, P.L. (1966). Behavioural correlates of the oestrous cycle in the rat. *Nature*, **209**, 223–4.

Butler, R.G. (1980). Population size, social behaviour, and dispersal in house mice: a quantitative investigation. *Animal Behaviour*, **28**, 78–85.

Byrne, D. (1964). Repression-sensitization as a dimension of personality. In *Progress in Experimental Personality Research*, vol. 1, ed. B. Maher, pp. 169–220. New York: Academic Press.

Caldwell, D.F. (1962). Effects of adrenal demedullation on retention of a conditioned avoidance response in the mouse. *Journal of Comparative and Physiological Psychology*, **55**, 1079–81.

Calhoun, J.B. (1952). The social aspects of population dynamics. *Journal of Mammalology,* **33**, 139–59.

Candland, D.K. & Campbell, B.A. (1962). Development of fear in the rat as measured by behaviour in the open field. *Journal of Comparative and Physiological Psychology,* **55**, 593–6.

Cannon, W.B. (1932). *The Wisdom of the Body.* New York: Norton.

Capaldi, E.J. (1967). A sequential hypothesis of instrumental learning. In *The Psychology of Learning and Motivation,* ed. K.W. Spence & J.T. Spence, pp. 67–156. New York & London: Academic Press.

Carew, T.J., Walters, E.T. & Kandel, E.R. (1981). Associative learning in *Aplysia*: cellular correlates supporting a conditioned fear hypothesis. *Science,* **211**, 501–4.

Cattell, R.B. (1950). *Personality.* New York: McGraw-Hill.

Cattell, R.B. (1965). *The Scientific Analysis of Personality.* Harmondsworth, England: Penguin.

Cattell, R.B. & Scheier, I.H. (1961). *The Meaning and Measurement of Neuroticism and Anxiety.* New York: Ronald Press.

Chamove, A.S. & Sanders, D.C. (1980). Emotional correlates of selection for avoidance learning in rats. *Biological Psychology,* **10**, 41–55.

Charney, D.S., Heininger, G.R. & Breier, A. (1984). Noradrenergic function in panic anxiety. *Archives of General Psychiatry,* **41**, 751–63.

Chen, J.S. & Amsel, A. (1982). Habituation to shock and learned persistence in pre-weanling, juvenile, and adult rats. *Journal of Experimental Psychology: Animal Behaviour Processes,* **8**, 113–30.

Chi, C.C. (1955). The effect of amobarbital sodium on conditioned fear as measured by the potentiated startle response in rats. *Psychopharmacologia,* **7**, 115–22.

Chitty, D. (1967). The natural selection of self-regulatory behaviour in animal populations. *Proceedings of the Ecological Society of Australia,* **2**, 51–78.

Chivers, S.M. & Einon D.F. (1982). Effects of early social experience on activity and object investigation in the ferret. *Developmental Psychobiology,* **15**, 75–80.

Chorazyna, H. (1962). Some properties of conditioned inhibition. *Acta Biologiae Experimentalis,* **22**, 5–13.

Christian, J.J. (1959). The roles of endocrine and behavioural factors in the growth of mammalian populations. In *Comparative Endocrinology,* ed. A. Gorbman, pp. 71–97. New York: Wiley.

Christian, J.J. (1963). Endocrine adaptive mechanisms and the physiologic regulation of population growth. In *Physiological Mammalology,* vol. 1, *Mammalian Populations,* ed. W.V. Mayer & R.G. van Gelder, pp. 189–353. New York: Academic Press.

Christian, J.J. (1970). Social subordination, population density, and mammalian evolution. *Science,* **168**, 84–90.

Christian, J.J. (1975). Hormonal control of population growth. In *Hormonal Correlates of Behaviour,* ed. B.E. Eleftheriou & R.L. Sprott, pp. 205–74. New York: Plenum.

Church, R.M., Wooten, C.L. & Matthews, T.J. (1970). Discriminative punishment and the conditioned emotional response. *Learning and Motivation,* **1**, 1–17.

Cohen, J.A. & Price, E.O. (1979). Grooming in the Norway rat: displacement activity or "boundary-shift"? *Behavioural and Neural Biology,* **26**, 177–88.

Collerain, I. & Ludvigson, H.W. (1972). Aversion of conspecific odor of frustrative nonreward in rats. *Psychonomic Science,* **27**, 54–6.

Collu, R., Gibb, W. & Ducharme, J.R. (1984). Effects of stress on the gonadal function. *Journal of Endocrinological Investigation,* **7**, 529–37.

Colombo, B. (1957). On the sex ratio in Man. *Cold Spring Harbor Symposia on Quantitative Biology*, **22**, 193–202.

Cooper, J.R., Bloom, F.E. & Roth, R.H. (1982). *The Biochemical Basis of Neuropharmacology*, 4th edn. Oxford University Press.

Coover, G.D. & Levine, S. (1976). Environmental control of suppression of the pituitary-adrenal system. *Physiology and Behaviour*, **17**, 35–7.

Crespi, L.P. (1942). Quantitative variation of incentive and performance in the white rat. *American Journal of Psychology*, **55**, 467–517.

Daly, H.B. (1969). Learning of a hurdle-jump response to escape cues paired with reduced reward or frustrative nonreward. *Journal of Experimental Psychology*, **79**, 146–57.

Dantzer, R. (1977). Behavioural effects of benzodiazepines: a review. *Neuroscience and Biobehavioural Reviews*, **1**, 71–86.

Dantzer, R., Arnove, M. & Mormede, P. (1980). Effects of frustration on behaviour and plasma corticosteroid level in pigs. Physiology and Behaviour, **24**, 1–4.

Darwin, C. (1872). *The Expression of the Emotions in Man and Animals*. London: John Murray.

Davenport J.W. & Thompson, C.I. (1965). The Amsel frustration effect in monkeys. *Psychonomic Science*, **3**, 481–2.

Davis, J.D. & Miller, N.E. (1963). Fear and pain: their effect on self-injection of amobarbital sodium by rats. *Science*, **141**, 1286–7.

Davis, M. (1979). Diazepam and flurazepam: effects on conditioned fear as measured with the potentiated startle paradigm. *Psychopharmacology*, **62**, 1–7.

Davis, N.M., Brookes, S., Gray, J.A. & Rawlins, J.N.P. (1981). Chlordiazepoxide and resistance to punishment. *Quarterly Journal of Experimental Psychology*, **33B**, 227–39.

Davis, N.M. & Gray, J.A. (1983). Brain 5-hydroxytryptamine and learned resistance to punishment. *Behavioural Brain Research*, **8**, 129–37.

De Molina, F.A. & Hunsperger, R.W. (1962). Organisation of the subcortical system governing defence and flight reactions in the cat. *Journal of Physiology*, **160**, 200–13.

De Vito, P.L. & Fowler, H. (1982). Transfer of conditioned appetitive stimuli to conditioned aversive excitatory and inhibitory stimuli. *Learning and Motivation*, **13**, 135–54.

DeVore, I. (ed.) (1965). *Primate Behaviour: Field Studies of Monkeys and Apes*. New York: Holt, Rinehart & Winston.

De Wied, D. (1964). The influence of the anterior pituitary on avoidance learning and escape behaviour. *American Journal of Physiology*, **207**, 255–9.

De Wied, D. (1966) Inhibitory effect of ACTH and related peptides on extinction of conditioned avoidance behaviour in rats. *Proceedings of the Society of Experimental Biology and Medicine*, **122**, 28–32.

De Wied, D. & Jolles, J. (1982). Neuropeptides derived from pro-opiocortin: behavioural, physiological, and neurochemical effects. *Physiological Reviews*, **62**, 976–1059.

Dearing, M.F. & Dickinson, A. (1979). Counterconditioning of shock by a water reinforcer in rabbits. *Animal Learning and Behavior*, **7**, 360–6.

Denenberg, V.H. (1964). Critical periods, stimulus input, and emotional reactivity: a theory of infantile stimulation. *Psychological Review*, **71**, 335–57.

Denenberg, V.H. & Whimbey, A.E. (1963). Behavior of adult rats is modified by the experiences their mothers had as infants. *Science*, **142**, 1192–3.

Deutsch, J.A. & Deutsch, D. (1966). *Physiological Psychology.* Homewood, Illinois: Dorsey Press.

Dickinson, A. (1977). Appetitive-aversive interactions: facilitation of aversive conditioning by prior appetitive training in the rat. *Animal Learning and Behavior,* 4, 416–20.

Dickinson, A. (1980). *Contemporary Animal Learning Theory.* Cambridge University Press.

Dickinson, A. & Dearing, M.F. (1979). Appetitive-aversive interactions and inhibitory processes. In *Mechanisms of Learning and Motivation,* ed. A. Dickinson & R.A. Boakes, pp. 203–31. Hillsdale, NJ: Erlbaum.

Dickinson, A. & Pearce, J.M. (1977). Inhibitory interactions between appetitive and aversive stimuli. *Psychological Bulletin,* 84, 690–711.

Dollard, J. & Miller, N.E. (1950). *Personality and Psychotherapy.* New York: McGraw-Hill.

Dörner, G., Geier, Th., Ahrens, L., Krell, L., Munx, G., Sieler, H., Kittner, E. & Muller, H. (1980). Prenatal stress as possible aetiogenetic factor of homosexuality in human males. *Endokrinologie* (Leipzig), Band 75 (Heft 3), 365–8.

Dörner, G., Gotz, F. & Docke, W.D. (1983). Prevention of demasculinization and feminization of the brain in prenatally stressed male rats by perinatal androgen treatment. *Experimental and Clinical Endocrinology,* 81, 88–90.

Douglas, R.J. (1967). The hippocampus and behavior. *Psychological Bulletin,* 67, 416–42.

Drago, F., Bohus, B., Scapagnini, U. & de Wied, O. (1980). Sexual dimorphism in passive avoidance behavior of rats: relation to body weight, age, shock intensity and retention interval. *Physiology and Behavior,* 24, 1161–4.

Drewek, K.J. & Broadhurst, P.L. (1979). Alcohol selection by strains of rats selectively bred for behavior. *Journal of Studies in Alcohol,* 40, 723–8.

Drewett, R.F. (1973). Oestrous and dioestrous components of the ovarian inhibition on hunger in the rat. *Animal Behaviour,* 21, 772–80.

Drewett, R.F., Gray, J.A., James, D.T.D., McNaughton, N., Valero, I. & Dudderidge, H.J. (1977). Sex and strain differences in septal driving of the hippocampal theta rhythm as a function of frequency: effects of gonadectomy and gonadal hormones. *Neuroscience,* 2, 1033–41.

Drugan, R.C. & Maier, S.F. (1983). Analgesic and opioid involvement in the shock-elicited activity and escape deficits produced by inescapable shock. *Learning and Motivation,* 14, 30–47.

Duffy, E. (1967). *Activation and Behavior.* New York: Wiley.

Eayrs, J.T., Glass, A. & Broadhurst, P.L. (1962). Thyroid function and central nervous activity. *Journal of Endocrinology,* 24, p. viii (Proceedings of the Society for Endocrinology).

Eccles, J. (1965). The synapse. *Scientific American,* 212 (1), 56–66.

Einon, D.F., Humphreys, A.P., Chivers, S.M., Field, S. & Naylor, V. (1981). Isolation has permanent effects upon the behavior of the rat, but not the mouse, gerbil, or guinea pig. *Developmental Psychobiology,* 14, 343–55.

Einon, D.F., Morgan, M.J. & Kibbler, C.C. (1978). Brief periods of socialization and later behavior in the rat. *Developmental Psychobiology,* 11, 213–25.

Ekman, P. *et al.* (1983). *Emotion in the Human Face,* 2nd edn. Cambridge University Press.

Elmadjian, F., Hope, J.M. & Lamson, E.T. (1958). Excretion of epinephrine and norepinephrine under stress. *Recent Progress in Hormone Research,* 4, 513–53.

Elmes, D.G., Jarrard, L.E. & Swart P.D. (1975). Helplessness in hippocampectomised rats: response perseveration? *Physiological Psychology*, **3**, 51–5.

Eriksen, C.W. (1966). Cognitive responses to internally cued anxiety. In *Anxiety and Behavior*, ed. C.D. Spielberger, pp. 327–60. New York: Academic Press.

Estes, W.K. & Skinner, B.F. (1941). Some quantitative properties of anxiety. *Journal of Experimental Psychology*, **29**, 390–400.

Eysenck, H.J. (1955). *Psychology and the Foundations of Psychiatry*. London: H.K. Lewis.

Eysenck, H.J. (1960a). Classification and the problem of diagnosis. In *Handbook of Abnormal Psychology*, ed. H.J. Eysenck, pp. 1–31. London: Pitman.

Eysenck, H.J. (1960b). The effects of psychotherapy. In *Handbook of Abnormal Psychology*, ed. H.J. Eysenck, pp. 697–725. London: Pitman.

Eysenck, H.J. (ed.) (1964). *Experiments in Behaviour Therapy*. Oxford: Pergamon Press.

Eysenck, H.J. (1967). *The Biological Basis of Personality*. Springfield, Illinois: C.C. Thomas.

Eysenck, H.J. (1977). *Crime and Personality*, 3rd edn. London: Routledge and Kegan Paul.

Eysenck H.J. & Eysenck, S.B.G. (1969). *Personality Structure and Measurement*. London: Routledge & Kegan Paul.

Eysenck, H.J. & Rachman, S. (1965). *The Causes and Cures of Neurosis*. London: Routledge & Kegan Paul.

Eysenck, S.B.G. & Eysenck, H.J. (1976). *Psychoticism as a Dimension of Personality*. London: Hodder and Stoughton.

Fanselow, M.S. (1979). Naloxone attenuates rat's preference for signaled shock. *Physiological Psychology*, **7**, 70–4.

Fanselow, M.S. (1985). Odors released by stressed rats produce opioid analgesia in unstressed rats. *Behavioral Neuroscience*, **99**, 589–92.

Fanselow, M.S. & Baackes, M.P. (1982). Conditioned fear-induced opiate analgesia on the formalin test: evidence for two aversive motivational systems. *Learning and Motivation*, **13**, 200–21.

Feldon, J. & Gray, J.A. (1981). The partial reinforcement extinction effect after treatment with chlordiazepoxide. *Psychopharmacology*, **73**, 269–75.

Feldon, J., Guillamon, A., Gray, J.A., DeWit, H. & McNaughton, N. (1979). Sodium amylobarbitone and responses to nonreward. *Quarterly Journal of Experimental Psychology*, **31**, 19–50.

Feldon, J., Rawlins, J.N.P. & Gray, J.A. (1985). Fornix-fimbria section and the partial reinforcement extinction effect. *Experimental Brain Research*, **58**, 435–9.

Ferrari, E.A., Todorov, J.C. & Graeff, F.G. (1973). Nondiscriminated avoidance of shock by pigeons pecking a key. *Journal of the Experimental Analysis of Behaviour*, **19**, 211–18.

Feuer, G. & Broadhurst, P.L. (1962). Thyroid function in rats selectively bred for emotional elimination. III. Behavioural differences. *Journal of Endocrinology*, **24**, 385–96.

File, S.E. (1980). The use of social interaction as a method for detecting anxiolytic activity of chlordiazepoxide-like drugs. *Journal of Neuroscience Methods*, **2**, 219–38.

File, S.E. & Peet, L.A. (1980). The sensitivity of the rat corticosterone response to environmental manipulations and to chronic chlordiazepoxide treatment. *Physiology and Behavior*, **25**, 753–8.

Floderus-Myrhed, B., Pedersen, N. & Rasmuson, I. (1980). Assessment of heritability for personality, based on a short-form of the Eysenck Personality Inventory: a study of 12,898 twin pairs. *Behavior Genetics*, **10**, 153–62.

Fowler, H. & Miller, N.E. (1963). Facilitation and inhibition of runway performance by hind- and forepaw shock of various intensities. *Journal of Comparative and Physiological Psychology*, **56**, 801–5.

Fowles, D. (1980). The three arousal model: implications of Gray's two-factor learning theory for heart rate, electrodermal activity and psychopathy. *Psychophysiology*, **17**, 87–104.

Fox, M.W. & Stelzner, D. (1966). Behavioural effects of differential early experience in the dog. *Animal Behaviour*, **14**, 273–81.

Frankenhaeuser, M. (1975). Experimental approaches to the study of catecholamines and emotion. In *Emotions: Their Parameters and Measurement*, ed. L. Levi, pp. 209–34. New York: Raven Press.

Freedman, J.L. (1973). The effects of population density on humans. In *Psychological Perspectives on Population*, ed. J.T. Fawcett, pp. 209–38. New York: Basic Books.

Freedman, P.E. & Rosen, A.J. (1969). The effects of psychotropic drugs on the double alley frustration effect. *Psychopharmacologia*, **15**, 39–47

Freeman, G.L. (1948). *The Energetics of Human Behavior*. Ithaca: Cornell University Press.

Freeman, H.L. & Kendrick, D.C. (1964). A case of cat phobia (treatment by a method derived from experimental psychology). In *Experiments in Behaviour Therapy*, ed. H.J. Eysenck, pp. 51–61. Oxford: Pergamon Press.

Fuller, J.L. & Clark, L.D. (1966). Genetic and treatment factors modifying the post-isolation syndrome in dogs. *Journal of Comparative and Physiological Psychology*, **61**, 251–7.

Funkenstein, D.H. (1955). Genetic and treatment factors modifying the post-isolation syndrome in dogs. *Scientific American*, **192** (5), 74–80.

Fuster, J.M. (1958). Effects of stimulation of brain stem on tachistoscopic perception. *Science*, **127**, 150.

Fuster, J.M. (1980). *The Prefrontal Cortex: Anatomy, Physiology, and Neuropsychology of the Frontal Lobe*. New York: Raven Press.

Galey, D., Durkin, T., Sitakis, G., Kempf, E. & Jaffard, R. (1985). Facilitation of spontaneous and learned spatial behaviours following 6-hydroxydopamine lesions of the lateral septum: a cholinergic hypothesis. *Brain Research*, **340**, 171–4.

Galle, O.R., Gove, W.R. & McPherson, J.M. (1972). Population density and pathology: what are the relations for Man? *Science*, **176**, 23–30.

Gallup, G.G., Jr (1965). Aggression in rats as a function of frustrative nonreward in a straight alley. *Psychonomic Science*, **3**, 99–100.

Gardner, E.T. & Lewis, P. (1976). Negative reinforcement with shock-frequency increase. *Journal of the Experimental Analysis of Behavior*, **25**, 3–14.

Gauthreaux, S.A., Jr (1978). The ecological significance of behavioral dominance. In *Perspectives in Ethology*, vol. 3, *Social Behavior*, ed. P.P.G. Bateson & P.K. Klopfer, pp. 17–54. New York: Plenum Press.

Geer, J.H. (1965). The development of a scale to measure fear. *Behaviour Research and Therapy*, **3**, 45–53.

Gold, M.S., Redmond, D.E., Jr & Kleber, H.D. (1978). Clonidine blocks acute opiate-withdrawal symptoms. *Lancet*, **ii**, 599–602.

Goldman, L., Coover, G.D. & Levine, S. (1973). Bidirectional effects of reinforcement shifts on pituitary adrenal activity. *Physiology and Behavior*, **10**, 209–14.

Gonzalez, C.A., Roe, C.L. & Levine, S. (1982). Cortisol responses under different housing conditions in female squirrel monkey. *Psychoneuroendocrinology, 7,* 209–16.

Goodman, J.H. & Fowler, H. (1983). Blocking and enhancement of fear conditioning by appetitive CSs. *Animal Learning and Behavior,* **11,** 75–82.

Gorenstein, E.E. & Newman, J.P. (1980). Disinhibitory psychopathology: a new perspective and a model for research. *Psychological Review,* **87,** 301–15.

Graeff, F.G. (1981). Minor tranquilizers and brain defense systems. *Brazilian Journal of Medical and Biological Research,* **14,** 239–65.

Graeff, F.G. (1987). The anti-aversive action of drugs. In *Advances in Behavioural Pharmacology,* vol. 6, ed. T. Thompson, P.B. Dews & J. Barrett. Hillsdale, NJ: Erlbaum.

Graeff, F.G., Quintero, S. & Gray, J.A. (1980). Median raphe stimulation, hippocampal theta rhythm and threat-induced behavioural inhibition. *Physiology and Behaviour,* **25,** 253–61.

Graeff, F.G. & Schoenfeld, R.I. (1970). Tryptaminergic mechanisms in punished and nonpunished behaviour. *Journal of Pharmacology and Experimental Therapeutics,* **173,** 277–83.

Graeff, F.G. & Silveira Filho, N.G. (1978). Behavioural inhibition induced by electrical stimulation of the median raphe nucleus of the rat. *Physiology and Behaviour,* **21,** 477–84.

Graeff, F.G., Zuardi, A.W., Giglio, J.S., Lima Filho, E.C. & Karniol, I.G. (1985). Effect of metergoline on human anxiety. *Psychopharmacology,* **86,** 334–8.

Graham-Jones, S., Holt, L., Gray, J.A. & Fillenz, M. (1985). Low-frequency septal stimulation increases tyrosine hydroxylase activity in the hippocampus. *Pharmacology, Biochemistry and Behavior,* **23,** 489–93.

Gray, J.A. (1964). Strength of the nervous system and levels of arousal: A reinterpretation. In *Pavlov's Typology,* ed. J.A. Gray, pp. 289–366. Oxford: Pergamon Press.

Gray, J.A. (1967). Disappointment and drugs in the rat. *Advancement of Science,* **23,** 595–605.

Gray, J.A. (1969). Sodium amobarbital and effects of frustrative non-reward. *Journal of Comparative and Physiological Psychology,* **69,** 55–64.

Gray, J.A. (1970*a*). The psychophysiological basis of introversion-extraversion. *Behaviour Research and Therapy,* **8,** 249–66.

Gray, J.A. (1970*b*). Sodium amobarbital, the hippocampal theta rhythm and the partial reinforcement extinction effect. *Psychological Review,* **77,** 465–80.

Gray, J.A. (1971). Sex differences in emotional behaviour in mammals including Man: endocrine bases. *Acta Psychologica,* **35,** 29–46.

Gray, J.A. (1972*a*). Causal theories of personality and how to test them. In *Multivariate Analysis and Psychological Theory,* ed. J.R. Royce, pp. 409–63. London: Academic Press.

Gray, J.A. (1972*b*). Learning theory, the conceptual nervous system and personality. In *The Biological Bases of Individual Behaviour,* ed. V.D. Nebylitsyn & J.A. Gray, pp. 372–99. New York: Academic Press.

Gray, J.A. (1972*c*). The psychophysiological basis of introversion–extraversion: a modification of Eysenck's theory. In *The Biological Bases of Individual Behaviour,* ed. V.D. Nebylitsyn & J.A. Gray, pp. 182–205. New York: Academic Press.

Gray, J.A. (1975). *Elements of a Two-Process Theory of Learning.* London: Academic Press.

Gray, J.A. (1976). The behavioural inhibition system: a possible substrate for anxiety. In *Theoretical and Experimental Bases of Behaviour Modification,* ed. M.P. Feldman & A.M. Broadhurst, pp. 3–41. London: Wiley.

Gray, J.A. (1977). Drug effects on fear and frustration: possible limbic site of action of minor tranquillizers. In *Handbook of Psychopharmacology*, vol. 8, ed. L.L. Iversen, S.D. Iversen & S.H. Snyder, pp. 433–529. New York: Plenum.

Gray, J.A. (1979). Emotionality in male and female rodents: a reply to Archer. *British Journal of Psychology*, **70**, 425–40.

Gray, J.A. (1981). A critique of Eysenck's theory of personality. In *A Model for Personality*, ed. H.J. Eysenck, pp. 246–76. Berlin: Springer.

Gray, J.A. (1982). *The Neuropsychology of Anxiety: An Enquiry into the Functions of the Septo-Hippocampal System*. Oxford University Press.

Gray, J.A. (1984). The hippocampus as an interface between cognition and emotion. In *Animal Cognition*, ed. H.L. Roitblat, T.G. Bever & H.S. Terrace, pp. 607–26. Hillsdale, NJ: Erlbaum.

Gray, J.A. (1987). Interactions between drugs and behaviour therapy. In *Theoretical Foundations of Behaviour Therapy*, ed. H.J. Eysenck & I. Martin. New York: Plenum Press.

Gray, J.A. *et al.* (1982). Précis and multiple peer review of 'The Neuropsychology of Anxiety: An Enquiry into the Functions of the Septo-hippocampal System'. *Behavioural and Brain Sciences*, **5**, 469–525.

Gray, J.A. & Ball G.G. (1970). Frequency-specific relation between hippocampal theta rhythm, behaviour and amobarbital action. *Science*, **168**, 1246–8.

Gray, J.A. & Buffery, A.W.H. (1971). Sex differences in emotional and cognitive behaviour in mammals including Man: Adaptive and neural bases. *Acta Psychologica*, **35**, 89–111.

Gray, J.A. & Dudderidge, H. (1971). Sodium amylobarbitone, the partial reinforcement extinction effect and the frustration effect in the double runway. *Neuropharmacology*, **10**, 217–22.

Gray, J.A., Lean, J. & Keynes, A. (1969). Infant androgen treatment and adult open-field behaviour: direct effects and effects of injections to siblings. *Physiology and Behavior*, **4**, 171–81.

Gray, J.A. & Levine, S. (1964). Effects of induced oestrus on emotional behaviour in selected strains of rats. *Nature*, **201**, 1198–200.

Gray, J.A. Levine, S. & Broadhurst, P.L. (1965). Gonadal hormone injection in infancy and adult emotional behaviour. *Animal Behaviour*, **13**, 33–45.

Gray, J.A. & McNaughton, N. (1983). Comparison between the behavioural effects of septal and hippocampal lesions: a review. *Neuroscience and Biobehavioral Reviews*, **7**, 119–88.

Gray, J.A., McNaughton, N., James, D.T.D. & Kelly, P.H. (1975). Effect of minor tranquillizers on hippocampal theta rhythm mimicked by depletion of forebrain noradrenaline. *Nature*, **258**, 424–5.

Gray, J.A. & Rawlins, J.N.P. (1986). Comparator and buffer memory: an attempt to integrate two models of hippocampal function. In *The Hippocampus*, vol. 4, ed. R.L. Isaacson & K.H. Pribram, pp. 159–201. New York: Plenum.

Gray, J.A. & Smith, P.T. (1969). An arousal-decision model for partial reinforcement and discrimination learning. In *Animal Discrimination Learning*, ed. R. Gilbert & N.S. Sutherland, pp. 243–72. London: Academic Press.

Gray, P. (1977). Effect of the estrous cycle on conditioned avoidance in mice. *Hormones and Behavior*, **8**, 235–41.

Graziano, A.M., De Giovanni, I.S. & Garcia, K.A. (1979). Behavioural treatment of children's fears: a review. *Psychological Bulletin*, **86**, 806–30.

Green, R.D., III, & Miller, J.W. (1966). Catecholamine concentrations: changes in plasma of rats during estrous cycle and pregnancy. *Science*, **151**, 825–6.

Grossen, N.E., Kostansek, D.J. & Bolles, R.W. (1969). Effects of appetitive discriminative stimuli on avoidance behaviour. *Journal of Experimental Psychology*, **81**, 340–3.

Grossman, S.P. (1967). *A Textbook of Physiological Psychology*. New York: Wiley.

Grzanna, R. & Molliver, M.E. (1980). The locus coeruleus in the rat: an immunohistochemical delineation. *Neuroscience*, **5**, 21–40.

Guhl, A.M., Collias, N.E. & Allee, W.C. (1945). Mating behavior and the social hierarchy in small flocks of white leghorns. *Physiological Zoology*, **18**, 365–90.

Guillamon, A., Gray, J.A. & Broadhurst, P.L. (1977). "Patterning effect": diferencias de sexo y raza. *Revista de Psicologia General y Aplicada*, **32**, 799–816.

Gupta, B.S. & Nagpal, M. (1978). Impulsivity/sociability and reinforcement in verbal operant conditioning. *British Journal of Psychology*, **69**, 203–6.

Guttman, N. & Kalish, H.I. (1956). Discriminability and stimulus generalization. *Journal of Experimental Psychology*, **51**, 79–88.

Gwinn, G.T. (1949). The effects of punishment on acts motivated by fear. *Journal of Experimental Psychology*, **39**, 260–9.

Haefely, W. (1984). Actions and interactions of benzodiazepine agonists and antagonists at GABAergic synapses. In *Actions and Interactions of GABA and Benzodiazepines*, ed. N.G. Bowery, pp. 263–85. New York: Raven Press.

Haefely, W., Kyburz, E., Gerecke, M. & Mohler, H. (1985). Recent advances in the molecular biology of benzodiazepine receptors and in the structure-activity relationships of their agonists and antagonists. In *Advances in Drug Research*, ed. B. Testa, pp. 165–322. London: Academic Press.

Haefely, W., Pieri, L., Polc, P. & Schaffner, R. (1981). General pharmacology and neuropharmacology of benzodiazepine derivatives. In *Handbook of Experimental Pharmacology*, vol. 55 (II), ed. F. Hoffmeister & G. Stille, pp. 10–262. Berlin: Springer-Verlag.

Halevy, G., Feldon, J. & Weiner, I. (1987). Resistance to extinction and punishment following training with shock and nonreinforcement: failure to obtain cross-tolerance. *Quarterly Journal of Experimental Psychology*, **39B**, 147–60.

Hall, C.S. (1941). Temperament: a survey of animal studies. *Psychological Bulletin*, **38**, 909–43.

Hall, C.S. (1951). The genetics of behavior. In *Handbook of Experimental Psychology*, ed. S.S. Stevens, pp. 304–29. London: Chapman & Hall.

Halliday, J.L. (1945). The incidence of psychosomatic affections in Britain. *Psychosomatic Medicine*, **7**, 135–46.

Halliday, M.S. (1966). Exploration and fear in the rat. In *Play, Exploration and Territory in Mammals*, ed. P.A. Jewell & C. Loizos, pp. 45–59. London: Academic Press.

Haltmeyer, G.C., Denenberg, V.H. & Zarrow, M.X. (1967). Modification of the plasma corticosterone response as a function of infantile stimulation and electric shock parameters. *Physiology and Behavior*, **2**, 61–3.

Hamilton, W.D. (1964). The genetical evolution of social behaviour. *Journal of Theoretical Biology*, **7**, 1–52.

Hare, R.D. & Schalling, D. (eds) (1978). *Psychopathic Behaviour: Approaches to Research*. Chichester: Wiley.

Harlow, H.F. & Harlow, M.K. (1965). The affectional systems. In *Behavior of Nonhuman Primates*, vol. 2, ed. A.M. Schrier, H.F. Harlow & F. Stollnitz. London: Academic Press.

Harman, H. (1960). *Modern Factor Analysis*. Chicago University Press.

Harris, G.W. (1964). Sex hormones, brain development and brain function. *Endocrinology*, **75**, 627–48.

Hartley, P.H.T. (1950). An experimental analysis of interspecific recognition. *Symposia of the Society for Experimental Biology*, **4**, 313–36.

Hearst, E. (1965). Approach, avoidance, and stimulus generalization. In *Stimulus Generalization*, ed. D.J. Mostofsky, pp. 331–55. Stanford University Press.

Hebb, D.O. (1946a). Emotion in Man and animal: an analysis of the intuitive processes of recognition. *Psychological Review*, **53**, 88–106.

Hebb, D.O. (1946b). On the nature of fear. *Psychological Review*, **53**, 259–76.

Hebb, D.O. (1949). *The Organization of Behavior*. New York: Wiley. London: Chapman and Hall.

Henke, P.G. (1974). Persistence of runway performance after septal lesions in rats. *Journal of Comparative and Physiological Psychology*, **86**, 760–7.

Henke, P.G. (1977). Dissociation of the frustration effect and the partial reinforcement extinction effect after limbic lesions in rats. *Journal of Comparative and Physiological Psychology*, **91**, 1032–8.

Hennessy, M.B., Vogt, J. & Levine, S. (1982). Strain of foster mother determines long-term effects of early handling: evidence for maternal mediation. *Physiological Psychology*, **10**, 153–7.

Herrnstein, R.J. (1969). Method and theory in the study of avoidance. *Psychological Review*, **76**, 49–69.

Herrnstein, R.J. & Hineline, P.N. (1966). Negative reinforcement as shock-frequency reduction. *Journal of the Experimental Analysis of Behavior*, **9**, 421–30.

Hilgard, E.R. & Marquis, D.G. (1961). *Conditioning and Learning*. New York: Appleton-Century-Crofts (revised edition by G.A. Kimble).

Hilton, S.M. & Zbrozyna, A. (1963). Defence reaction from the amygdala and its afferent and efferent connections. *Journal of Physiology*, **165**, 160–73.

Hinde, R.A. (1954). Factors governing the changes in strength of a partially inborn response as shown by the mobbing behaviour of the chaffinch (*Fringilla coelebs*). I. The nature of the response and an examination of its course. *Proceedings of the Royal Society of London* (Series B), **142**, 306–31.

Hinde, R.A. (1966). *Animal Behaviour*. London: McGraw-Hill.

Hineline, P.N. (1970). Negative reinforcement without shock reduction. *Journal of the Experimental Analysis of Behavior*, **14**, 259–68.

Holt, L. & Gray, J.A. (1983a). Septal driving of the hippocampal theta rhythm produces a long-term, proactive and non-associative increase in resistance to extinction. *Quarterly Journal of Experimental Psychology*, **35B**, 97–118.

Holt, L. & Gray, J.A. (1983b). Proactive behavioural effects of theta-blocking septal stimulation in the rat. *Behavioural and Neural Biology*, **39**, 7–21.

Holt, L. & Gray, J.A. (1985). Proactive behavioural effects of theta-driving septal stimulation on conditioned suppression and punishment in the rat. *Behavioural Neuroscience*, **99**, 60–74.

Homskaya, E.D. (1964). Verbal regulation of the vegetative components of the orienting reflex in focal brain lesions. *Cortex*, **1**, 63–76.

Hull, C.L. (1943). *Principles of Behavior*. New York: Appleton-Century.

Hunt, H.F. & Otis, L.S. (1953). Conditioned and unconditioned emotional defecation in the rat. *Journal of Comparative and Physiological Psychology*, **46**, 378–82.

Hutchinson, R.R. & Renfrew, J.W. (1966). Stalking attack and eating behaviors elicited from the same sites in the hypothalamus. *Journal of Comparative and Physiological Psychology*, **61**, 360–7.

Imada, H. (1972). Emotional reactivity and conditionability in four strains of rats. *Journal of Comparative and Physiological Psychology*, **79**, 474–80.

Imada, H., Mino, T., Sugioka, K. & Ohki, Y. (1981). Measurement of current flow through the rat under signaled and unsignaled grid-shock conditions. *Animal Learning and Behavior*, **9**, 75–9.

Isaac, W. (1960). Arousal and reaction times in cats. *Journal of Comparative and Physiological Psychology*, **53**, 234–6.

Isaacson, R.L. (1982). *The Limbic System*, 2nd edn. New York: Plenum Press.

Ison, J.R., Daley, H.B. & Glass, D.H. (1967). Amobarbital sodium and the effects of reward and nonreward in the Amsel double runway. *Psychological Reports*, **20**, 491–6.

Ison, J.R., Glass, D.H. & Bohmer, H.M. (1966). Effects of sodium amytal on the approach to stimulus change. *Proceedings of the American Psychological Association*, **2**, 5–6.

Ison, J.R. & Pennes, E.S. (1969). Interaction of amobarbital sodium and reinforcement schedule in determining resistance to extinction of an instrumental running response. *Journal of Comparative and Physiological Psychology*, **68**, 215–19.

Ison, J.R. & Rosen, A.J. (1967). The effects of amobarbital sodium on differential instrumental conditioning and subsequent extinction. *Psychopharmacologia*, **10**, 417–25.

Izard, C.E. (1971). *The Face of Emotion*. New York: Appleton-Century-Crofts.

Izard, C.E. (1977). The emotions and emotion concepts in personality and culture research. In *Handbook of Modern Personality Theory*, ed. R.B. Cattell & R.M. Dreger, pp. 496–510. Washington: Hemisphere.

Izard, C.E., Hembree, E.A., Dougherty, L.M. & Coss, C.L. (1983). Changes in two- to nineteen-month-old infants' facial expressions following acute pain. *Developmental Psychology*, **19**, 418–26.

Jacklin, C.N., Maccoby, E.E. & Doering, C.H. (1983). Neonatal sex-steroid hormones and timidity in 6- to 18-month-old boys and girls. *Developmental Psychobiology*, **16**, 163–8.

Jackson, R.L., Alexander, J.H. & Maier, S.F. (1980). Learned helplessness, inactivity, and associative deficits: effects of inescapable shock on response choice escape learning. *Journal of Experimental Psychology: Animal Behavior Processes*, **6**, 1–20.

Jackson, R.L., Maier, S.F. & Coon, D.J. (1979). Long-term analgesic effects of inescapable shock and learned helplessness. *Science*, **206**, 91–3.

Jackson, R.L., Maier, S.F. & Rapoport, P.M. (1978). Exposure to inescapable shock produces both activity and associative deficits in the rat. *Learning and Motivation*, **9**, 69–98.

Jacobson, E. (1962). *You Must Relax*, 4th edn. New York: McGraw-Hill.

James, D.T.D., McNaughton, N., Rawlins, J.N.P., Feldon, J. & Gray, J.A. (1977). Septal driving of hippocampal theta rhythm as a function of frequency in the free-moving male rat. *Neuroscience*, **2**, 1007–17.

James, W. (1890). *The Principles of Psychology*. New York: Henry Holt.

Jersild, A.T. (1955). *Child Psychology*, 4th edn. London: Staples Press.

Jersild, A.T. & Holmes, F.B. (1935). *Children's Fears*. New York: Teachers College Bureau of Publications.

Joffe, J.M. (1969). *Prenatal Determinants of Behavior*. Oxford: Pergamon Press.

Jones, H.E. & Jones, M.C. (1928). A study of fear. *Childhood Education*, **5**, 136–43.

Joseph, R. & Gallagher, R.E. (1980). Gender and early environmental influences on activity, overresponsiveness, and exploration. *Developmental Psychobiology*, **13**, 527–44.

Kamin, L.J. (1956). The effects of termination of the CS and avoidance of the US on avoidance learning. *Journal of Comparative and Physiological Psychology,* **49**, 420–3.

Kamin, L.J. (1957). The gradient of delay of secondary reward in avoidance learning. *Journal of Comparative and Physiological Psychology,* **50**, 445–9.

Kamin, L.J. (1969). Predictability, surprise, attention and conditioning. In *Punishment and Aversive Behavior,* ed. R. Church & B. Campbell, pp. 279–96. New York: Appleton-Century-Crofts.

Kamin, L.J., Brimer, C.J. & Black, A.H. (1963). Conditioned suppression as a monitor of fear of the CS in the course of avoidance training. *Journal of Comparative and Physiological Psychology,* **56**, 497–501.

Kantorowitz, D.A. (1978). Personality and conditioning of tumescence and detumescence. *Behaviour Research and Therapy,* **16**, 117–23.

Karpicke, J. & Dout, D. (1980). Withdrawal from signals for imminent inescapable electric shock. *Psychological Record,* **30**, 511–23.

Katz, R. (1984). Unconfounded electrodermal measures in assessing the aversiveness of predictable and unpredictable shocks. *Psychophysiology,* **21**, 452–8.

Kelley, A.E. & Domesick, V.B. (1982). The distribution of the projection from the hippocampal formation to the nucleus accumbens in the rat: an anterograde- and retrograde-horseradish peroxidase study. *Neuroscience,* **7**, 2321–35.

Kelsey, J.E. (1984). Ventromedial septal lesions in rats reduce stomach erosions produced by inescapable shock. *Physiological Psychology,* **11**, 283–6.

Kelsey, J.E. & Baker, M.D. (1983). Ventromedial septal lesions in rats reduce the effects of inescapable shocks on escape performance and analgesia. *Behavioral Neuroscience,* **97**, 945–61.

Kenny, A. (1963). *Action, Emotion and Will.* London: Routledge & Kegan Paul. New York: Humanities Press.

King, F.A. & Meyer, P.M. (1958). Effects of amygdaloid lesions upon septal hyper-emotionality in the rat. *Science,* **128**, 655–6.

Kinsey, A.C., Pomeroy, W.B., Martin, C.E. & Gebhard, P.H. (1953). *Sexual Behavior in the Human Female.* Philadelphia & London: Saunders.

Kinsman, R.A. & Bixenstine, V.E. (1967). Secondary reinforcement and shock termination. *Journal of Experimental Psychology,* **76**, 62–8.

Kitay, J.I. (1963). Pituitary-adrenal function in the rat after gonadectomy and gonadal hormone replacement. *Endocrinology,* **73**, 253–60.

Klein, D.F. (1981). Anxiety reconceptualized. In *Anxiety: New Research and Changing Concepts,* ed. D.F. Klein & J. Rabkin, pp. 235–63. New York: Raven Press.

Kling, A. & Hutt, P.J. (1958). Effect of hypothalamic lesions on the amygdala syndrome in the cat. *American Medical Association Archives of Neurology and Psychiatry,* **79**, 511–17.

Klüver, H. & Bucy, P.C. (1937). "Psychic blindness" and other symptoms following bilateral temporal lobectomy in rhesus monkeys. *American Journal of Physiology,* **119**, 352–3.

Knobil, E. (1974). On the control of gonadotropin secretion in the rhesus monkey. *Recent Progress in Hormorne Research,* **30**, 1–36.

Köhler, W. (1925). *The Mentality of Apes.* New York: Harcourt, Brace.

Konorski, J. (1967). *Integrative Activity of the Brain.* University of Chicago Press.

Konorski, J. & Szwejkowska, G. (1952). Chronic extinction and restoration of conditioned reflexes. IV. The dependence of the course of extinction and restoration of conditioned reflexes on the 'history' of the conditioned stimulus (the principle of the primacy of first training). *Acta Biologiae Experimentalis,* **17**, 141–65.

Kraemer, G.W. & McKinney,W.T. (1985). Social separation increases alcohol consumption in rhesus monkeys. *Psychopharmacology,* **86**, 182–9.

Krank, M.D. (1985). Asymmetrical effects of Pavlovian excitatory and inhibitory transfer on Pavlovian appetitive responding and acquisition. *Learning and Motivation,* **16**, 35–62.

Krebs, C.J., Gaines, M.S, Keller, B.L., Myers, J.H. &Tamarin, R.H. (1973). Population cycles in small rodents. *Science,* **179**, 35–41.

Lack, D. (1966). *Population Studies of Birds.* Oxford University Press.

Laird, J.D. (1984). The real role of facial response in the experience of emotion: a reply to Tourangeau and Ellsworth, and Others. *Journal of Personality and Social Psychology,* **47**, 909–17.

Langley, L.L. (1965). *Outline of Physiology.* New York & London: McGraw-Hill.

LaTorre, R.A. (1980). Devaluation of the human love object: heterosexual rejection as a possible antecedent to fetishism. *Journal of Abnormal Psychology,* **89**, 295–8.

Lawler, E.E., III (1965). Secondary reinforcement value of stimuli associated with shock reduction. *Quarterly Journal of Experimental Psychology,* **17**, 57–62.

Leclerc, R. (1985). Sign-tracking behavior in aversive conditioning: its acquisition via a Pavlovian mechanism and its suppression by operant contingencies. *Learning and Motivation,* **16**, 63–82.

Leshner, A.I. (1978). *An Introduction to Behavioural Endocrinology.* NewYork: Oxford University Press.

Levine, S. (1962a). Psychophysiological effects of infantile stimulation. In *Roots of Behavior,* ed. E.L. Bliss, pp. 246–53. New York: Harper.

Levine, S. (1962b).The effects of infantile experience on adult behavior. In *Experimental Foundations of Clinical Psychology,* ed. A.J. Bachrach, pp. 139–69. New York: Basic Books.

Levine, S. (1966). Sex differences in the brain. *Scientific American,* **214** (4), 84–90.

Levine, S., Haltmeyer, G.C., Karas, G.C. & Denenberg, V.H. (1967). Physiological and behavioral effects of infantile stimulation. *Physiology and Behavior,* **2**, 55–9.

Levine, S. & Mullins, R.F., Jr (1966). Hormonal influences on brain organization in infant rats. *Science,* **152**, 1585–92.

Levine, S. & Soliday, S. (1962). An effect of adrenal demedullation on the acquisition of a conditioned avoidance response. *Journal of Comparative and Physiological Psychology,* **55**, 214–16.

Levis, D.J. & Boyd,T.L. (1979). Symptom maintenance: an infrahuman analysis and extension of the conservation of anxiety principle. *Journal of Abnormal Psychology,* **88**, 107–20.

Liang, B. & Blizard, D.A. (1978). Central and peripheral norepinephrine concentrations in rat strains selectively bred for differences in response to stress: confirmation and extension. *Pharmacology, Biochemistry and Behavior,* **8**, 75–80.

Liang, B., Dunlap, C.E., III, Freedman, L.S. & Blizard, D.A. (1982). Cardiac β-receptor variation in rat strains selectively bred for difference in susceptibility to stress. *Life Sciences,* **31**, 533–9.

Lidbrink, P., Corrodi, H., Fuxe, K. & Olson, L. (1973).The effects of benzodiazepines, meprobamate, and barbiturates on central monoamine neurons. In *The Benzodiazepines,* ed. S. Garattini, E. Mussini & L.O. Randall, pp. 203–23. New York: Raven Press.

Livett, B.G. (1973). Histochemical visualization of peripheral and central adrenergic neurones. In *Catecholamines,* ed. L.L. Iversen, pp. 93–9. *British Medical Bulletin,* Suppl. 29.

Lockard, J.S. (1963). Choice of a warning signal or no warning signal in an unavoidable shock situation. *Journal of Comparative and Physiological Psychology*, **56**, 526–30.

Lord, B.J., King, M.G. & Pfister, H.P. (1976). Chemical sympathectomy and two-way escape and avoidance learning in the rat. *Journal of Comparative and Physiological Psychology*, **90**, 303–16.

Lorenz, K. (1956). Plays and vacuum activities. In *L'Instinct dans le Comportement des Animaux et de l'Homme*, ed. M. Autuori et al., pp. 633–8. Paris: Masson.

Lovibond, P.F. & Dickinson, A. (1982). Counterconditioning of appetitive and defensive CRs in rabbits. *Quarterly Journal of Experimental Psychology*, **34B**, 115–26.

Lynch, M.A., Lindsay, J. & Ounsted, C. (1975). Tranquillizers causing aggression. *British Medical Journal*, no. 5952, p. 260.

Lyons, W. (1980). *Emotion*. Cambridge University Press.

Lysle, D.T. & Fowler, H. (1985). Inhibition as a 'slave' process: deactivation of conditioned inhibition through extinction of conditioned excitation. *Journal of Experimental Psychology: Animal Behavior Processes*, **11**, 71–94.

Maas, J.W. (1963). Neurochemical differences between two strains of mice. *Nature*, **197**, 255–7.

McCord, P.R. & Wakefield, D.A. (1981). Arithmetic achievement as a function of introversion-extraversion and teacher-presented reward and punishment. *Personality and Individual Differences*, **2**, 145–52.

McFarland, D.J. (1966). On the causal and functional significance of displacement activities. *Zeitschrift fur Tierpsychologie*, **23**, 217–35.

McFie, J. (1972). Factors of the brain. *Bulletin of the British Psychological Society*, **25**, 11–14.

McGonigle, B., McFarland, D.J. & Collier, P. (1967). Rapid extinction following drug-inhibited incidental learning. *Nature*, **214**, 531–2.

Mackintosh, N.J. (1974). *The Psychology of Animal Learning*. New York: Academic Press.

Mackintosh, N.J. (1983). *Conditioning and Associative Learning*. Oxford University Press.

MacLennan, A.J., Drugan, R.C., Hyson, R.L., Maier, S.F., Maden, J. & Barchas, J.D. (1982). Dissociation of long-term analgesia and the shuttle box escape deficit caused by inescapable shock. *Journal of Comparative and Physiological Psychology*, **96**, 904–12.

McNaughton, N. & Gray, J.A. (1983). Pavlovian counterconditioning is unchanged by chlordiazepoxide or by septal lesions. *Quarterly Journal of Experimental Psychology*, **35B**, 221–33.

McNaughton, N. & Mason, S.T. (1980). The neuropsychology and neuropharmacology of the dorsal ascending noradrenergic bundle – a review. *Progress in Neurobiology*, **14**, 157–219.

McNaughton, N., James, D.T.D., Stewart, J., Gray, J.A., Valero, I. & Drewnowski, A. (1977). Septal driving of hippocampal theta rhythm as a function of frequency in the male rat: effects of drugs. *Neuroscience*, **2**, 1019–27.

McNiven, M.A. (1960). 'Social-releaser mechanisms' in birds: a controlled replication of Tinbergen's study. *Psychological Record*, **10**, 259–65.

Mah, C., Suissa, A. & Anisman, H. (1980). Dissociation of antinociception and escape deficits induced by stress in mice. *Journal of Comparative and Physiological Psychology*, **94**, 1160–71.

Maier, S.F., Coon, D.J., McDaniel, M.A., Jackson, R.L. & Grau, J. (1979). The time course of learned helplessness, inactivity, and nociceptive deficits in rats. *Learning and Motivation,* **10**, 467–87.

Margulies, S., (1961). Response duration and operant level, regular reinforcement, and extinction. *Journal of the Experimental Analysis of Behavior,* **4**, 317–21.

Marks, I.M. (1969). *Fears and Phobias.* London: Heinemann.

Marler, P.R. & Hamilton, W.J. (1966). *Mechanisms of Animal Behavior.* New York: Wiley.

Marshall, G.D. & Zimbardo, P.G. (1979). Affective consequences of inadequately explained physiological arousal. *Journal of Personality and Social Psychology,* **37**, 970–88.

Martin, J.R. & Bättig, K. (1980). Exploratory behaviour of rats at oestrus. *Animal Behaviour,* **28**, 900–5.

Martin, R.C. & Melvin, K.B. (1964). Fear responses of bobwhite quail (*Colinus virginianus*) to a model and a live red-tailed hawk (*Buteo janaicensis*). *Psychologische Forschung,* **27**, 323–36.

Maslach, C. (1979). Negative emotional biasing of unexplained arousal. *Journal of Personality and Social Psychology,* **37**, 953–69.

Mason, J.W. (1959). Psychological influences on the pituitary-adrenal cortical system. *Recent Progress in Hormone Research,* **15**, 345–78.

Masserman, J.H. & Yum, K.S. (1946). An analysis of the influence of alcohol on experimental neuroses in cats. *Psychosomatic Medicine,* **8**, 36–52.

Mathews, A.M. (1978). Fear-reduction research and clinical phobias. *Psychological Bulletin,* **85**, 390–404.

Mathews, A.M., Gelder, M.G. & Johnston, D.W. (1981). *Agoraphobia: Nature and Treatment.* New York: Guildford Press.

Mayer-Gross, W., Slater, E.T.O. & Roth, M. (1979). *Clinical Psychiatry.* Baltimore: Williams and Wilkins.

Meaney, M.J., Aitken, D.H., Bodnoff, S.R., Ing, C.J., Tatarewicz, J.E. & Sapolsky, R.M. (1985). Early postnatal handling alters glucocorticoid receptor concentrations in selected brain regions. *Behavioral Neuroscience,* **99**, 765–70.

Mellitz, M., Hineline, P.N., Whitehouse, W.G. & Laurence, M.T. (1983). Duration-reduction of avoidance sessions as negative reinforcement. *Journal of the Experimental Analysis of Behavior,* **40**, 57–67.

Melzack, R. (1954). The genessis of emotional behaviour: an experimental study of the dog. *Journal of Comparative and Physiological Psychology,* **47**, 166–8.

Melzack, R., Penick, E. & Beckett, A. (1959). The problem of 'innate fear' of the hawk shape: an experimental study with mallard ducks. *Journal of Comparative and Physiological Psychology,* **52**, 694–8.

Melzack, R. & Scott, T.H. (1957). The effects of early experience on the response to pain. *Journal of Comparative and Physiological Psychology,* **50**, 155–61.

Melzack, R. & Wall, P.D. (1983). *The Challenge of Pain.* New York: Basic Books.

Miczek, K.A., Thompson, M.L. & Shuster, L. (1985). Naloxone injections into the periaqueductal grey area and arcuate nucleus block analgesia in defeated mice. *Psychopharmacology,* **87**, 39–42.

Millenson, J.R. (1967). *Principles of Behavioral Analysis.* New York: Macmillan.

Miller, N.E. (1951). Learnable drives and rewards. In *Handbook of Experimental Psychology,* ed. S.S. Stevens, pp. 435–72. New York: Wiley.

Miller, N.E. (1959). Liberalization of basic S-R concepts: extensions to conflict behavior, motivation and social learning. In *Psychology: A Study of a Science,* study 1, vol. 2, ed. S. Koch, pp. 196–292. New York: McGraw-Hill.

Miller, N.E. (1960). Learning resistance to pain and fear: effects of overlearning, exposure and rewarded exposure in context. *Journal of Experimental Psychology,* **60**, 137–45.

Miller, N.E. (1964). The analysis of motivational effects illustrated by experiments on amylobarbitone sodium. In *Animal Behaviour and Drug Action*, ed. H. Steinberg, pp. 1–18. London: Churchill.

Miller, N.E. (1976). Learning, stress and psychosomatic symptoms. *Acta Neurobiologiae Experimentalis,* **36**, 141–56.

Miller, R.E. (1971). Experimental studies of communication in the monkey. In *Primate Behavior: Developments in Field and Laboratory Research*, vol. 2, ed. L.A. Rosenblum, pp. 139–75. New York: Academic Press.

Miller, R.E. (1974). Social and pharmacological influences on the nonverbal communication of monkeys and of Man. In *Nonverbal Communication*, ed. L. Krames, P. Pliner & T. Alloway, pp. 77–101. New York: Plenum.

Miller, R.E. & Ogawa, N. (1962). The effect of adrenocorticotrophic hormone (ACTH) on avoidance conditioning in the adrenalectomized rat. *Journal of Comparative and Physiological Psychology,* **55**, 211–13.

Mineka, S. (1979). The role of fear in theories of avoidance learning, flooding, and extinction. *Psychological Bulletin,* **86**, 985–1010.

Mineka, S. (1987). A primate model of phobic fears. In *Theoretical Foundations of Behavior Therapy*, ed. H.J. Eysenck & I. Martin. New York: Plenum.

Mineka, S., Cook, M. & Miller, S. (1984). Fear conditioned with escapable and inescapable shock: effects of a feedback stimulus. *Journal of Experimental Psychology: Animal Behavior Processes,* **10**, 307–23.

Mineka, S., Davidson, M., Cook, M. & Keir, R. (1984). Observational conditioning of snake fear in rhesus monkeys. *Journal of Abnormal Psychology,* **93**, 355–72.

Mineka, S. & Gino, A. (1979). Dissociative effects of different types and amounts of nonreinforced CS exposure on avoidance extinction and the CER. *Learning and Motivation,* **10**, 141–60.

Mineka, S. & Gino, A. (1980). Dissociation between conditioned emotional response and extended avoidance performance. *Learning and Motivation,* **11**, 476–502.

Mineka, S. & Kihlstrom, J.F. (1978). Unpredictable and uncontrollable events: a new perspective on experimental neurosis. *Journal of Abnormal Psychology,* **87**, 256–71.

Mischel, T. (ed.) (1969). *Human Action*. New York: Academic Press.

Money, J. & Ehrhardt, A.A. (1968). Prenatal hormone exposure: possible effects on behaviour in man. In *Endocrinology and Human Behaviour*, ed. R.P. Michael, pp. 32–48. London: Oxford University Press.

Morgan, C.T. (1965). *Physiological Psychology*, 3rd edn. New York: McGraw-Hill.

Morris, R.G.M. (1975). Preconditioning of reinforcing properties to an exteroceptive feedback stimulus. *Learning and Motivation*, 6, 289–98.

Mowrer, O.H. (1960). *Learning and Behavior*. New York: Wiley.

Moye, T.B., Hyson, R.L., Grau, J.W. & Maier, S.F. (1983). Immunization of opioid analgesia: effects of prior escapable shock on subsequent shock-induced and morphine-induced antinociception. *Learning and Motivation,* **14**, 238–51.

Moynihan, M. (1955). Some aspects of reproductive behaviour in the black-headed gull (*Larus ridibundus ridibundus L.*) and related species. *Behaviour*, Suppl. 4, 1–201.

Muenzinger, K.F., Brown, W.O., Crow, W.J. & Powlowski, R.F. (1952). Motivation in learning. XI. An analysis of electric shock for correct responses into its avoidance and accelerating components. *Journal of Experimental Psychology,* **43**, 115–19.

Munck, A., Guyre, P.M. & Holbrook, N.J. (1984). Physiological functions of glucocorticoids in stress and their relation to pharmacological actions. *Endocrine Reviews,* **5**, 25–44.

Myer, J.S. (1971). Some effects of noncontingent aversive stimulation. In *Aversive Conditioning and Learning*, ed. F.R. Brush, pp. 469–536. New York: Academic Press.

Myers, J.H. & Krebs, C.J. (1974). Population cycles in rodents. *Scientific American,* **230**, 38–46.

Nashold, B.S., Jr, Wilson, W.P. & Slaughter, G.S. (1969). Sensations evoked by stimulation in the midbrain of Man. *Journal of Neurosurgery,* **30**, 14–24.

Nation, J.R. & Cooney, J.B. (1982). The time course of extinction-induced aggressive behavior in humans: evidence for a stage model of extinction. *Learning and Motivation,* **13**, 95–112.

Nelson, P.B. & Wollen, K.A. (1965). Effects of ethanol and partial reinforcement upon runway acquisition. *Psychonomic Science,* **3**, 135–6.

Newman, J.P., Widom, C.S. & Nathan, S. (1985). Passive avoidance in syndromes of disinhibition: psychopathy and extraversion. *Journal of Personality and Social Psychology,* **48**, 1316–27.

Nicholson, J.N. & Gray, J.A. (1972). Peak shift, behavioural contrast and stimulus generalization as related to personality and development in children. *British Journal of Psychology,* **63**, 47–68.

Nieto, J. (1984). Transfer of conditioned inhibition across different aversive reinforcers in the rat. *Learning and Motivation,* **15**, 37–57.

Notterman, J.M. (1959). Force emission during bar pressing. *Journal of Experimental Psychology,* **58**, 341–7.

Numan, R, (1978). Cortical-limbic mechanisms and response control: a theoretical review. *Physiological Psychology,* **6**, 445–70.

Oei, T.P.S. & King, M.G. (1978). Central catecholamine and peripheral noradrenaline depletion by 6-hydroxydopamine and active avoidance learning in rats. *Journal of Comparative and Physiological Psychology,* **92**, 94–108.

Öhman, A. (1979). Fear relevance, autonomic conditioning, and phobias: a laboratory model. In *Trends in Behaviour Therapy*, ed. P.O. Sjoden, S. Bates & W.W. Dockens, pp. 107–33. New York: Academic Press.

Öhman, A. & Dimberg, U. (1978). Facial expressions as conditioned stimuli for electrodermal responses: a case of 'preparedness'? *Journal of Personality and Social Psychology,* **36**, 1251–8.

O'Keefe, J. & Nadel, L. (1978). *The Hippocampus as a Cognitive Map*. Oxford: Clarendon Press.

Olds, J. & Olds, M. (1965). Drives, rewards, and the brain. In *New Directions in Psychology*, vol. II, ed. F. Barron, W.C. Dement, W. Edwards, H. Lindmann, L.D. Phillips, J. Olds & M. Olds, pp. 329–410. New York: Holt, Rinehart and Winston.

Olds, N.E. & Fobes, J.C. (1981). The central basis of motivation: intracranial self-stimulation studies. In *Annual Review of Psychology*, vol. 32, ed. M.K. Rosenzweig & C.W. Porter, pp. 523–74. Palo Alto: Annual Reviews Inc.

Olsen, R.W. (1981). GABA-benzodiazepine-barbiturate receptor interactions. *Journal of Neurochemistry,* **37**, 1–13.

Olton D.S., Becker, J.T. & Handelmann, G.E. (1979). Hippocampus, space, and memory. *Behavioural and Brain Sciences,* **2**, 313–22.

Orr, S.P. & Lanzetta, J.T. (1984). Extinction of an emotional response in the presence of facial suppression of emotion. *Motivation and Emotion,* **8**, 55–66.

Overmier, J.B. & Seligman, M.E.P. (1976). Effects of inescapable shock upon subsequent escape and avoidance learning. *Journal of Comparative and Physiological Psychology,* **63**, 28–33.

Owen, J.W., Cicala, G.A. & Herdegen, R.T. (1978). Fear inhibition and species specific defense reaction termination may contribute independently to avoidance learning. *Learning and Motivation,* **9**, 297–313.

Owen, S., Boarder, M., Gray, J.A. & Fillenz, M. (1982). Acquisition and extinction of continuously and partially reinforced running in rats with lesions of the dorsal noradrenergic bundle. *Behavioral Brain Research,* **5**, 11–41.

Page, H.A. (1955). The facilitation of experimental extinction by response prevention as a function of the acquisition of a new response. *Journal of Comparative and Physiological Psychology,* **48**, 14–16.

Panksepp, J. (1982). Towards a general psychobiological theory of emotions. *Behavioral and Brain Sciences,* **5**, 407–67.

Passingham, R.E. (1972). Crime and personality: A review of Eysenck's theory. In *The Biological Bases of Individual Behaviour,* ed. V.D. Nebylitsyn & J.A. Gray, pp. 342–71. New York & London: Academic Press.

Patterson, M.L. (1973). Compensation in nonverbal immediacy behaviours: a review. *Sociometry,* **36**, 237–52.

Pavlov, I.P. (1927). *Conditioned Reflexes,* translated and edited by G.V. Anrep. Oxford University Press.

Perkins, C.C., Jr, Seymann, R.G., Levis, D.J. & Spencer, H.R., Jr (1966). Factors affecting preference for signal-shock over shock-signal. *Journal of Experimental Psychology,* **72**, 190-6

Petty, F. & Sherman, A.D. (1981). GABA-ergic modulation of learned helplessness. *Pharmacology, Biochemistry, and Behavior,* **15**, 567–70.

Pfaff, D.W. & Zigmond, R.E. (1971). Neonatal androgen effects on sexual and non-sexual behaviour of adult rats reared under various hormone regimes. *Neuroendocrinology,* **7**, 129–45.

Pfeifer, W.D. & Davis, L.C. (1974). Effect of handling in infancy on responsiveness of adrenal tyrosine hydroxylase in maturity. *Behavioral Biology,* **10**, 239–45.

Politch, J.A. & Herrenkohl, L.R. (1984). Prenatal ACTH and corticosterone: effects on reproduction in male mice. *Physiology and Behavior,* **32**, 135–7.

Powell, G.E. (1979). *Brain and Personality.* London: Saxon House.

Quintero, S., Buckland, C., Gray, J.A., McNaughton, N. & Mellanby, J. (1985a). The effects of compounds related to γ-aminobutyrate and benzodiazepine receptors on behavioural responses to anxiogenic stimuli in the rat: choice behaviour in the T-maze. *Psychopharmacology,* **86**, 328–33.

Quintero, S., Henney, S., Lawson, P., Mellanby, J. & Gray, J.A. (1985b). The effects of compounds related to γ-aminobutyrate and benzodiazepine receptors on behavioural responses to anxiogenic stimuli in the rat: punished bar-pressing. *Psychopharmacology,* **85**, 244–51.

Quintero, S., Mellanby, J., Thompson, M.R., Nordeen, H., Nutt, D., McNaughton, N. & Gray, J.A. (1985c). Septal driving of hippocampal theta rhythm: role of γ-aminobutyrate-benzodiazepine receptor complex in mediating effects of anxiolytics. *Neuroscience,* **16**, 875–84.

Rachman, S. (1967). Systematic desensitization. *Psychological Bulletin,* **67**, 93–103.

Rachman, S. & Hodgson, R. (1978). *Obsessions and Compulsions.* New York: Prentice Hall.

Rawlins, J.N.P. (1985). Associations across time: the hippocampus as a temporary memory store. *Behavioural and Brain Sciences,* **8**, 479–528.

Rawlins, J.N.P., Feldon, J. & Gray, J.A. (1980*a*). The effects of hippocampectomy and of fimbria section upon the partial reinforcement extinction effect in rats. *Experimental Brain Research*, **38**, 273–83.

Rawlins, J.N.P., Feldon, J. & Gray, J.A. (1980*b*). Discrimination of response-contingent and response-independent shock by rats: effects of chlordiazepoxide HCl and sodium amylobarbitone. *Quarterly Journal of Experimental Psychology*, **32**, 215–32.

Rawlins, J.N.P., Feldon, J., Salmon, P., Gray, J.A. & Garrud, P. (1980*c*). The effects of chlordiazepoxide HCl administration upon punishment and conditioned suppression in the rat. *Psychopharmacology*, **70**, 317–22.

Rawlins, J.N.P., Feldon, J., Ursin, H. & Gray, J.A. (1985). Resistance to extinction after schedules of partial delay or partial reinforcement in rats with hippocampal lesions. *Experimental Brain Research*, **59**, 273–81.

Redmond, D.E., Jr (1979). New and old evidence for the involvement of a brain norepinephrine system in anxiety. In *Phenomenology and Treatment of Anxiety*, ed. W.G. Fann, I. Karacan, A.D. Pokorny & R.L. Williams, pp. 153–203. New York: Spectrum.

Reiman, E.M., Raichle, M.E., Butler, F.K., Hersovitch, P. & Robins, E. (1984). A focal brain abnormality in panic disorder, a severe form of anxiety. *Nature*, **310**, 683–5.

Rescorla, R.A. & LoLordo, V.M. (1965). Inhibition of avoidance behavior. *Journal of Comparative and Physiological Psychology*, **59**, 406–12.

Rescorla, R.A. & Solomon, R.L. (1967). Two-process learning theory: relationships between Pavlovian conditioning and instrumental learning. *Psychological Review*, **74**, 151–82.

Revelle, W., Humphreys, M.S., Simon, L. & Gilliland, K. (1980). The interactive effect of personality, time of day and caffeine: a test of the arousal model. *Journal of Experimental Psychology: General*, **109**, 1–31.

Ricciuti, H.N. (1974). Fear and the development of social attachments in the first year of life. In *The Origins of Fear*, ed. M. Lewis & L.A. Rosenblum, pp. 73–106. New York: Wiley.

Richards, M.P.M. (1966). Infantile handling in rodents: a reassessment in the light of recent studies of maternal behaviour. *Animal Behavior*, **14**, 582.

Richter, C.P. (1927). Animal behavior and internal drives. *Quarterly Review of Biology*, **2**, 302–43.

Richter, C.P. (1959). Rats, Man, and the welfare state. *American Psychologist*, **14**, 18–28.

Ritter, S., Pelzer, N.L. & Ritter, R.C. (1978). Absence of glucoprivic feeding after stress suggests impairment of noradrenergic neuron function. *Brain Research*, **149**, 399–411.

Roberts, W.W. & Kiess, H.O. (1964). Motivational properties of hypothalamic aggression in cats. *Journal of Comparative and Physiological Psychology*, **58**, 187–93.

Robertson, H.A., Martin, I.L. & Candy, J.M. (1978). Differences in benzodiazepine receptor binding in Maudsley reactive and Maudsley nonreactive rats. *European Journal of Pharmacology*, **50**, 455–7.

Robinson, R. (1965). *Genetics of the Norway Rat*. Oxford: Pergamon Press.

Rodriguez, W.A. & Logan, F.A. (1980). Preference for punishment of the instrumental or the consummatory response. *Animal Learning and Behavior*. **8**, 116–19.

Rosellini, R.A., DeCola, J.P. & Shapiro, N.R. (1982). Cross-motivational effects of inescapable shock are associative in nature. *Journal of Experimental Psychology: Animal Behavior Processes*, **8**, 376–88.

Roth, M. (1979). A classification of affective disorders based on a synthesis of new and old concepts. In *Research in the Psychobiology of Human Behavior*, ed. E. Meyer III & J.V. Brady, pp. 75–114. Baltimore: Johns Hopkins University Press.

Roth, M., Gurney, C., Mountjoy, G.Q., Kerr,T.A. & Schapira, K. (1976).The relationship between classification and response to drugs in affective disorder – problems posed by drug response in affective disorders. In *Monoamine Oxidase and its Inhibition*, Ciba Foundation Symposium 39 (NS), pp. 297–325. Amsterdam: Elsevier.

Royce, J.R. (1977). On the construct validity of open-field measures. *Psychological Bulletin*, **84**, 1098–106.

Russell, P.A. (1971). Infantile stimulation in rodents: a consideration of possible mechanisms. *Psychological Bulletin*, **75**, 192–202.

Ryan, T.J. & Watson, P. (1968). Frustrative nonreward theory applied to children's behaviour. *Psychological Bulletin*, **69**, 111–25.

Sackett, G.P. (1966). Monkeys reared in isolation with pictures as visual input: evidence for an innate releasing mechanism. *Science*, **154**, 1468–70.

Salmon, P. (1982). 'The Effect of Propranolol on Emotional Behaviour in Rats.' Unpublished Ph.D. thesis, University of Oxford.

Sanger, D.J. & Joly, D. (1985). Anxiolytic drugs and the acquisition of conditioned fear in mice. *Psychopharmacology*, **85**, 284–8.

Savage, R.D. & Eysenck, H.J. (1964).The definition and measurement of emotionality. In *Experiments in Motivation*, ed. H.J. Eysenck, pp. 292–314. Oxford: Pergamon Press.

Sawrey, W.L., Conger, J.J. & Turrell, E.S. (1956). An experimental investigation of the role of psychological factors in the production of gastric ulcers in rats. *Journal of Comparative and Physiological Psychology*, **49**, 457–61.

Sawrey,W.L. & Long, D.H. (1962). Strain and sex differences in ulceration in the rat. *Journal of Comparative and Physiological Psychology*, **55**, 603–5.

Scavio, M.J., Jr (1974). Classical-classical transfer: effects of prior aversive conditioning upon appetitive conditioning in rabbits. *Journal of Comparative and Physiological Psychology*, **86**, 107–15.

Scavio, M.J., Jr & Gormezano, I. (1980). Classical-classical transfer: effects of prior appetitive conditioning upon aversive conditioning in rabbits. *Animal Learning and Behavior*, **8**, 218–24.

Schachter, S. & Singer, J.E. (1962). Cognitive, social, and physiological determinants of emotional state. *Psychological Review*, **69**, 379–99.

Schleidt, W.M. (1961). Reaktionen von Truthuhnern auf fliegende Raubvogel und Versuche zur Analyse ihrer AAM's. *Zeitschrift für Tierpsychologie*, **18**, 534–60.

Schwartsbaum, J.S. (1960). Changes in reinforcing properties of stimuli following ablation of the amygdaloid complex in monkeys. *Journal of Comparative and Physiological Psychology*, **53**, 388–95.

Sechenov, I.M. (1935). Reflexes of the brain. In *Selected Works of I.M. Sechenov*, pp. 263–336. Moscow: State Publishing House for Biological and Medical Literature.

Segal, M. (1977*a*). The effects of brainstem priming stimulation on interhemispheric hippocampal responses in the awake rat. *Experimental Brain Research*, **28**, 529–41.

Segal, M. (1977*b*). Excitability changes in rat hippocampus during conditioning. *Experimental Brain Research*, **55**, 67–73.

Seitz, P.F.D. (1954).The effects of infantile experiences upon adult behavior in animal subjects. I. Effects of litter size during infancy upon adult behavior in the rat. *American Journal of Psychiatry*, **110**, 916–27.

Seligman, M.E.P. (1971). Phobias and preparedness. *Behavioral Therapy*, **2**, 307–20.

Seligman, M.E.P. (1975). *Helplessness*. San Francisco: Freeman.

Seligman, M.E.P. & Maier, S.F. (1967). Failure to escape traumatic shock. *Journal of Experimental Psychology*, **74**, 1–9.

Selye, H. (1952). *The Story of the Adaptation Syndrome*. Montreal: Acta Inc.

Seunath, O.M. (1975). Personality, reinforcement and learning. *Perceptual and Motor Skills*, **41**, 459–63.

Seyfarth, R.M., Cheney, D.L. & Marler, P. (1980). Vervet monkey alarm calls: semantic communication in a free-ranging primate. *Animal Behaviour*, **28**, 1070–94.

Shagass, C. & Naiman, J. (1956). The sedation threshold as an objective index of manifest anxiety in psychoneurosis. *Journal of Psychosomatic Research*, **1**, 49–57.

Sheard, M.H. & Flynn, J.P. (1967). Facilitation of attack behavior by stimulation of the midbrain of cats. *Brain Research*, **4**, 324–33.

Sherman, A.D. & Petty, F. (1982). Specificity of the learned helplessness animal model of depression. *Pharmacology, Biochemistry and Behavior*, **16**, 449–54.

Shipley, R.H. (1974). Extinction of conditioned fear in rats as a function of several parameters of CS exposure. *Journal of Comparative and Physiological Psychology*, **87**, 699–707.

Shipley, R.H., Mock, L.A. & Levis, D.J. (1971). Effects of several response prevention procedures on activity, avoidance responding, and conditioned fear in rats. *Journal of Comparative and Physiological Psychology*, **77**, 256–70.

Sidman, M. (1966). Avoidance behavior. In *Operant Behavior: Areas of Research and Application*, ed. W.K. Honig, pp. 448–98. New York: Appleton-Century-Crofts.

Sidman, M., Mason, J.W., Brady, J.V. & Thack, J., Jr (1962). Quantitative relations between avoidance behavior and pituitary-adrenal cortical activity. *Journal of the Experimental Analysis of Behavior*, **5**, 353–62.

Siegel, P.S. & Brantley, J.J. (1951). The relationship of emotionality to the consummatory response of eating. *Journal of Experimental Psychology*, **42**, 304–6.

Simon, N.G., Gandelman, R. & Gray, J.L. (1984). Endocrine induction of intermale aggression in mice: a comparison of hormonal regimens and their relationship to naturally occurring behavior. *Physiology and Behavior*, **33**, 379–83.

Sines, J.O. (1961). Behavioral correlates of genetically enhanced susceptibility to stomach lesion development. *Journal of Psychosomatic Research*, **5**, 120–6.

Sklar, L.S. & Anisman, H. (1981). Stress and cancer. *Psychological Bulletin*, **89**, 369–406.

Slater, J. & Blizard, D.A. (1976). A re-evaluation of the relation between estrogen and emotionality in female rats. *Journal of Comparative and Physiological Psychology*, **90**, 755–64.

Smith, J.M. (1964). Group selection and kin selection. *Nature*, **201**, 1145–7.

Smythies, J.R. (1966). *The Neurological Foundations of Psychiatry*. Oxford: Blackwell.

Sokolov, E.N. (1960). Neuronal models and the orienting reflex. In *The Central Nervous System and Behaviour*, Transactions of the Third Conference, ed. M.A.B. Brazier. pp. 187–276. New York: Josiah Macy Jr. Foundation.

Sokolov, E.N. (1966). Neuronal mechanisms of the orienting reflex. In *Eighteenth International Congress of Psychology, Moscow, Symposium No. 5, 'Orienting Reflex, Alertness and Attention'*, pp. 31–6.

Solomon, R.L. (1964). Punishment. *American Psychologist*, **19**, 239–53.

Solomon, R.L. Kamin, L.J. & Wynne, L.C. (1953). Traumatic avoidance learning: the outcome of several extinction procedures with dogs. *Journal of Abnormal and Social Psychology*, **48**, 291–302.

Solomon, R.L., Turner, L.H. & Lessac, M.S. (1968). Some effects of delay of punishment on resistance to temptation in dogs. *Journal of Personality and Social Psychology,* **8**, 233–8.

Solomon, R.L. & Wynne, L.C. (1953). Traumatic avoidance learning: acquisition in normal dogs. *Psychological Monographs,* **67**, 4 (whole no. 354), pp. 1–19.

Soltysik, S. (1960). Studies on avoidance conditioning. III. Differentiation and extinction of avoidance reflexes. *Acta Biologiae Experimentalis,* **20**, 171–82.

Soltysik, S. (1964). Inhibitory feedback in avoidance conditioning. In *Feedback Systems Controlling Nervous Activity* (First Conference on Neurobiology), ed. A. Escobar, pp. 316–31. Mexico: Sociedad Mexicana de Ciencias Fisiologicas.

Soltysik, S.S., Wolfe, G.E., Nicholas, T., Wilson, W.J. & Garcia-Sanchez, J.L. (1983). Blocking of inhibitory conditioning within a serial conditioned stimulus-conditioned inhibitor compound: maintenance of acquired behaviour without an unconditioned stimulus. *Learning and Motivation,* **14**, 1–29.

Soubrié, P. (1986). Reconciling the role of central serotonin neurons in human and animal behavior. *Behavioral and Brain Sciences,* **9**, 319–63.

Southwick, C.H. (1959). Eosinophil response of C57BR mice to behavioural disturbance. *Ecology,* **40**, 156–7.

Spence, J.T. & Spence, K.W. (1966). The motivational components of manifest anxiety: drive and drive stimuli. In *Anxiety and Behavior,* ed. C.D. Spielberger, pp. 291–326. New York & London: Academic Press.

Spielberger, C.D. ed. (1966). *Anxiety and Behavior.* New York and London: Academic Press.

Staddon, J.E.F. & Innis, N.K. (1966). An effect analogous to 'frustration' on interval reinforcement schedules. *Psychonomic Science,* **4**, 287–8.

Stanley, M., Virgilio, J. & Gershon, S. (1982). Initiated imipramine binding, sites are decreased in the frontal cortex of suicides. *Science,* **216**, 1337–9.

Starr, M.D. & Mineka, S. (1977). Determinants of fear over the course of avoidance learning. *Learning and Motivation,* **8**, 332–50.

Starzl, T.E., Taylor, C.W. & Magoun, H.W. (1951). Collateral afferent excitation of reticular formation of brain stem. *Journal of Neurophysiology,* **14**, 479–96.

Stein, L., Wise, C.D. & Berger, B.D. (1973). Anti-anxiety action of benzodiazepines: decrease in activity of serotonin neurons in the punishment system. In *The Benzodiazepines,* ed. S. Garattini, E. Mussini & L.O. Randall, pp. 299–326. New York: Raven Press.

Sterman, M.B. & Fairchild, M.D. (1966). Modification of locomotor performance by reticular formation and basal forebrain stimulation in the cat: evidence for reciprocal systems. *Brain Research,* **2**, 205–17.

Stevens, R. & Goldstein, R. (1981). Effects of neonatal testosterone and estrogen on open-field behaviour in rats. *Physiology and Behavior,* **26**, 551–3.

Strongman, K.T. (1965). The effect of anxiety on food intake in the rat. *Quarterly Journal of Experimental Psychology,* **17**, 255–60.

Sudak, H.S. & Maas, J.W. (1964). Behavioural-neurochemical correlation in reactive and nonreactive strains of rats. *Science,* **146**, 418–20.

Swanson, H.H. (1967). Alteration of sex-typical behaviour of hamsters in open field and emergence tests by neonatal administration of androgen or oestrogen. *Animal Behaviour,* **15**, 209–16.

Taub, E. & Berman, A.J. (1968). Movement and learning in the absence of sensory feedback. In *The Neuropsychology of Spatially Oriented Behaviour,* ed. S.J. Freedman, pp. 173–92. Homewood, Illinois: Dorsey Press.

Thiébot, M.-H., Hamon, M. & Soubrié, P. (1983). The involvement of nigral serotonin innervation in the control of punishment-induced behavioral inhibition in rats.

Pharmacology, Biochemistry and Behavior, **19**, 225–9.

Thiébot, M.-H., Jobert, A. & Soubrié, P. (1980). Conditioned suppression of behavior: its reversal by intra-raphe microinjection of chlordiazepoxide and GABA. *Neuroscience Letters*, **16**, 213–17.

Thiessen, D.D., Zolman, J.F. & Rodgers, D.A. (1962). Relation between adrenal weight, brain cholinesterase activity, and hole-in-wall behavior of mice under different living conditions. *Journal of Comparative and Physiological Psychology*, **55**, 186–90.

Thomas, E. & DeWold, L. (1977). Experimental neurosis: neuropsychological analysis. In *Psychopathology: Experimental Models*. ed. M.E.P. Seligman & J. Maser, pp. 214–31. San Francisco: Freeman.

Thompson, R.F., Barchas, J.D., Clark, G.A., Donegan, N., Kettner, R.E., Lavond, D.G., Madden, J., IV, Monk, M.D. & McCormick, D.A. (1984). Neuronal substrates of associative learning in the mammalian brain. In *Primary Neural Substrates of Learning and Behavioral Change*. ed. D. Alkon & J. Farley, pp. 71–7. Cambridge University Press.

Thompson, T. & Bloom, W. (1966). Aggressive behavior and extinction-induced response-rate increase. *Psychonomic Science*, **5**, 335–6.

Thompson, W.R. (1957). Influence of prenatal maternal anxiety on emotionality in young rats. *Science*, **125**, 698–9.

Tinbergen, N. (1948). Social releasers and the experimental method required for their study. *Wilson Bulletin*, **60**, 6–51.

Tinbergen, N. (1951). *The Study of Instinct*. Oxford University Press.

Tsaltas, E., Gray, J.A. & Fillenz, M. (1984). Alleviation of response suppression to conditioned aversive stimuli by lesions of the dorsal noradrenergic bundle. *Behavioral Brain Research*, **13**, 115–27.

Tsuda, A., Tanaka, M., Nishikawa, T. & Hirai, H. (1983). Effects of coping behavior on gastric lesions in rats as a function of the complexity of coping tasks. *Physiology and Behavior*, **30**, 805–8.

Tugendhat, B. (1960*a*). The normal feeding behavior of the three-spined stickleback (*Gasterosteus aculeatus L.*). *Behaviour*, **15**, 284–318.

Tugendhat, B. (1960*b*). The disturbed feeding behavior of the three-spined stickleback. I. Electric shock is administered in the food area. *Behaviour*, **16**, 159–87.

Turner, C.D. & Bagnara, J.T. (1976). *General Endocrinology*, 6th edn. New York: Holt-Saunders.

Tye, N.C., Everitt, B.J. & Iversen, S.D. (1977). 5-Hydroxytryptamine and punishment. *Nature*, **268**, 741-2.

Valentine, C.W. (1930). The innate bases of fear. *Journal of Genetic Psychology*, **37**, 394–419.

Valins, S. (1966). Cognitive effects of false heart-rate feedback. *Journal of Personality and Social Psychology*, **4**, 400–8.

Van Abeelen, J.H.F. (1966). Effects of genotype on mouse behaviour. *Animal Behaviour*, **14**, 218–25.

van de Poll, N.E., Smeet, J., van Oyen, H.G. & van der Zwan, S.M. (1982). Behavioral consequences of agonistic experience in rats: sex differences and the effects of testosterone. *Journal of Comparative and Physiological Psychology*, **96**, 893–903.

Vanderwolf, C.H., Kramis, R. & Robinson, T.E. (1978). Hippocampal electrical activity during waking behaviour and sleep: analyses using centrally acting drugs. In *Functions of the Septo-Hippocampal System*, ed. K. Elliott & J. Whelan, Ciba Foundation Symposium 58, pp. 199–221. Amsterdam: Elsevier.

van Oyen, H.G., van der Zuzen, S.M., van de Poll, N.E. & Walg, H. (1981a). Punishment of food rewarded lever holding in male and female rats. *Physiology and
Behavior,* **26**, 1037–40.

van Oyen, H.G., Walg, H. & van de Poll, N.E. (1981b). Discriminated lever press
avoidance conditioning in male and female rats. *Physiology and Behavior,* **26**,
313–17.

Van-Toller, C. & Tarpy, R.M. (1974). Immunosympathectomy and avoidance behavior.
Psychological Bulletin, **81**, 132–7.

Vieth, W., Curio, E. & Ernst, U. (1980). The adaptive significance of avian mobbing.
III. Cultural transmission of enemy recognition in blackbirds: cross-species
tutoring and properties of learning. *Animal Behaviour,* **28**, 1217–29.

Vinogradova, O.S. (1966). Investigation of habituation in single neurons of different
brain structure with special reference to the hippocampus. *Eighteenth International Congress of Psychology Moscow, Symposium No. 5, 'Orienting Reflex,
Alertness and Attention'*, pp. 55–8.

Wagner, A.R. (1959). The role of reinforcement and nonreinforcement in an apparent
frustration effect. *Journal of Experimental Psychology,* **57**, 130–6.

Wagner, A.R. (1966). Frustration and punishment. In *Current Research on Motivation*,
ed. R.N. Haber, pp. 229–39. New York: Holt, Rinehart & Winston.

Wagner, A.R. & Rescorla, R.A. (1972). Inhibition in Pavlovian conditioning: application of a theory. In *Inhibition and Learning*, ed. R.A. Boakes & M.S. Halliday.
London: Academic Press.

Walters, E.T., Carew, T.H. & Kandel, E.R. (1981). Associative learning in *Aplysia*:
evidence for conditioned fear in an invertebrate. *Science*, **211**, 504–6.

Wang, G.H. (1923). The relation between 'spontaneous' activity and oestrous cycle in
the white rat. *Comparative Psychology Monographs* 2 (no. 6), 1–27.

Ward, I.L. & Ward, O.B. (1985). Sexual behavior differentiation: effects of prenatal
manipulations in rats. In *Handbook of Behavioral Neurobiology*, vol. 7, ed. N.
Adler, D. Pfaff & R.W. Goy, pp. 77–98. New York: Plenum Publishing Corpo-
ration.

Ward, I.L. & Weisz, J. (1980). Maternal stress alters plasma testosterone in fetal males.
Science, **207**, 328–9.

Warren, J.M. & Levy, S.J. (1979). Fearfulness in female and male cats. *Animal Learning
and Behavior,* **7**, 521–4.

Watson, J.B. (1913). Psychology as the behaviorist views it. *Psychological Review*, **20**,
158–77.

Watson, J.B. (1924). *Behaviorism*. New York: Norton.

Watson, R.H.J. (1960). Constitutional differences between two strains of rats with
different behavioural characteristics. In *Advances in Psychosomatic Medicine*,
vol. 1, ed. A. Jones & B. Stokvis, pp. 160–5. New York: Karger.

Weinberg, J., Erskine, M. & Levine, S. (1980). Shock induced fighting attenuates the
effects of prior shock experience in rats. *Physiology and Behavior,* **25**, 9–16.

Weisman, R.G. & Litner, J.S. (1969). Positive conditioned reinforcement of Sidman
avoidance behavior in rats. *Journal of Comparative and Physiological Psychology,* **68**, 597–603.

Weiss, J.M. (1968). Effects of coping responses on stress. *Journal of Comparative and
Physiological Psychology,* **65**, 251–60.

Weiss, J.M. (1977). Ulcers. In *Psychopathology: Experimental Models*, ed. M.E.P.
Seligman & J. Maser, pp. 232–69. San Francisco: Freeman.

Weiss, J.M., Bailey, W.H., Goodman, P.A., Hoffman, L.J., Ambrose, M.J., Salman,

S. & Charry, J.M. (1982). A model for neurochemical study of depression. In *Behavioral Models and the Analysis of Drug Action*, ed. M.Y. Spiegelstein & A. Levy. Amsterdam: Elsevier.

Weiss, J.M., Glazer, H.I. & Pohorecky, L.A. (1976*a*). Coping behavior and neurochemical changes: an alternative explanation for the original 'learned helplessness' experiments. In *Animal Models in Human Psychobiology*, ed. A. Serban & A. Kling, pp. 141–73. New York: Plenum Press.

Weiss, J.M., Pohorecky, L.A., Salman, S. & Gruenthal, M. (1976*b*). Attenuation of gastric lesions by psychological aspects of aggression in rats. *Journal of Comparative and Physiological Psychology*, **90**, 252–9.

Weisz, J. & Ward, I.L. (1980). Plasma testosterone and progesterone titers of pregnant rats, their male and female fetuses, and neonatal offspring. *Endocrinology*, **106**, 306–16.

Whimbey, A.E. & Denenberg, V.H. (1967). Two independent behavioral dimensions in open-field performance. *Journal of Comparative and Physiological Psychology*, **63**, 500–4.

Whishaw, I.Q. & Vanderwolf, C.H. (1973). Hippocampal EEG and behavior: changes in amplitude and frequency of RSA (theta rhythm) associated with spontaneous and learned movement patterns in rats and cats. *Behavioral Biology*, **8**, 461–84.

Whiting, J.W.M. & Mowrer, O.H. (1943). Habit progression and regression – a laboratory study of some factors relevant to human socialization. *Journal of Comparative Psychology*, **36**, 229–53.

Wilcock, J. & Fulker, D.W. (1973). Avoidance learning in rats: genetic evidence for two distinct behavioral processes in the shuttle-box. *Journal of Comparative and Physiological Psychology*, **82**, 247–53.

Willett, R. (1960). The effects of psychosurgical procedures on behaviour. In *Handbook of Abnormal Psychology*, ed. H.J. Eysenck, pp. 566–610. London: Pitman.

Williams, J.H. & Azmitia, E.C. (1981). Hippocampal serotonin reuptake and nocturnal locomotor activity after microinjection of 5,7-DHT in the fornix-fimbria. *Brain Research*, **207**, 95–107.

Willingham, W.W. (1956). The organization of emotional behavior in mice. *Journal of Comparative and Physiological Psychology*, **49**, 345–8.

Willner, P. (1985). *Depression: A Psychobiological Synthesis*. New York: Wiley.

Wilson, E.H. & Dinsmoor, J.A. (1970). Effect of feeding on 'fear' as measured by passive avoidance in rats. *Journal of Comparative and Physiological Psychology*, **70**, 431–6.

Wilson, E.O. (1975). *Sociobiology: The New Synthesis*. Cambridge, Mass.: Belknap.

Wolf. S.G. (1965). *The Stomach*. New York: Oxford University Press.

Wolf. S.G. & Wolff, H.G. (1943). *Human Gastric Function: An Experimental Study of a Man and his Stomach*. Oxford Medical Publications.

Wolpe, J. (1958). *Psychotherapy by Reciprocal Inhibition*. Stanford University Press.

Wynne, L.C. & Solomon, R.L. (1955). Traumatic avoidance learning: acquisition and extinction in dogs deprived of normal peripheral autonomic function. *Genetic Psychology Monographs*, **52**, 241–84.

Wynne-Edwards, V.C. (1962). *Animal Dispersion in Relation to Social Behaviour*. Edinburgh: Oliver & Boyd.

Wynne-Edwards, V.C. (1964). Population control in animals. *Scientific American,* **211** (2), 68–74.

Yeo, C.H., Hardiman, M.T. & Glickstein, M. (1985). Classical conditioning of the nictitating membrane response of the rabbit. *Experimental Brain Research*, **66**, 87–98.

Yerkes, R.M. & Yerkes, A.W. (1936). Nature and conditions of avoidance (fear) response in chimpanzee. *Journal of Comparative Psychology,* **21**, 53–66.

Young, W.G., Goy, R.W. & Phoenix, C.H. (1964). Hormones and sexual behavior. *Science,* **143**, 212–18.

Zeigler, H.P. (1964). Displacement activity and motivational theory: a case study in the history of ethology. *Psychological Bulletin,* **61**, 362–76.

Zimmer-Hart, C.L. & Rescorla, R.A. (1974). Extinction of Pavlovian conditioned inhibition. *Journal of Comparative and Physiological Psychology,* **86**, 837–45.

Author index

Subject index